Nuclear Physics 1

Nuclear Physics 1

Nuclear Deexcitations, Spontaneous Nuclear Reactions

Ibrahima Sakho

ISTE

WILEY

First published 2021 in Great Britain and the United States by ISTE Ltd and John Wiley & Sons, Inc.

ISTE Ltd
27-37 St George's Road
London SW19 4EU
UK

www.iste.co.uk

John Wiley & Sons, Inc.
111 River Street
Hoboken, NJ 07030
USA

www.wiley.com

Library of Congress Control Number: 2021945496

British Library Cataloguing-in-Publication Data
A CIP record for this book is available from the British Library
ISBN 978-1-78630-641-8

MIX
Paper from responsible sources
FSC® C013604

Contents

Preface

Nuclear physics is devoted to the study of the properties of atomic nuclei. These properties relate to the internal structure of the nucleus which facilitate the understanding of the properties of nucleons (neutrons and protons), the mechanisms of nuclear reactions (spontaneous or induced), in order to describe the different processes of elastic and inelastic nucleus-nucleus interactions, the fields of application of nuclear physics and, finally, the impact of nuclear radiation on human health and the environment.

In general, nuclear physics is the physics of low energies, ranging from 250 eV to 10 GeV [SAO 04, GER 07, LAL 11]. The range of energies above 10 GeV [SAO 04, GER 07, LAL 11] relate to the physics of high energies whose purpose is to study the constituent particles of matter and the fundamental interactions between them. In this field, experimenters use particle accelerators that operate at very high energies or deliver very large beam intensities, thus allowing access to the fundamental laws of subatomic physics at very short distances. The most spectacular achievement to date is of course the Large Hadron Collider (LHC), launched in September 2008 at CERN.

Nuclear physics is an area that has experienced considerable growth since the discovery of radioactivity in 1896 by Henri Becquerel [HAL 11], well before the discovery of the atomic nucleus in 1911 by Ernest Rutherford [RUT 11]. Research in nuclear physics covers several topics ranging from subatomic particles to stars. It thus constitutes a fundamental component of physics, allowing the exploration of the infinitely large and the infinitely small [ARN 10]. In addition, nuclear physics makes it possible to understand many astrophysical phenomena such as nucleosynthesis processes (primordial, stellar and explosive) within the framework of the Big Bang model. The study of these processes allows us to understand the origin of chemical elements and to describe the evolution of supernova and neutron stars [SUR 98].

This book is the fruit of a 25-year long teaching career. Initially this was teaching final-year high-school S1 and S2 science students at Alpha Molo Baldé High School in Kolda, from 1996 to 2002. It was then at Bambey High School from 2002 to 2008 and at Maurice Delafosse Technical High School from 2008 to 2010. This was followed by 9 years at Assane Seck University, Ziguinchor, teaching final-year Physics undergraduates and, since February 2019, teaching final-year Physics and Chemistry undergraduates at the University of Thiès.

Nuclear Physics 1 consists of four chapters, as follows.

Chapter 1 is reserved for general information regarding the *atomic nucleus* with a view to establishing the general properties of nuclei. It begins with a presentation of the experimental facts that led to the *discovery of the electron* (β^- particle), the *proton*, the *neutron* and the *nucleus* itself. It then focuses on the study of the *composition* and *dimensions* of the nucleus. Next, the *nomenclature of nuclides* and the *stability of nuclei* are studied. The chapter culminates with a series of exercises with answers.

Chapter 2 is dedicated to the study of nuclear deexcitation processes. The *nuclear shell model*, which offers an understanding of the discrete structure of nuclear levels, is studied in detail. Subsequently, the study examines the properties of *angular momentum* and *parity*, the processes of *gamma deexcitation* and *internal conversion* and the phenomenon of deexcitation by nuclear emission. A detailed study of the Bethe–Weizsäcker *semi-empirical mass formula* via the liquid-drop model and of the *mass parabola equation* for odd A completes the chapter and is followed by a series of exercises complete with answers.

Chapter 3 is devoted to the study of *alpha (α) radioactivity*. It begins with the experimental facts that led to the discovery of *radioactivity* itself, the discovery of α radioactivity and β^- radioactivity, the discovery of the *positron* (β^+ particle), *neutrino* and experiments highlighting α, β and γ *radiation*. The chapter goes on to focus on the study of *radioactive disintegration* and the properties of *α decay*. A series of exercises complete with answers is at the end of the chapter.

Chapter 4 is reserved for the study of β^- and β^+ *decay modes* and for the study of *radioactive family trees*. At the beginning of the chapter, we present the experimental facts that led to the discovery of *artificial radioactivity*. We then focus the development on the study of the properties of *β decay* and the link between β decay and decay by *electron capture*. In addition, *double β decay* and the process of *atomic deexcitation* by *Auger effect* are studied in this chapter. The study subsequently focuses on the presentation of *radioactive series*, enabling the introduction of the *Bateman equations*. The mechanism for *radionuclide production* by nuclear bombardment features prominently in the chapter, which is rounded off with a series of exercises complete with answers.

Two appendices follow the above chapters. The first appendix is dedicated to the determination of the quantified expression of the energy of the three-dimensional quantum harmonic oscillator, in relation to the harmonic potential nuclear shell model. Two approaches are adopted to achieve this. The first approach integrates the Schrödinger equation applied to a quantum harmonic oscillator. In the second approach, a more flexible operative approach is adopted using creation and annihilation operators. The second appendix provides a listing, in table form, of the atomic masses of isotopes of atomic numbers $Z = 1$–93.

This book is written for Physical Science teachers in high schools, for final-year Physics undergraduate students (*Licence 3* under the French LMD system) and for university lecturers responsible for the Nuclear Physics module in their programs. It is written using clear and concise language, underpinned by a very original pedagogical style. Each chapter begins with an overview of the general objective, the specific objectives and the prerequisites for understanding the chapter as it unfolds. In addition, each concept or law introduced follows a direct application for sound understanding of the nuclear phenomena and properties studied. The chapters are interspersed with succinct biographies of all the great thinkers who have contributed to the development of nuclear physics in relation to the topics developed.

This book does not attempt to cover all aspects relating to understanding nuclear deexcitation processes and the properties of spontaneous nuclear reactions. Nevertheless, it contains the fundamental basics of nuclear physics relating to the topics studied here. As with all human endeavors, there is always room for improvement. We therefore remain open to our readers for any suggestions, comments or criticisms that could be used to improve the scientific quality of this work.

September 2021

1

Overview of the Nucleus

Overall objective	
To know the general properties of nuclei	
Specific objectives	
To compare the atomic models of Thomson, Perrin and Rutherford	To define the effective scattering cross-section
To compare Rutherford's and Blackett's observations on the first nuclear transmutation reaction	To define the effective elementary scattering cross-section
To know the properties of the isospin operator	To define the separation energy of a nucleon
To know the properties of the spin angular momentum of a nucleus	To determine the separation energy of a neutron from a proton for a given nuclide
To know the fundamental properties of nucleons	To determine the total isospin corresponding to the ground state of a nucleus
To know the expression for the radius of a nucleus assumed to be spherical	To deduce, from the binding energy, the nuclear charge of the most stable isobar
To know the expression for the Sakho unit nuclear radius	To establish the relationship between binding energy and mass defect
To know the principle of a mass spectrograph	To establish the relationship between skin thickness and the diffusivity parameter
To compare the stability of nuclei from their nuclear binding energy	To write the balanced equation of the first nuclear transmutation reaction
To know the effect of nuclear forces on the stability of nuclei	To establish Rutherford's differential effective cross-section

For a color version of all of the figures in this chapter, see www.iste.co.uk/sakho/nuclear1.zip.

To differentiate between *u* and *d* quarks according to the Gell-Mann and Zweig model	To make the analogy between electron gyromagnetic ratio and nuclear gyromagnetic ratio
To differentiate between isospin and nuclear spin	To make the analogy between Bohr magneton and Bohr nuclear magneton
To differentiate between unit nuclear radius and electromagnetic unit radius	To make the analogy between spin multiplicity and isospin multiplicity
To differentiate between isotopes, isobars and isotones	To interpret Chadwick's experiment
To differentiate between mirror nucleus and magic nucleus	To interpret Geiger and Marsden's experiment
To distinguish between valley of stability and line of stability	To interpret Rutherford's scattering experiment
To define a monoisotopic element	To interpret Rutherford's nuclear transmutation experiment
To define a nuclear isomer	To interpret the shape of the Woods–Saxon charge distribution density
To define the nuclear dipole magnetic moment	To interpret the Segrè diagram
To define the nuclear Landé factor	To interpret the Aston curve
To define the skin thickness of a nucleus	To situate the nuclear energy surface or stability valley in the Segrè diagram
To define the atomic mass unit	
Prerequisites	
Material structure	Motion of a charged particle in a uniform magnetic field
Atomic models	Vector product properties
Shell model of electron configurations	Fundamental theorems of the dynamics of the material point
Quantum numbers of the electron	

1.1. Discovery of the electron

1.1.1. *Hittorf and Crookes experiments*

In around 1869, Johann Wilhelm Hittorf studied electric discharge in rarefied gases using a vacuum tube. With the help of a Sprengel pump, Hittorf managed to obtain pressures below 0.001 mbar and found that electric discharges were accompanied by the emission of glow rays, which he called "cathode rays" [LEP 56, CAR 79, ROU 60, PER 95, SAK 11]. Subsequently, he observed that cathode rays are gifted with the property of being deflected by a magnetic field. But Hittorf stopped at these observations, without providing any physical interpretation.

Still in 1869, Sir William Crookes invented the electronic tube bearing his name [CRO 79, ROU 60, SAK 11]. The *Crookes tube* was a cold cathode tube, that is, a tube that did not have a heating filament as in the case of cathode tubes, designed to generate electrons.

Essentially, a Crookes tube is a bulb containing a gas and is equipped with two electrodes (Figure 1.1). When a voltage of approximately 50,000 volts is applied between the two electrodes and the gas pressure is gradually reduced, a dark space (called Crookes space) fills the tube at around 0.01 millimeters of mercury. Today it is now possible to interpret Crookes' observations.

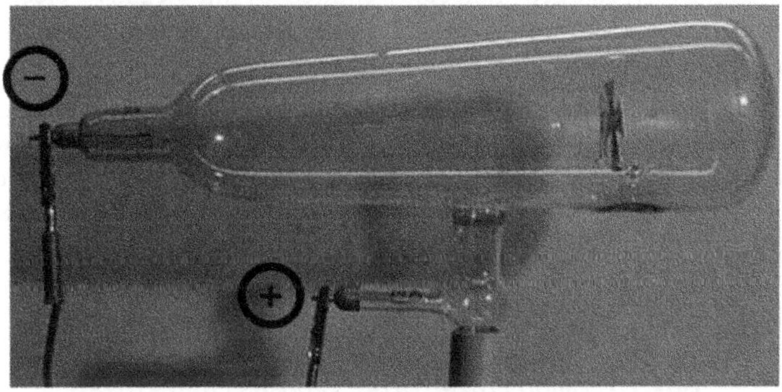

Figure 1.1. *Crookes tube*

Inside the Crookes tube, electrons are generated by ionization of gas molecules excited by the applied continuous voltage. Under the action of the established electric field, the ions created in the tube are accelerated. They collide with gas molecules, knocking electrons off them. The positive ions thus formed are attracted by the cathode. As they strike the latter, they eject a large number of electrons, called cathode rays. As they strike the glass, the electrons excite the atoms in the walls of the tube, thus causing its fluorescence usually in the yellow-green range. The electrons flow in a straight line from the cathode to the anode. This motion is highlighted by the shadow cast by the cross on the fluorescent wall (Figure 1.2). Moreover, cathode radiation has the property of being deflected by a magnetic field. Crookes considered cathode radiation to be the fourth state of matter which he called the radiant state. But what is the nature of this radiation?

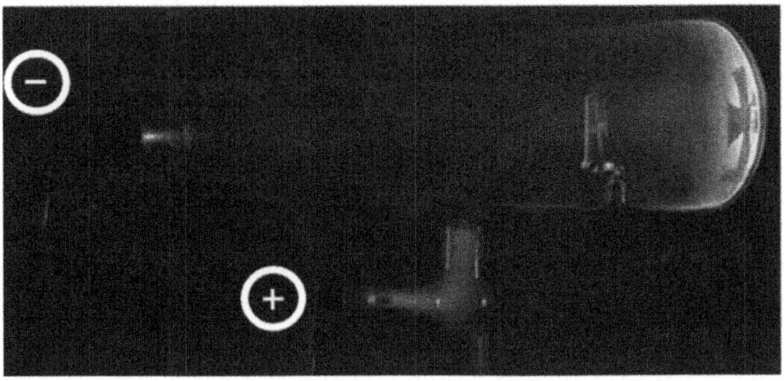

Figure 1.2. *Crookes tube wall fluorescence*

Crookes gave the following response to this question in 1885: "cathode radiation consists of negatively charged molecules emitted by the negative electrode". The proof that they are indeed molecules, he cried, is that the action of a magnet makes them deviate from their trajectory, just like particles of iron filings.

Johann Wilhelm Hittorf was a German physicist. He is also known for his work on the interpretation of electrical conductivity in electrolytic solutions in 1859, on the quantitative study of metal ion allotropy in 1865 and on the demonstration of cathode rays in 1869.

Sir William Crookes was a British physicist and chemist. He is most famous for having invented the "Crookes tube" in 1869 which allowed him to highlight cathode rays.

Box 1.1. *Hittorf (1824–1914); Crookes (1832–1919)*

1.1.2. *Perrin and Thomson experiments*

While cathode radiation was an experimental reality that no-one could question, the molecular nature of the famous radiation puzzled many physicists, including Sir Joseph John Thomson [THO 97, ROU 60, CAR 79, PER 95, FAL 87, SAK 11]. The latter, as early as 1881, boldly took the opposite view to Crookes' conception of the nature of cathode radiation. For Thomson, "cathode radiation consists not of molecules but of *particles of pure negative electricity*," which he would later call electrons (a term introduced in 1891 by Stoney [CAR 79]). However, Thomson argued that "particles of pure electricity are not matter". Moreover, he noted, even if experimental observations require particles of pure electricity to be given a mass, then it is only necessary to understand that "this mass is nothing other than their inertia induced by their motion under the action of the magnetic field". However,

Thomson did not say anything regarding the dimension of particles of pure negative electricity.

Challenging the ideas of Crookes, Thomson and others, Perrin suggested that cathode rays should not be considered as being made up of molecules but of "smaller particles charged with pure negative electricity". In 1895, he sought to verify this proposal through experimentation. Perrin then used a Faraday box (a small cylinder capable of trapping cathode rays), in contact with the plate of a positively-charged electroscope (Figure 1.3), to gather the cathode radiation as it exited a Crookes tube.

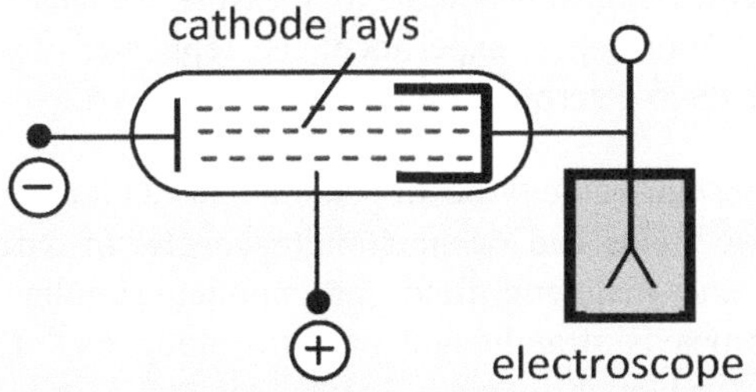

Figure 1.3. *Perrin's simplified experimental set-up*

Perrin found that the famous particles of pure electricity suggested by Thomson roughly neutralized the positive charge of the electroscope. These observations confirmed that the electricity transported by cathode radiation is negative in nature. But it should be noted that, until that date, there was no information on the size of the particles of pure negative electricity. Were they smaller than atoms? This question could be answered by measuring their mass, m. In chronological terms, the measurement of this mass was preceded by the measurement of the electron mass-to-charge ratio, e/m, by Thomson, and by the measurement of the elementary electrical charge, e, by Robert Millikan. Thomson's experiments form the subject of the study that follows. Millikan's experiment is discussed in section 1.1.3.

In 1897, Thomson conducted a series of historical experiments that measured the mass-to-charge ratio of the electron.

First experiment: Thomson studied the possibility of separating the negative electrical charge from the cathode rays by a magnetic field. He built a cathode-ray tube that ends in a pair of cylinders with slots connected to an electrometer. This first experiment showed that the cathode rays are not deflected under the action of a magnetic field. Thomson concluded that the negative charge cannot be separated from the rays (which tacitly proves that the cathode rays are composed of identical charges).

Second experiment: Thomson studied the action of an electric field on cathode rays. To do this, he built a cathode-ray tube with a deeper vacuum, within which is an electric field created by a voltage applied between two conductive metal plates. He placed a coat of phosphorescent paint at the end of the tube to detect incident rays. Thomson observed that rays are attracted by the positive plate. He thus proved that the electrical charge of cathode rays is negative, in accordance with Perrin's experimental observations.

Third experiment: Thomson sought to measure the *electron mass-to-charge ratio, e/m*. Today, Thomson's experiment is replicated by performing more meticulous experiments to determine the value of e/m.

Electrons in a constant velocity beam penetrate at O into a space where uniform electric and magnetic fields can occur simultaneously. In a first experiment, the orthogonal electric and magnetic fields are applied simultaneously so that the motion of the electrons is straight and uniform along OO' (Figure 1.4(a)). The electric field is vertical and directed downwards. In a second experiment, the magnetic field is removed; the characteristics of the electric field and the electron velocity vector remain unchanged (Figure 1.4(b)). The set-ups of the two experiments conducted can be schematically presented in parallel, as shown in Figure 1.4.

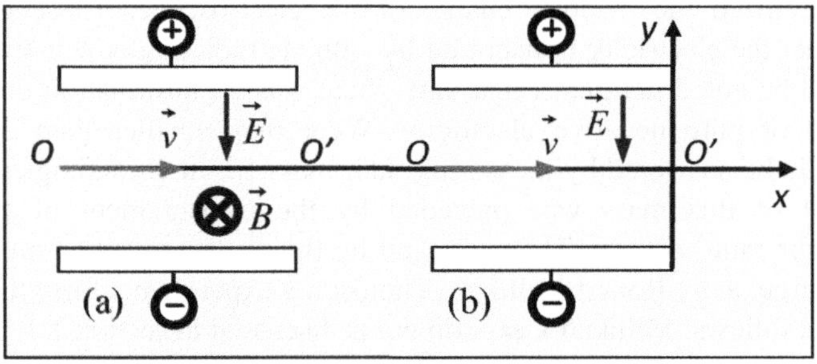

Figure 1.4. *Simplified set-up for measuring the electron mass-to-charge ratio: (a) simultaneous action of the electric field and the magnetic field; (b) action of the electric field alone*

It is then shown that the electron mass-to-charge ratio has the following value:

$$\frac{e}{m} = 1.76 \times 10^{11}\,\mathrm{C \cdot kg^{-1}} \qquad [1.1]$$

APPLICATION 1.1.– Using the device shown in Figure 1.4, show that if Y designates the vertical deviation along the $O'y$ axis, the electron mass-to-charge ratio is given by the expression:

$$\frac{e}{m} = 2\frac{E}{B^2 l^2} Y \qquad [1.2]$$

Then find the result [1.1].

Given data: $E = 50$ kV · m^{-1}; $B = 1$ mT; $OO' = l = 10.0$ cm, $Y = 1.76$ cm.

ANSWER.– Considering Figure 1.4(a), the electrical force is compensated by the magnetic force since the motion is uniform and rectilinear (it is a velocity filter). Let:

$$q\vec{E} = -q\vec{v} \wedge \vec{B} \Rightarrow v = \frac{E}{B} \qquad [1.3]$$

Let us thus determine the equation of the trajectory of an electron in the electric field (Figure 1.4(b)). Using the theorem of the center of inertia gives:

$$\begin{cases} q\vec{E} = m\vec{a} \\ \vec{a} = \dfrac{q\vec{E}}{m} \end{cases} \Rightarrow \begin{cases} a_x = 0 \Rightarrow x = vt \\ a_y = -\dfrac{qE}{m} \Rightarrow y = -\dfrac{1}{2}\dfrac{qE}{m}t^2 \end{cases}$$

Using the last equation system, the equation for the trajectory of an electron within the electric field is written ($q = -e$):

$$y = \frac{1}{2}\frac{eE}{mv^2}x^2 \qquad [1.4]$$

The vertical deviation, Y, along the $O'y$ axis is obtained for $x = OO' = l$. Using [1.4] and taking account of [1.3], we then obtain the following:

$$Y = \frac{e}{m}\frac{B^2}{2E}l^2 \qquad [1.5]$$

Using [1.5], result [1.2] is found.

NOTE.–

$$\frac{e}{m} = 2\frac{5\times10^{4}}{10^{-6}\times10^{-2}} \times 1.76 \times 10^{-2} = 1.76 \times 10^{11}\ \mathrm{C \cdot kg^{-1}}$$

The result [1.1] is indeed found.

The CODATA (Committee on Data and Technology) recommended value is $1.75882001076(53)\times 10^{11}\ \mathrm{C \cdot kg^{-1}}$.

George Johnstone Stoney was an Irish physicist. He introduced the concept of an "electricity atom" or "electricity particle," for which he invented the term "electron" in 1891.

Sir Joseph John Thomson was a British physicist and chemist. He is famous for his work on the study of the structure of matter. He is known primarily for inventing the mass spectrograph which is very useful in the separation of isotopes. Thomson confirmed the existence of the electron, and measured its mass-to-charge ratio in 1897. He was awarded the Nobel Prize in Physics 1906 for his theoretical and experimental research on electrical conductivity in gases.

Jean Perrin was a French physicist. He is known for demonstrating that cathode rays are negatively charged in 1895. In 1901, Perrin envisioned the atom as a solar system formed at the center of a quasi-point-like concentration of positive matter surrounded by a procession of electrons in motion. In addition, in 1908 he measured Avogadro's number. Moreover, in 1919, Perrin was the first to put forward the hypothesis that the transformation of hydrogen into helium was at the origin of energy radiated by the Sun. He was awarded the Nobel Prize in Physics 1926 for his work on the discontinuity of matter and, more specifically, for his discovery of sedimentation equilibrium.

Box 1.2. *Stoney (1826–1911); Thomson (1856–1940); Perrin (1870–1942)*

1.1.3. *Millikan experiment*

In 1909 [MIL 10, ROU 60, CAR 79, PER 95, SAK 10, SAK 11], Robert Andrews Millikan began measuring the *elementary electrical charge, e.* The experimental set-up used by Millikan can be schematically presented as indicated in Figure 1.5(a) [SAK 10].

Droplets assimilated to small, homogeneous spheres of radius r, mass m and, in small numbers, are obtained by spraying oil between two plates, P_1 and P_2, through a small orifice in the upper plate, P_1. Between P_1 and P_2, the droplets encounter ions produced by radiation from a source, S (X-ray tube, radium bulb, etc.). Millikan's simplified experimental device is illustrated in Figure 1.5(a).

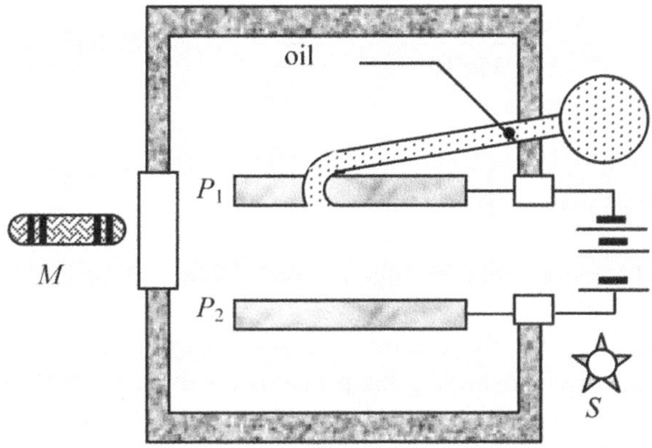

Figure 1.5a. *Millikan's simplified experimental device*

From time to time, an ion attaches itself to one of the droplets. The latter is then electrified and is subjected to the double action of the antagonistic gravity field and electric field reigning between the two plates. The velocity assumed by an electrified droplet depends on its charge, q, its mass, m, its radius, r, and the viscosity coefficient, η, of the air. The motion of the droplets is observed by means of a microscope, M. When the velocity of a droplet changes abruptly, it means it has attached an ion. Measuring this velocity for various electric field values (including zero field) and knowing η then enables the absolute value $|q|$ of the charge carried by a droplet to be measured, and the value of the elementary electrical charge, e, to be deduced. The value of the elementary electrical charge measured by Millikan is $(4.77 \pm 0.009)\ 10^{-10}$ uemcgs [BIS 19]. This corresponds to the international system value, 1.592×10^{-19} coulombs [BIS 19].

Today, more meticulous experiments produce a far more accurate value than that obtained by Millikan. Using glycerin as oil, the experiment shows that for several observations in an electric field of intensity E, the velocities, v, acquired by the droplets are distributed according to an arithmetic progression of difference $\Delta v \approx 3 \times 10^{-5}$ m $\cdot$ s $^{-1}$ [SAK 10]. When a droplet attaches the smallest charge, e, its velocity varies by Δv. The elementary electrical charge is then given by the expression:

$$e = \frac{6\pi\eta\, r\Delta v}{E} \qquad [1.6]$$

Numerically, for $\eta = 1.83 \times 10^{-5}$ SI; $E = 93.5$ kV $\cdot$ m^{-1} 1; $r = 1.45$ µm [SAK 10], we obtain, knowing that $\Delta v \approx 3 \times 10^{-5}$ m $\cdot$ s^{-1}:

$$e = \frac{6\pi \times 1.83 \times 10^{-5} \times 1.45 \times 10^{-6} \times 3 \times 10^{-5}}{9.35 \times 10^{4}} = 1.604833047 \times 10^{-19}$$

Thus:

$$e = 1.604833047 \times 10^{-19} \text{ C}$$

The CODATA recommended value is $1.602179487(40)10^{-19}$ C.

Using value [1.1] for the electron mass-to-charge ratio, the mass, m, of the electron is deduced by considering the previous result: $E = 1.604833047 \times 10^{-19}$ C, which gives $m = 9.11836958 \times 10^{-31}$ kg. The CODATA recommended value is $9.10938215(45)\ 10^{-31}$ kg.

Robert Andrew Millikan was an American physicist. He is best known for his experimental work on the drop of oil sprayed by X-rays, which allowed him to measure elementary electrical charge in 1909. He was awarded the Nobel Prize in Physics 1923 for his work on the "elementary charge of electricity and the photoelectric effect".

Box 1.3. *Millikan (1868–1953)*

APPLICATION 1.2.– Let us consider Figure 1.5(b). When a droplet of radius r and charge q is in motion in the electric field space, demonstrate relationship [1.6]. We will make use of the principle of inertia and will take account of the fact that when a droplet attaches the lowest charge, e, its velocity varies by Δv. The intensity of the *Stokes force* acting on a droplet of velocity v is given: $f = 6\pi\eta r v$, with η the viscosity of the air and r the radius of a droplet. We will produce a diagram showing the forces acting on the droplet concerned.

ANSWER.– Let us consider a droplet of charge q and mass m. It is subjected to its weight, $\vec{P}$, the electrical force, $\vec{F}$, and the Stokes force, $\vec{f}$. With the gravity field and the electric field between the two plates being antagonistic, the electric field is directed upwards (the droplets move from P_1 to P_2 and in the same direction as the Stokes force (Figure 1.5(b)).

With the principle of inertia being verified, the motion of a droplet is uniform and rectilinear along the vertical, that is:

$$\vec{P}+\vec{F}+\vec{f}=m\vec{g}+q\vec{E}-6\pi\eta r\vec{v}=\vec{0} \qquad [1.7]$$

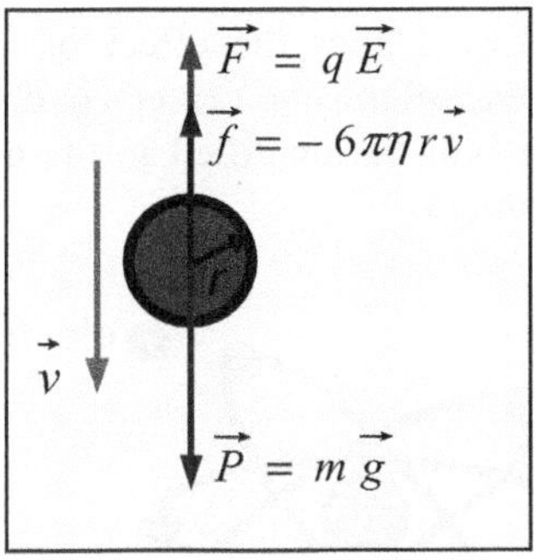

Figure 1.5b. *Uniform drop motion of a droplet of charge q, mass m and radius r. $\vec{P}$: weight, $\vec{F}$: electrical force, $\vec{f}$: Stokes force*

By projecting [1.7] in the direction of the velocity, we obtain:

$$mg - qE - 6\pi\eta rv = 0$$

Which then gives:

$$qE = mg - 6\pi\eta rv \qquad [1.8]$$

From a physical point of view, the discontinuous variations, Δv, in the velocity limit of a droplet correspond to discontinuous variations, Δq, in the charge, q, carried by the droplet. Equation [1.8] then gives:

$$(q + \Delta q)E = mg - 6\pi\eta r\,(v + \Delta v) \Rightarrow |\Delta q| = e = 6\pi\eta r\Delta v/E$$

[1.6] is indeed found.

NOTE.– The following reasoning can be adopted. When the velocity of a droplet varies from v to v', then equation [1.8] is written $qE = 6\pi\eta rv' - 6\pi\eta r\ rv' = 6\pi\eta r\,(v' - v)$. Let: $|q|E = 6\pi\eta r\Delta v$, with $\Delta v = |v' - v|$. Yet the amount of electricity $|q| = ne$. Hence $neE = 6\pi\eta r\Delta v$. Given that the variation in velocity, Δv, is due to the attaching of the elementary charge, then $n = 1$. We thus obtain: $e = 6\pi\eta r\Delta v/E$.

1.2. The birth of the nucleus

1.2.1. *Perrin and Thomson atomic model*

In 1901, Perrin envisioned the atom as a miniature solar system where electrons acting like planets gravitate freely around a positively charged, quasi-point-like center of matter (Figure 1.6). Under the effect of attractive electrostatic force, electrons turn around the quasi-point-like center via elliptic trajectories [DUM 15]. This quasi-point-like center will be identified in the atomic nucleus following the Rutherford scattering experiment.

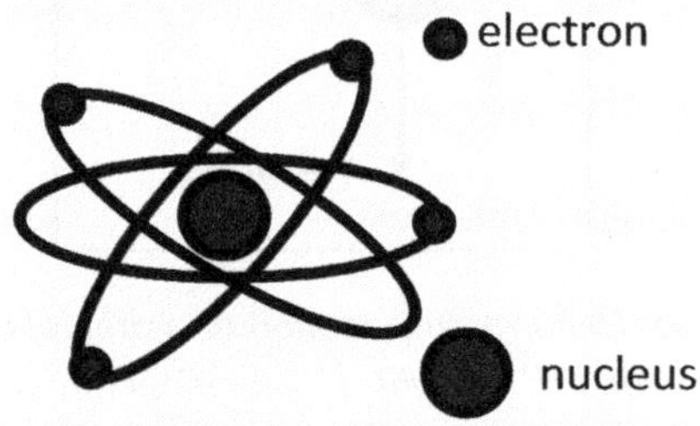

Figure 1.6. *Planetary model of the atom, envisioned by Perrin in 1901*

Through the experiments of Thomson and Millikan, it became known that in addition to atoms, there existed a much smaller particle, called an electron. The internal structure of the quasi-point center in the Perrin atomic model had not yet been elucidated.

In 1902, Thomson envisioned an atomic model that was different to that of Perrin by developing the "raisin bread" theory on the atomic structure [ROU 60, SAK 11]. In this model, electrons are considered to be negative raisins distributed throughout bread in a positive matter, hence this model is known as the plum pudding atom, as shown in Figure 1.7 [SAK 11].

But because of the positive electron–substance attraction and Coulomb electron–electron repulsion forces, the atomic matter as described is unstable. The Thomson atomic model is therefore not suitable, as confirmed by Rutherford in 1911 (see section 1.2.3).

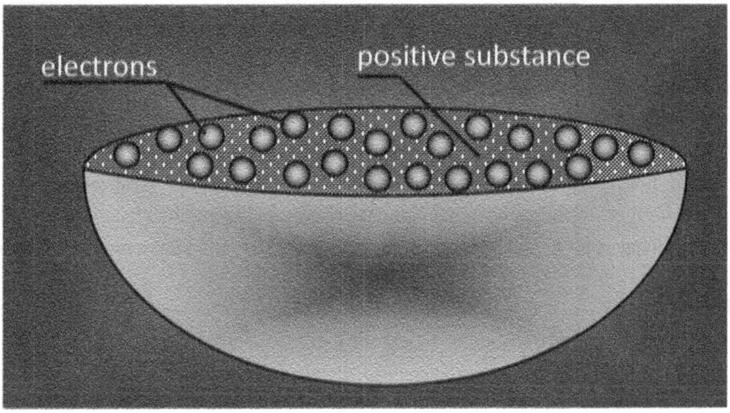

1.2.2. G

In 19 … carried
out *parti* … ils. The
simplifie … resented
below (F … ingle, θ,
of partic

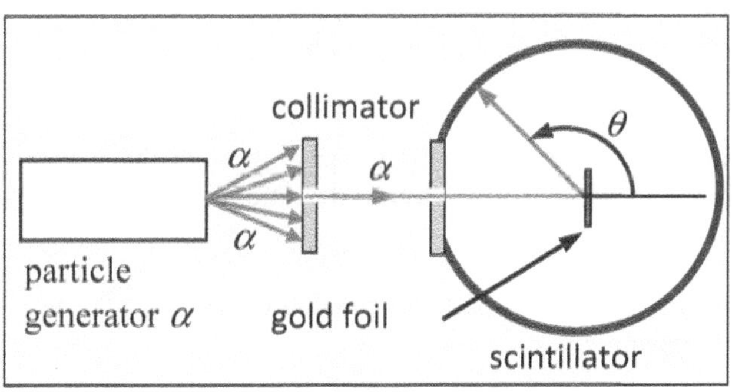

Figure 1.8. *Simplified experimental device by Geiger and Marsden*

The Geiger and Marsden experiments showed that one in 8,000 α particles is scattered at an angle θ greater than 90°. These observations contradicted the Thomson model (Figure 1.7), which predicted small angular deflections by simple scattering, and a very low probability of large multiple scattering deflection.

Johannes Wilhelm "Hans" Geiger was a German physicist. In 1928, together with German physicist Walther Müller (1905-1979), he invented the particle detector known as the Geiger–Müller counter, the operating principle of which he devised in 1913. Geiger is also famous for having established, together with English physicist John Mitchell Nuttall (1890-1958), the law giving the rate of decay as a function of time, known as the Geiger–Nuttall law (see Chapter 3). In addition, in 1908 under Rutherford's supervision (see Box 1.5), Geiger, together with Marsden, carried out the α particle scattering experiment with thin gold strips, which enabled Rutherford to devise the planetary model of the atom.

Sir Ernest Marsden was an English-New Zealand physicist. He is known for his experimental work on α particle scattering by thin gold blades.

Box 1.4. *Geiger (1882–1945); Marsden (1889–1970)*

1.2.3. *Rutherford scattering: Planetary atomic model*

In 1911, Ernest Rutherford [RUT 11, SIV 86, EVA 61, GUY 03, STÖ 07, SAK 11, DUM 15, SAK 19] used radioactive radiation consisting of α particles or helion from a radium source to bombard thin metal foils. This experiment, known as *Rutherford scattering*, enabled him to explain Geiger and Marsden's experimental observations, and find that the atomic structure could not be represented in the static form that Thomson had envisaged in 1902. The experimental set-up created by Rutherford is schematically presented as indicated in Figure 1.9 [SAK 19].

Rutherford's experience makes it possible to make at least two important observations:

First observation: numerous particles pass through the matter without being deflected (although theoretically several deviations should have been observed according to Thomson's model).

Second observation: alpha particles passing close to the "center" of matter are deflected at a large angle.

The first observation allows us to conclude that the positive particles scattered in the Thomson model are concentrated at the "center" of the matter. The second observation, meanwhile, proves that the "center" of the matter pushing alpha particles is a positively charged point-like particle. Rutherford demonstrated that an α particle can be obtained by twice ionizing a helium atom: $He \rightarrow He^{2+} + 2e^{-}$. The α particle is a He^{2+} helium nucleus. The positive "center" of matter was identified as the atomic nucleus. Drawing on the astronomical model of Johannes Kepler

(1571-1630) of the solar system, Rutherford proposed the model of the planetary atom in which electrons gravitate around the nucleus, as shown in Figure 1.10 [SAK 11].

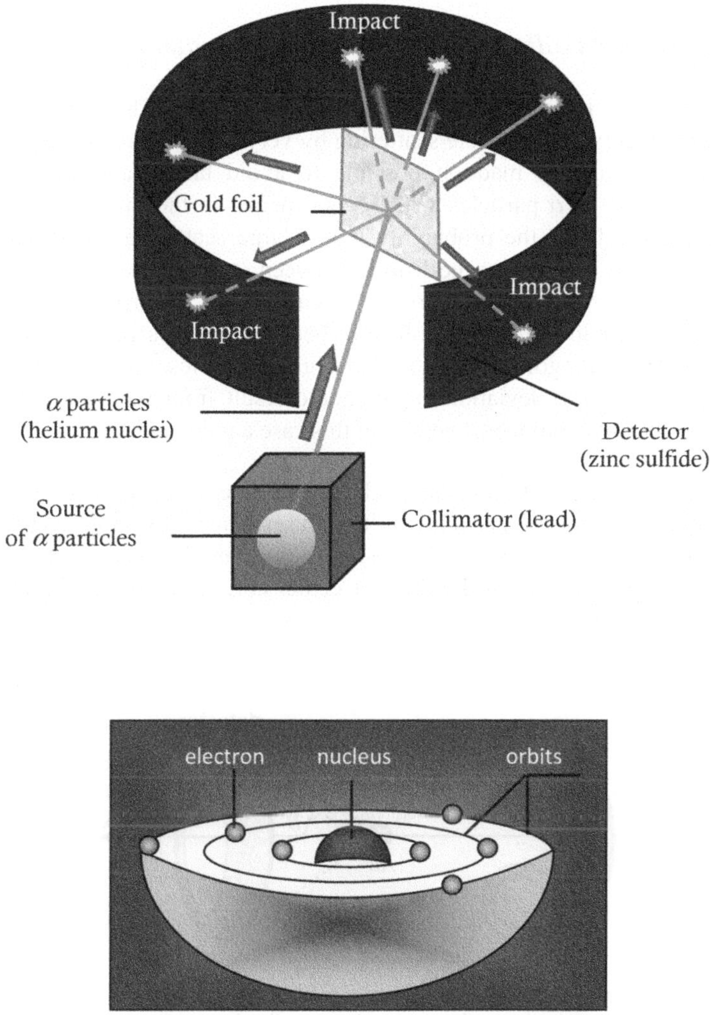

Figure 1.10. *Planetary atomic model envisaged by Rutherford*

Note that the planetary model is unstable. The electrons, as charged particles, are subjected to a centripetal acceleration due to their orbital motion. Yet, according to predictions of classical electrodynamics, any charged particle subjected to acceleration loses energy by radiation. In the planetary model, electrons ought to lose energy through radiation and eventually fall onto the nucleus in around 10^{-11} s (see exercise 1.7.1). Yet that does not happen.

1.2.4. *Rutherford's differential effective cross-section*

To theoretically interpret these experimental observations, Rutherford developed a quantitative theory of α particle scattering by very thin gold foils, 10^{-5} to 10^{-4} cm thick. These thicknesses made it possible to avoid taking account of potential multiple collisions of α particles with several nuclei in the interpretation of large scattering angles. Thus, the probability that a large scattering angle would result from two or more successive collisions between α particles and gold nuclei is entirely negligible. Moreover, due to low electron mass, the probability that a large scattering angle would result from collisions between the α particles and electrons in the electron cloud of a gold nucleus is also negligible. Thus, under the conditions of the experiment, a large deviation angle should result from a collision between an α particle and a single point-like center, in this case a gold nucleus.

In the general case, let us consider the elastic interaction process of a constant velocity beam of identical particles by a diffuser center (example target nucleus) placed at the origin, *O*, of the coordinates. By definition, the *impact parameter*, *b*, of a particle is the distance from the diffuser center core to the initial direction of the particle [SIV 86, STÖ 07, DUM 15]. The geometry of Rutherford's theoretical model for studying classical scattering is shown in Figure 1.11.

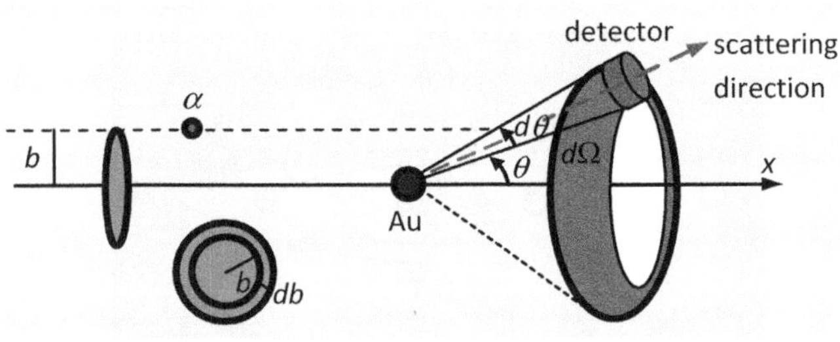

Figure 1.11. *Scattering of a beam of α particles of an angle, θ, in the solid angle, dΩ, b is the impact parameter*

Rutherford's differential effective cross-section is the theoretical expression subject to experimental verification. Let us briefly review some useful definitions.

The *solid angle, d*Ω, between the two cones of revolution, with the same vertex, *O*, the same axis, *Ox,* and with semi-vertex angles, θ and $\theta + d\theta$, is:

$$d\Omega = 2\pi \sin\theta \, d\theta \qquad [1.9]$$

The particles scattered at a larger angle are those that are deflected under the angles between θ and $\theta + d\theta$ with an impact parameter between b and $b + db$ (the scattering angle is maximum for $b = 0$). With the effective cross-section having the dimension of a surface, let us consider a disk of radius b and axis of revolution *Ox*. The surface of this disk is equal to the *effective scattering cross section* in the direction of angle θ, noted σ, that is: $\sigma = \pi b^2$. The *elementary effective scattering cross-section*, $d\sigma$, is then given by the relationship:

$$d\sigma = 2\pi b db \qquad [1.10]$$

By definition, the *differential effective cross-section* is equal to $\dfrac{d\sigma}{d\Omega}$. Using the definition relationships [1.9] and [1.10], we then obtain:

$$\frac{d\sigma}{d\Omega} = \left| \frac{b}{\sin\theta} \frac{db}{d\theta} \right| \qquad [1.11]$$

If we use ze to designate the *charge of the incident particle of kinetic energy*, E, and Ze for the charge of the target nucleus, then the impact parameter is equal to:

$$b = \frac{1}{4\pi\varepsilon_0} \frac{zZe^2}{2E} \frac{1}{\tan(\theta/2)} \qquad [1.12]$$

Using [1.12], Rutherford's differential effective cross-section is then written according to [1.11]:

$$\frac{d\sigma}{d\Omega} = \left(\frac{1}{4\pi\varepsilon_0} \right)^2 \left(\frac{zZe^2}{4E} \right)^2 \frac{1}{\sin^4(\theta/2)} \qquad [1.13]$$

The effective cross-section is expressed in *barn* (b) with 1 barn = 10^{-28} m^2 and the solid angle in steradians (sr). The differential cross section [1.13] is then expressed in barn per steradian ($\text{b} \cdot \text{sr}^{-1}$).

APPLICATION 1.3.– Demonstrate expression [1.13]. Perform the numerical application for α particles of total energy $E = 5.5$ MeV, scattered under an angle of 90° by a gold foil with atomic number $Z = 79$. We will take $e = 1.6 \times 10^{-19}$ C; 1 eV = 1.6×10^{-19} J and $(1/4\pi\varepsilon_0) = 9.0 \times 10^9$ SI.

ANSWER.– Using [1.12], we obtain:

$$\frac{db}{d\theta} = \frac{1}{4\pi\varepsilon_0}\frac{zZe^2}{2E}\frac{d}{d\theta}\left(\frac{\cos(\theta/2)}{\sin(\theta/2)}\right) = -\frac{1}{4\pi\varepsilon_0}\frac{zZe^2}{4E}\frac{1}{\sin^2(\theta/2)} \quad [1.14]$$

Considering that $\sin\theta = 2\sin(\theta/2)\cos(\theta/2)$, we obtain, by considering [1.12]:

$$\frac{b}{\sin\theta} = \frac{1}{4\pi\varepsilon_0}\frac{zZe^2}{2E}\frac{\cos(\theta/2)}{\sin(\theta/2)} \times \frac{1}{2\sin(\theta/2)\cos(\theta/2)}$$

Let:

$$\frac{b}{\sin\theta} = \frac{1}{4\pi\varepsilon_0}\frac{zZe^2}{4E}\frac{1}{\sin^2(\theta/2)} \quad [1.15]$$

Using [1.14] and [1.15] gives:

$$\frac{b}{\sin\theta} \times \frac{db}{d\theta} = -\frac{1}{4\pi\varepsilon_0}\frac{zZe^2}{4E}\frac{1}{\sin^2(\theta/2)} \times \frac{1}{4\pi\varepsilon_0}\frac{zZe^2}{4E}\frac{1}{\sin^2(\theta/2)}$$

Considering the absolute value of this equation, we find [1.13].

For α particles, $z = 2$, we find, numerically:

$$\frac{d\sigma}{d\Omega} = (9\times10^9)^2\left(\frac{2\times79\times1.6\times10^{-19}}{4\times5.5\times10^6}\right)^2 \frac{1}{\sin^4(\pi/4)}$$

$$= 4.278 \times 10^{-28} = \text{b} \cdot \text{sr}^{-1}$$

We then find:

$$\frac{d\sigma}{d\Omega} = 4.28 \text{ b} \cdot \text{sr}^{-1}$$

APPLICATION 1.4.– The elastic scattering of an α particle of velocity $\vec{v}$ and kinetic energy E = 6.0 MeV is achieved with a nucleus of gold, Au, that is assumed to be immobile. For a frontal collision, a *minimum approach distance* of $a_{\min}$ = 3.8 × 10^{-12} cm is found. Produce a diagram and then deduct the atomic number of the gold nucleus from the data. We will assume that the α particle comes from the infinite where the potential energy is zero. We will also assume that the $\{\alpha$ particle – Au nucleus$\}$ system is conservative. The data given for application 1.3 will be used.

ANSWER.– The frontal collision corresponds to a zero-impact parameter (b = 0). We thus obtain the diagram shown below (Figure 1.12).

Rutherford scattering is known to be studied in the context of classical mechanics (the resting energy, m_0c^2, of the particles in the system studied is therefore not taken into account). At infinity, the mechanical energy of the $\{\alpha - \text{Au}\}$ system is equal to the kinetic energy, E, of the α particle. At the distance $r = a_{\min}$, the α particle stops then turns back.

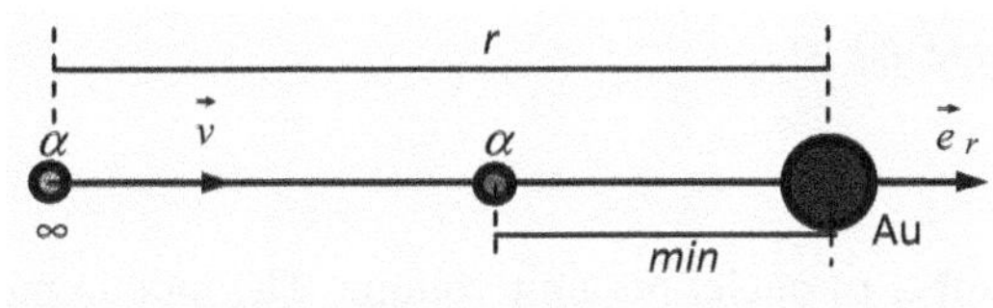

Figure 1.12. *Frontal collision between an α particle and an immobile nucleus of gold, Au*

The mechanical energy of the $\{\alpha - \text{Au}\}$ system is reduced to its potential energy:

$$V(r) = k\frac{zZe^2}{r} \qquad [1.16]$$

In relationship [1.16], the electrical constant $k = 1/4\pi\varepsilon_0 = 9 \times 10^9$ SI. With the $\{\alpha - \text{Au}\}$ system being conservative, we obtain:

$$E = V(r_{\min}) = k\frac{zZe^2}{a_{\min}} \Rightarrow Z = \frac{a_{\min}E}{kze^2} \qquad [1.17]$$

Numerically we obtain from [1.17]:

$$Z = \frac{3.8\times10^{-14}\times6\times10^6}{9\times10^9\times2\times1.6\times10^{-19}} = 79.17 \approx 79$$

Let us now seek to establish an expression deduced from Rutherford's formula [1.13], subject to experimental verification.

If ϕ designates the *α particle flow*, the number, dN_1, of particles scattered by a nucleus per unit time in the solid angle $d\Omega$ is:

$$dN_1 = \phi d.\sigma \tag{1.18}$$

If I is the intensity of a parallel plane beam of α particles falling on the nucleus, or the number of particles passing per unit time through a surface, S, normal to the beam, then:

$$\phi = \frac{I}{S} \tag{1.19}$$

According to [1.19], the flow of ϕ particles is expressed in $s^{-1} \cdot cm^{-2}$.

Using [1.19], the number, dN_1, of particles scattered per unit time and per unit area in the solid angle $d\Omega$ is:

$$dN_1 = Id\sigma \tag{1.20}$$

Let n be the number of nuclei per unit volume. The total number of target nuclei in volume V is then equal to Vn. The total number, dN, of particles scattered per unit time and per unit area in the solid angle $d\Omega$ is then equal to $VndN_1$. Thus, using [1.20]:

$$dN = IVnd\sigma \tag{1.21}$$

By considering Rutherford's formula [1.13], we then obtain:

$$dN = IVn\left(\frac{1}{4\pi\varepsilon_0}\right)^2\left(\frac{zZe^2}{4E}\right)^2\frac{d\Omega}{\sin^4(\theta/2)} \tag{1.22}$$

Let us compare the theoretical predictions according to the Rutherford model [1.22] with the experimental observations of Geiger and Marsden for α particles of energy 5.5 MeV, scattered by gold nuclei. The resulting curves are shown in Figure 1.13. Overall, it is noted that for scattering angles less than $140°$, the number of α particles scattered as a function of the scattering angle is compatible with the model of the nucleus of highly-concentrated charge density at a point in space. This constitutes experimental evidence of the validity of Rutherford's point-like nucleus model. Moreover, this concordance between theory and experience confirms the planetary atomic model according to Rutherford (Figure 1.10) and overturns the

Thomson atomic model (Figure 1.7). However, the nucleus is not point-like. It has a skin thickness indicating that the mass distribution is not uniform in a nuclear volume of spherical form (see section 1.4.2).

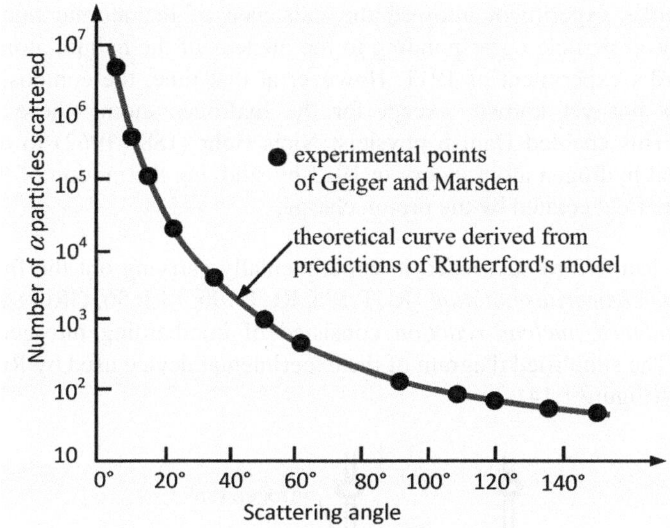

Figure 1.13. *Comparisons of theoretical predictions according to the Rutherford model with the experimental observations of Geiger and Marsden*

Lord Ernest Rutherford of Nelson was a New Zealand-born British physicist and chemist. Rutherford is considered the father of nuclear physics for his notable discoveries in this field. He discovered α radiation (helium nuclei) and β^- radiation (electron). Rutherford also discovered that radioactivity is accompanied by the decay of chemical elements. This won him the Nobel Prize in Chemistry 1908. In 1909, he established, along with British radiochemist Frederick Soddy (1877-1956), the experimental law of radioactive decay of radioelements. Drawing on his past experience of the gold-leaf α particle scattering, achieved by Geiger and Marsden, Rutherford revealed the existence of the atomic nucleus in 1911. This allowed him to put forward the planetary atomic model. Moreover, in 1919, he carried out the first artificial nuclear transmutation reaction, thus paving the way for the study of the properties of the nuclear reactions induced.

Box 1.5. *Rutherford (1871–1937)*

1.3. Composition of the nucleus

1.3.1. *Discovery of the proton*

Rutherford's experiment allowed the existence of the atomic nucleus to be revealed. The α particle corresponding to the nucleus of the helium atom was used in Rutherford's experiment of 1911. However at that time, the composition of the nucleus was not yet known, except for the hydrogen atom whose nucleus is the proton. This enabled Danish physicist Niels Bohr (1885-1962) to develop his semi-classical hydrogen atom theory in 1913 by studying the motion of the electron in the electric field created by the proton charge.

In 1919, Rutherford succeeded in experimentally carrying out the first *reaction of artificial nuclear transmutation* [RUT 19a, RUT 19b, LEP 56, GRO 85, NES 17]. This first *induced nuclear reaction* consisted of bombarding nitrogen gas with α particles. The simplified diagram of the experimental device used by Rutherford is given below (Figure 1.14).

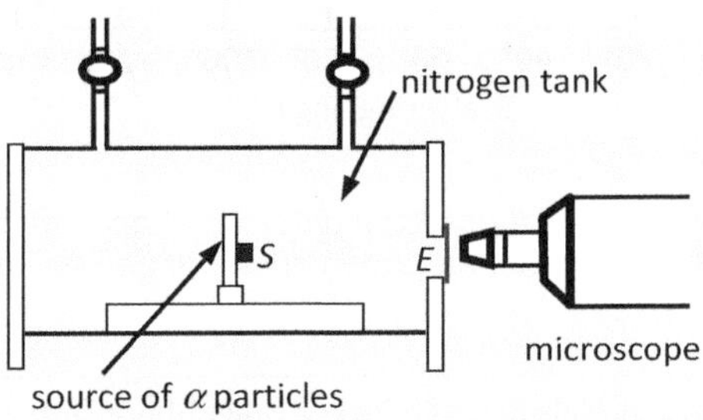

Figure 1.14. *Rutherford's simplified experimental device for conducting the first induced nuclear reaction*

Using an oscilloscope, Rutherford observed scintillations on the zinc sulfide-coated E-screen due to particles emitted as a result of the collision between the α particles and the nitrogen nuclei. He deduced from his observations that in striking nitrogen atoms, alpha particles produced a proton. In 1920, at a conference of the British Association for the Advancement of Science, Rutherford suggested that the nucleus of the hydrogen atom be termed the proton [KRI 19a]. However, Rutherford did not know what had become of the residual nucleus. Thus, he tasked Patrick Blackett (1897-1974), a researcher working under his direction, with identifying the residual nucleus [KRI 19b]. Blackett proceeded to conduct a series of experiments with a Wilson chamber (or fog chamber) and confirmed in 1925 that the particles emitted in collisions between α particles and nitrogen nuclei are ^1_1H protons [BLA 25]. The nuclear transmutation reaction is thus written:

$$^4_2He + ^{14}_7N \rightarrow ^1_1H + ^{17}_8O \qquad [1.23a]$$

The reaction shows the transmutation of the nitrogen nucleus 14 to the oxygen nucleus 17 with production of a proton. For this reason, Rutherford is credited with the discovery of the proton. Between 1930 and 1932, a series of experiments conducted on the study of nuclear transmutation reactions led to the birth of the neutron, the proton's natural companion in the nucleus.

In 1930, German physicist Walther Bothe and his student Herbert Becker conducted transmutation experiments by bombarding atomic targets with α particles [BOT 30, NES 17]. At that time, the available sources of alpha radiation were radium 226 (α of 4.9 MeV) and its descendant, radon 222 (α of 5.6 MeV) or, for polonium 210, emitting α particles of 5.4 MeV.

Using beryllium bombarded by α particles from a source of polonium-210 as a target, the reaction produced penetrating radiation. The latter was interpreted by Bothe and Becker as electromagnetic radiation like the γ radiation known at this time. Thus, they opposed that this radiation was produced by the deexcitation of radiocarbon 13 formed during the transmutation, according to the equation:

$$^4{}_2He + ^9{}_4Be \rightarrow ^{13}{}_6C^* \rightarrow ^{13}{}_6C + \gamma \qquad [1.23b]$$

Bothe and Becker evaluated γ-ray energy by interposing lead plates of varying thicknesses between the target and the counter. By measuring the attenuation of the gamma-ray flux of known energy, they estimated the energy of the γ photons at around 5 MeV. Unfortunately, this energy was higher than the usual γ photons produced by radioactivity, which are of the order of 1 to 2 MeV.

Intrigued by these results, Irène and Frédéric Joliot-Curie [CUR 32b, CUR 32c] conducted transmutation experiments of boron and beryllium bombarded by α particles, with a view to understanding the real nature of what is known as Bothe–Becker radiation. These experiments showed that the penetrating radiation observed in the Bothe and Becker experiments could eject protons from paraffin. Yet Irène and Frédéric Joliot-Curie did not interpret their observations correctly as they believed that the Compton effect phenomenon had occurred between the light element and gamma rays, whose energy they estimated. At the time, there was no knowledge of any penetrating radiation capable of ejecting particles other than electrons by photoelectric or Compton effect.

In 1932, Sir James Chadwick offered the correct interpretation of Bothe and Becker's observations by conducting a decisive experiment that would solve the enigma of the penetrating radiation observed in the transmutation reactions of boron and beryllium.

Patrick Maynard Stewart Blackett was a British experimental physicist. He is famous for his experiments on the induced nuclear reactions that made it possible, in 1925, to identify the residual nucleus (oxygen) in the transmutation reaction initiated by Rutherford in 1919. He was awarded the Nobel Prize in Physics 1948 for his development of the Wilson cloud chamber and for his discoveries in the fields of nuclear physics and cosmic rays.

Walter Wilhelm Georg Bothe was a German physicist. He is most famous for his contributions to nuclear transmutation reactions. He won half of the Nobel Prize in Physics 1954 for designing the coincidence method (with the other half awarded to the English physicist, Max Born (1882-1970). In nuclear physics, the principle of the coincidence method consists of using sensitive devices to detect the simultaneous (or near-simultaneous) emission of two particles, each collected in a different counter. For more details on this subject see [VUC 55].

Box 1.6. *Blackett (1897–1974); Bothe (1891–1954)*

1.3.2. *Discovery of the neutron*

As noted above, the aim of the Chadwick experiment was to identify radiation penetrating the boron and beryllium transmutation reactions. The experimental set-up used by Chadwick is schematically presented in Figure 1.15.

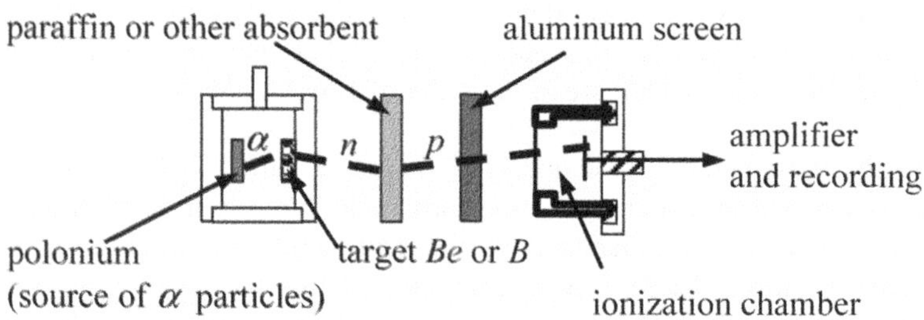

Figure 1.15. *Chadwick's experimental set-up*

Polonium, a source of α particles of 5.3 MeV is introduced into a chamber with a high vacuum. The collision between the α particles and the beryllium producing penetrating radiation (here neutron n to be identified) in all directions. This radiation then bombards a paraffin screen, inducing the ejection of protons (p). These protons then pass through thin aluminum screens to access an ionization chamber filled with nitrogen gas. The pulse produced by protons and recoil nitrogen nuclei is amplified and recorded. The observed pulses corresponding to protons penetrating into the ionization chamber. The track of the recoil protons is then measured in the aluminum screens interposed between the detector and the paraffin.

In 1920, the concept of the neutron consisting of an intimate combination of a proton and an electron was discussed by Rutherford. But this combination was never confirmed theoretically or experimentally. However, interpreting these observations, Chadwick showed that the measured recoil energies were compatible with the neutron n hypothesis [CHA 32a, CHA 32b, NES 17]. On the basis of this hypothesis, the bombardment of beryllium or boron target nuclei with α particles produced two types of induced nuclear reactions: the ^{9}Be $(\alpha, n)^{12}$C reaction and the ^{11}B $(\alpha, n)^{14}$N reaction. These reactions are explicitly written:

$$^{4}_{2}He + {}^{9}_{4}Be \rightarrow {}^{1}_{0}n + {}^{12}_{6}C \qquad [1.24]$$

$$^{4}_{2}He + {}^{11}_{5}B \rightarrow {}^{1}_{0}n + {}^{14}_{7}N \qquad [1.25]$$

Since the nucleus mass of beryllium 9 was not known in 1932, reaction [1.25] was used by Chadwick. By measuring the energy released by this reaction and using the available masses of ^{11}B, ^{14}N, and ^{4}He, Chadwick found a neutron mass, m_n, between 1.005 u and 1.008 u, i.e. [EVA 61]:

$$1.005 \text{ u} \leq m_n \leq 1.008 \text{ u} \qquad [1.26a]$$

The CODATA recommended value is 1.00866491597(43) u. Although not precise, the Chadwick result is acceptable seeing as the upper limit of neutron mass only differs from the CODATA recommended value by around 0.0007 u.

Chadwick found the neutron he had been seeking for 12 years. He announced his results in a brief note, "Possible Existence of a Neutron", in the journal *Nature* on February 17 (published on February 27) and detailed his experiments in a longer – and more affirmative – article, "The Existence of a Neutron" on May 10, published on July 1 in *Proceedings of the Royal Society*. Note that he does not speak of "the" neutron, but "a" neutron. The text published by Chadwick is presented in box 1.7. However, it was not immediately accepted that Chadwick had discovered the neutron. Was it a real particle, like the proton or the electron? The problem was solved in 1933 by Chadwick and the American physicist, Maurice Goldhaber (1911-2011).

In 1931, American chemist Harold Urey (1893-1981) discovered deuterium by demonstrating with his colleagues the existence of heavy water (deuterium oxide, D_2O, a discovery that saw him awarded the Noble Prize in Chemistry 1934). In 1933, Chadwick and Goldhaber succeeded in achieving deuteron photodissociation (deuterium nucleus). By bombarding the deuteron with γ photons of 2.6 MeV, they measured the energy of the protons released. Chadwick and Goldhaber thus determined that the deuteron binding energy was 2.2 MeV. Subsequently, they estimated the neutron mass by drawing on the principle of energy conservation. Let:

$$m_{\mathrm{n}}c^2 = m_{\mathrm{d}}c^2 - m_{\mathrm{p}}c^2 + E_{\mathrm{l}} \qquad [1.26b]$$

In equation [1.26b] m_{n} is the neutron mass, m_{d} designates the deuteron mass ($m_{\mathrm{d}}c^2 = 1876.1$), m_{p} represents the proton mass ($m_{\mathrm{p}}c^2 = 938.3$), and E_{l} designates the deuteron binding energy ($E_{\mathrm{l}} = 2.2$ MeV). Numerically we find:

$$m_{\mathrm{n}}c^2 = 1876.1 - 938.3 + 2.2 = 940 \text{ MeV}. \qquad [1.26c]$$

The accepted value of the precise neutron mass equals 939.565346 MeV/c^2 (see Table 1.3). This value is in close accordance with result [1.26c] by Chadwick and Goldhaber, since the mass difference is 0.434654 MeV/c^2.

In 1933, the neutron was accepted as a fundamental particle constituting atomic nuclei.

Possible Existence of a Neutron

It has been shown by Bothe and others that beryllium when bombarded by α-particles of polonium emits a radiation of great penetrating power, which has been an absorption coefficient in lead of about 0.3 (cm)$^{-1}$. Recently Mme. Curie-Joliot and M. Joliot found, when measuring the ionization produced by this beryllium radiation in a vessel with a thin window, that the ionization increased when matter-containing hydrogen was placed in front of the window. The effect appeared to be due to the ejection of protons with velocities up to a maximum of nearly 3 x 109 cm. per sec. They suggested that the transference of energy to the proton was by a process similar to the Compton effect, and estimated that the beryllium radiation had a quantum energy of 50 x 106 electron volts. I have made some experiments using the valve counter to examine the properties of this radiation excited in beryllium. The *valve counter consists of a small ionization chamber connected to an amplifier, and the sudden production of ions by the entry of a particle, such as a proton or α-particle, is recorded by the deflexion of an oscillograph. These experiments have shown that the radiation ejects particles from hydrogen, helium, lithium, beryllium, carbon, air, and argon. The particles ejected from hydrogen behave, as regards range and ionizing power, like protons with speeds up to about 3.2 x 10^9 cm. per sec. The particles from the other elements have a large ionizing power, and appear to be in each case recoil atoms of the elements.*

If we ascribe the ejection of the proton to a Compton recoil from a quantum of 52 x 10^6 electron volts, then the nitrogen recoil atom arising by a similar process should have an energy not greater than about 400,000 volts, should produce not more than about 10,000 ions, and have a range in air at N.T.P. of about 1.3 mm. Actually, some of the recoil atoms in nitrogen produce at least 30,000 ions. In collaboration with Dr. Feather, I have observed the recoil atoms in an expansion chamber, and their range, estimated visually, was sometimes as much as 3 mm at N.T.P. These results, and others I have obtained in the course of the work, are very difficult to explain on the assumption that the radiation from beryllium is a quantum radiation, if energy and momentum are to be conserved in the collisions.

The difficulties disappear, however, if it be assumed that the radiation consists of particles of mass 1 and charge 0, or neutrons. The capture of the α particle by the Be^9 nucleus may be supposed to result in the formation of a C^{12} nucleus and the emission of the neutron. From the energy relations of this process the velocity of the neutron emitted in the forward direction may well be about 3 x 10^9 cm. per sec. The collisions of the neutron with the atoms through which it passes give rise to the recoil atoms, and the observed energies of the recoil atoms are in fair agreement with this view. Moreover, I have observed that the protons ejected from hydrogen by the radiation emitted in the opposite direction to that of the exciting α-particle appear to have a much smaller range than those ejected by the forward radiation. This again receives a simple explanation of the neutron hypothesis.

If it be supposed that the radiation consists of quanta, then the capture of the α-particle by the Be9 nucleus will form a C^{13} nucleus. The mass defect of C^{13} is known with sufficient accuracy to show that the energy of the quantum emitted in this process cannot be greater than about 14 x 10^6 volts. It is difficult to make such a quantum responsible for the effects observed. It is to be expected that many of the effects of a neutron in passing through matter should resemble those of a quantum of high energy, and it is not easy to reach the final decision between the two hypotheses. Up to the present, all the evidence is in favor of the neutron, while the quantum hypothesis can only be upheld if the conservation of energy and momentum be relinquished at some point.

J. Chadwick, Cavendish Laboratory, Cambridge, Feb. 17. [CHA 32b]

Sir James Chadwick was a British physicist. He is particularly well known for his historical experiment that enabled the discovery of the neutron in 1932. This discovery saw him receive the Nobel Prize in Physics 1935. This discovery permitted an understanding of the internal structure of nuclei and paved the way for the study of various induced nuclear reactions, such as fission.

Box 1.7. *Chadwick (1891–1974)*

1.3.3. *Internal structure of nucleons:* u *and* d *quarks*

Aside from the electrical charge, the proton and neutron are spin-1/2 fermions like the electron. Their mass difference is equal to 0.00138 u. Thus the idea of German physicist Werner Heisenberg (1901-1976) was accepted, namely that proton and neutron are the two possible states of the same heavy particle, called the *nucleon*. Thus, the atomic nucleus consists of nucleons, i.e. the protons and neutrons. The nucleus of the hydrogen atom consists of a single nucleon, the proton. The α particle corresponding to the nucleus of the helium atom consists of four nucleons: two protons and two neutrons.

With Chadwick's discovery of the neutron in 1932, the internal structure of the nucleus was established. But the internal structure of nucleons remained a mystery. What do they consist of? This question would remain unanswered until 1964.

Developing their theory on elementary particles, in 1964 Murray Gell-Mann and George Zweig independently assumed the existence of particles that would be the constituent elements of all others [LEP 56, GRO 85, PAG 85, STÖ 07, SEC 10]. Gell-Mann called these elements *quarks*. This concept is derived from the literary work of Irish novelist and poet James Joyce (1882-1941) in his book, *Finnegans Wake*, which includes the line, "Three quarks for Muster Mark," where half-real, half-imaginary objects are presented three by three. In the Gell-Mann and Zweig model, the existence of the *u* ("up") *quark* with a charge of (2/3) *e* and the *d* ("down") *quark* with a charge of (–1/3) *e*, to explain the internal structure of the nucleons is assumed. Associated with these quarks, each of spin-1/2, are two *antiquarks*, noted $\bar{u}$ and $\bar{d}$, respectively.

In order to satisfy the *principle of electric charge conservation*, the proton has a *udu*-type structure, three-quark internal configuration and the neutron has a *udd*-type structure (Figure 1.16).

In both the neutron and the proton, the strong interaction between quarks is due to an exchange of particles called *gluons*, noted *g* (spin $s = 1$, mass $m = 0$). These exchanged particles thus ensure the cohesion of the nucleons.

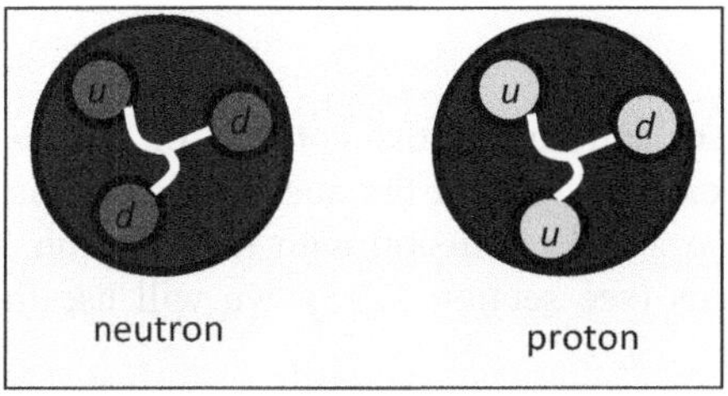

Figure 1.16. *Internal proton and neutron structures according to the Gell-Mann and Zweig quark model*

In particle physics, two types of particle systems can be distinguished:

– the *elementary system*: a quantum object fully characterized by its mass and spin;

– the *elementary particle*: an elementary system in which one cannot distinguish between the elementary constituents of given mass and spin.

The proton and neutron are thus elementary systems (they are considered elementary particles while ignoring their internal structure) whereas quarks are elementary particles (their internal structure not being known at this time). Whereas the proton is a stable particle with an estimated lifetime of more than 1.2×10^9 years, the neutron is an unstable particle with a radioactive period of 885.7 s [STÖ 07].

For a given nucleus, a mass number or nucleon number is defined and noted A. By definition, $A = Z + N$, with Z the proton number and N the neutron number of the nucleus considered. If X designates the symbol of the chemical element considered, then a nucleus is symbolized by writing ${}_{Z}^{A}X$. This conventional notation of nuclei then enables understanding of the writing of the reagents and products of the induced nuclear reactions, [1.23] and [1.24].

Murray Gell-Mann was an American physicist. He is famous most notably for having formulated the theory of quarks which he considered to be the constituent elementary particles of all the other particles. He was awarded the Nobel Prize in Physics 1969.

George Zweig was a Russian-American physicist. In particle physics he developed quark theory, independently from Gell-Mann. We thus speak of the Murray–Zweig quark model. He later went on to study neurobiology.

Box 1.8. *Gell-Mann (1929–2019); Zweig (born 1937)*

1.3.4. *Isospin*

In 1932, Heisenberg introduced the concept of *isospin*, or *isobaric spin*, to characterize the two possible states of the nucleon. Isospin is noted I [STÖ 15] or T [EVA 61, SEC 10]. To avoid confusion with nuclear spin, I [EVA 61, BRÖ 01], reserved for the nucleus (see section 1.3.5), we will use the notation T below to designate isospin.

In a "hypothetical isospin space", the nucleon is characterized by the isospin vector, $\vec{T}$, which is an operator of components T_x, T_y, and T_z. By making the analogy with electronic spin, taking the two possible values, $m_s = \pm 1/2$, it is agreed that the projection, Tz, of the *isospin operator* along a preferred direction, Oz, in an isospin space takes the values:

$$T_z = +1/2 \text{ for the proton; } T_z = -1/2 \text{ for the neutron} \qquad [1.27]$$

According to the *rule of addition of angular momenta*, the total isospin, $\vec{T}$, resulting from the coupling of the respective individual isospins, $\vec{T}_1$ and $\vec{T}_2$, of two nucleons, (1) and (2), and its projection, T_z along a reference direction, Oz, in the *isospin space* satisfy the relationships:

$$\vec{T} = \vec{T}_1 + \vec{T}_2 \; ; T_z = T_{z1} + T_{z2} \qquad [1.28]$$

For the nucleon, the introduction of isospin is based on the fundamental assumption that *pp* (proton–proton), *pn* (proton–neutron), or *nn* (neutron–neutron) nuclear interaction is the same because of the *charge independence* of nuclear forces. By analogy to the spin multiplicity $(2S + 1)$ leading to the singlet state, $S = 0$, and triplet state, $S = 1$, the number of states in the isospin space is equal to the *isospin multiplicity* $(2T + 1)$. The following nucleon states and systems are thus obtained.

– For $T = 0$, $T_z = 0$: singlet state corresponding to the *pn* nucleonic system.

– For $T = 1$: triplet state. Three nucleon systems are obtained: *pp* ($T_z = +1$); *pn* ($T_z = 0$) and *nn* ($T_z = -1$).

APPLICATION 1.5.– Is the *ground state of the deuteron* a singlet state or a triplet state? We will substantiate the answer.

ANSWER.– The deuteron (deuterium nucleus) consists of a proton ($T_{z1} = +1/2$) and a neutron ($T_{z2} = -1/2$). This gives $T_z = 0$ and $T = 0$: the ground state of the deuteron is a singlet state.

Generally, for an arbitrary nucleus of mass number $A = N + Z$, the projection, Tz, of the total isospin quantum number is given by the relationship:

$$T_z = \frac{1}{2}(N - Z) \quad [1.29]$$

Relationship [1.29] shows that T_z depends on the excess neutrons ($N - Z$). Therefore, the ground state of even-even nuclei is a singlet state.

APPLICATION 1.6.– Determine the *total isospin*, T, corresponding to the ground state of the following isobar nuclei: ^{14}C ($Z = 6$), ^{14}N ($Z = 7$) and ^{14}O ($Z = 8$).

ANSWER.– Using formula [1.29], we obtain the following for the nuclei, respectively: ^{14}C, ^{14}N and ^{14}O: $T_z = +1$, $T_z = 0$ and $T_z = -1$. The ground state of the carbon-14 and oxygen-14 nuclei has isospin $T = 1$, while $T = 0$ for the nitrogen-14 nucleus.

1.3.5. *Nuclear spin*

By analogy to the electron's own rotational motion characterized by the spin angular momentum, the atomic nucleus is associated with a *nuclear spin angular momentum* noted I. Similarly, by analogy to the magnetic quantum number of spin, m_s, taking ($2s + 1$) values with s the spin of the electron, the magnetic quantum number of nuclear spin, m_I, taking ($2I + 1$) values with I the *nuclear spin*.

Nuclear spin, I, is the sum of the nucleon spins and their possible orbital angular momenta in the nucleus. The proton and neutron are *fermions* of spin $I = 1/2$. According to *Pauli's exclusion principle*, two nucleons belonging to the same nucleus can have only parallel spin orientations if they are different (*pn*), or antiparallel if they are identical (*PP* or *nn*). For a given nucleus with mass number A, I is integer if A is even and half-integer if A is odd. For example, for deuteron $^2_1$H (*pn*), $I = 1$, for lithium $^7_3$Li (*pp nn pn n*), $I = 3/2$ and for cobalt $^{59}_{27}$Co, $I = 7/2$.

1.3.6. *Nuclear magnetic moment*

As a general rule, it is assumed that any charged particle following a closed curve is a source of magnetic field. At a very far distance from the particle, the magnetic field can be considered as that created by a *magnetic dipole* placed at the location of the charged particle [SAK 20].

As with electrons, the nucleus is associated with a *spin angular momentum,* $\vec{I}$, which is an observable. The square $\vec{I}^2$ and one of these projections, for example I_z, following a preferred direction (here *Oz*) have determined values equaling, respectively:

$$\hbar^2 I(I+1) \text{ and } M_I\hbar .$$

The *nuclear magnetic quantum number,* M_I, then takes the values:

$$-I, -(I+1), \ldots\ldots, +(I-1), +I$$

This is a total of $(2I + 1)$ values. By definition, the *nuclear dipole magnetic moment*, noted $\overrightarrow{M}_N$, is given by the relationship:

$$\overrightarrow{M}_N = g_I \gamma_I \vec{I} \qquad [1.30]$$

In relationship [1.30], g_I is the *nuclear Landé factor* and the magnitude γ_I is the *nuclear gyromagnetic ratio* [BIÉ 06, SAK 20] given by the following expression (this is the analogue of the *electron gyromagnetic ratio* $\gamma = -e/2m$):

$$\gamma_I = \frac{e}{2m_p} \qquad [1.31]$$

In relationship [1.31], m_p is the proton mass.

Likewise, we define the *nuclear magneton*, also called the *nuclear Bohr magneton*, noted μ_N (by analogy to the *Bohr magneton* $\mu_B = -\gamma\hbar = e\hbar/2m$). The nuclear magneton is given by the expression:

$$\mu_N = \gamma_I \hbar = \frac{e\hbar}{2m_p} \qquad [1.32]$$

Using [1.31] and [1.32], the dipole magnetic moment [1.32] is written:

$$\overrightarrow{M}_N = g_I \frac{\mu_N}{\hbar} \vec{I} \qquad [1.33]$$

Noting that $\mu_N/\mu_B = m/m_p$ and $m_p/m = 1836.1526715$, then we obtain:

$$\mu_N = 5.05078317 \times 10^{-27}\ \text{J} \cdot \text{T}^{-1}$$

For the proton:

$$I = 1/2;\ g_I = 5.5883;\ \mu_p = 2.7928473375\ \mu_N \qquad [1.34]$$

Although electrically neutral, a dipole magnetic moment, μ_n, is associated with the *neutron's spin angular momentum*. Thus, for the neutron:

$$I = 1/2;\ g_I = -3.8263;\ \ \mu_n = -1.91304272\ \mu_N \qquad [1.35]$$

Table 1.1 summarizes the fundamental properties of nucleons.

Properties	Proton	Neutron
Mass	$1.672621637(83)\times 10^{-27}$ kg	$1.674927211(84)\times 10^{-27}$ kg
Charge	$1.602179487(40)\times 10^{-19}$ C	0
Spin	½	½
Isospin	+ ½	– ½
Radioactive period	$\sim 2.1\times 10^{29}$ years	(885.7 ± 08) s
Nuclear Landé factor	5.5883	– 3.8263
Dipole magnetic moment	$2.7928473375\ \mu_N$	$-1.91304272\ \mu_N$

Table 1.1. *Fundamental properties of nucleons*

1.4. Nucleus dimensions

1.4.1. *Nuclear radius*

Most nuclei are considered to be spherical. In this approximation, the nuclear radius, R, given by the relationship [EVA 61, MIS 96, BRÖ 01, STÖ 07, SEC 10, DUM 15, SAK 18a]:

$$R = r_0\, A^{1/3} \qquad [1.36]$$

In relationship [1.36], r_0 is a nuclear radius whose value depends on the nucleus model adopted:

– for the *nuclear model with constant nucleon density*, the nucleus charge is considered to be uniformly distributed within the nuclear volume, r_0 is called the *unit nuclear radius* within the framework of classical mechanics and worth:

$$r_0 = 1.45\ \times 10^{-15}\ \text{m} = 1.45\ \text{fm} \qquad [1.37]$$

– for the nuclear model, it is considered that the nucleus charge is distributed uniformly at the surface of the nucleus; r_0 is called the *electromagnetic unit radius* within the framework of quantum mechanics and is worth:

$$r_0 = 1.2 \times 10^{-15} \text{ m} = 1.2 \text{ fm} \quad [1.38]$$

The first theoretical expression for the unit nuclear radius was established in 2018 by the author as part of the nuclear model with constant nucleon density. The *Sakho unit nuclear radius* is given by the expression [SAK 18a]:

$$r_0 = \frac{\alpha^2}{(1+N/Z)} a_0 \quad [1.39]$$

In relationship [1.39], α is the *fine-structure constant* and a_0 represents the *Bohr radius.*

For most nuclei, $A \approx 2Z$ ($N = Z$). The unit nuclear radius is then written:

$$r_0 = \frac{\alpha^2}{2} a_0 \quad [1.40]$$

Based on $\alpha^2 = 5.325 \times 10^{-5}$ and $a_0 = 5.2917 \times 10^{-11}$ m, relationship [1.40] gives:

$$r_0 = 1.41 \times 10^{-15} \text{ m} = 1.41 \text{ fm} \quad [1.41]$$

The theoretical result [1.41] is very much consistent with the experimental results, $r_{0exp} = 1.40 \times 10^{-15}$ m $= 1.40$ fm obtained on the scattering of α particles by light nuclei Li, Be, Mg and Al [POL 35] and on the scattering of fast neutrons by nuclei [DAY 53].

NOTE.– The radii of exotic nuclei do not confirm relationship [1.36]. An *exotic nucleus* is a nucleus created artificially in particle accelerators.

APPLICATION 1.7.– Using [1.36], give a framing of the nuclear radii, R_X, corresponding to natural nuclei. Let $r_0 = 1.2$ fm.

ANSWER.– The lightest nucleus is that of the hydrogen atom: $R_H = 1.20$ fm. The heaviest natural nucleus is uranium 238: $R_U = 1.2 \times 10^{-15} \times (238)^{1/3} = 7.4$ fm. Thus, for a natural nucleus of radius R_X, we have:

$$(1.0 < R_X < 8.0) \text{ fm} \quad [1.42]$$

APPLICATION 1.8.– Estimate the *density of nuclear matter*. We will compare with the mass of 1 m^3 of water and then draw a conclusion. The mass of a nucleus, $m \approx$ Au, with u the atomic mass unit.

Let: $r_0 = 1.2$ fm; $u = 1.66 \times 10^{-27}$ kg.

ANSWER.– The density of a nucleus assumed to be spherical of radius $R = r_0 A^{1/3}$ is:

$$\rho = \frac{m}{V} = \frac{Au}{(4/3)\pi R^3} = \frac{3u}{4\pi r_0^3} \qquad [1.43]$$

NOTE.– $\rho \approx 2.3 \times 10^{17}$ kg · m^{-3}

Given that 1 m^3 of water weighs 1,000 kg, 1 m^3 of nuclear matter weighs around 230 million billion kilograms.

CONCLUSION.– nuclear material is extremely dense.

1.4.2. *Nuclear density, skin thickness*

As noted in section 1.4.1, the value of the radius, r_0, depends on the adopted nucleus model according to whether the nuclear mass is uniformly distributed in volume or surface. As a result, the nuclear density is not constant for a given nucleus.

With nuclear rays varying approximately between 1 and 8 fm [1.42], to probe the inside of the nuclei, the probe particles must have a de Broglie wavelength of the order of 1 fm. To achieve this, it is possible to envisage conducting experiments of, for example, photon, electron or neutron scattering by target nuclei. The information obtained on the inside of the nucleus will then depend on the nature of the probe particle used and the probe-nucleon particle interaction.

Scattering experiments use probe particles with very high energies. These are therefore relativistic particles of total energy:

$$E = \sqrt{p^2c^2 + m_0^2c^4} \qquad [1.44]$$

By definition, the *de Broglie wavelength* $\lambda = h/p$. Using relationship [1.44], the relativistic expression for the energy of the probe particles is obtained:

$$E = \sqrt{\frac{h^2c^2}{\lambda^2} + m_0^2c^4} \qquad [1.45]$$

For an electromagnetic probe, the experiment would require a photonic energy of more than 1,200 MeV. In photon scattering experiments, *X* photons with energies in the keV range are generally used. From a technical point of view, it proves very complex to create a photon beam with 1,200 MeV energy.

APPLICATION 1.9.– Show that the use of an electromagnetic probe would require energy of more than 1,200 MeV.

Given data: $c = 3.0 \times 10^8$ m · s $^{-1}$ and $h = 6.63 \times 10^{-34}$ J · s; 1 eV = 1.6×10^{-19} J.

ANSWER.– For the photon, the resting mass, $m_0 = 0$, expression [1.45] gives:

$$E = \frac{hc}{\lambda} \quad [1.46]$$

NOTE.– (de Broglie wavelength of the order of 1 fm: $\lambda = 1$ fm).

$$E = 1.243 \times 10^{9} \text{ eV} = 1243 \text{ MeV}$$

The scattering of high-energy electrons (approximately 1 GeV) on nuclei was the first source of information on nuclear charge distribution. At these energies, electrons have a wavelength comparable to or less than the dimensions of the nucleus, and can thus probe the spatial distribution of protons within the nucleus [DEL 03]. As neutrons are neutral, neutron scattering experiments offer access to the mass distribution of the nucleus.

Let us thus consider a nucleus of *nuclear charge distribution density* $\rho\,(r)$. For the electron, the resting energy $m_0c^2 = 0.511$ MeV. For $\lambda = 1$ fm, expression [1.45] then shows that the energy of the electron beam to be used is equal to 1.24 GeV. However, the experiment reveals that electrons of energy between 100 and 1,000 MeV enables probing of the inside of the nuclei.

Using electrons as probe particles, the experiment shows that the density $\rho\,(r)$ presents saturation in a certain range of the nuclear volume and then gradually decreases at the surface of the nucleus. The study of several nuclei has established the form of $\rho\,(r)$, the simplest of which, accounting satisfactorily for the charge distribution in the nucleus assumed to be spherical, is that of Woods–Saxon, given by [CAR 79, DEL 03, STÖ 07, SEC 10, DUM 15, MAY 17]:

$$\rho(r) = \frac{\rho_0}{1 + \exp\left(\dfrac{r - R}{a}\right)} \quad [1.47]$$

In relationship [1.47], $\rho_0 = \rho(0)$ designates the *saturation density*, the quantity, a, called the *diffusivity parameter* characterizes the extent of the nuclear surface, r is the distance measured in spherical coordinates from the center of the nucleus, and R is the radius of the nucleus assumed to be spherical.

The set of experimental results for fairly heavy nuclei ($A \geq 30$) agrees with the formula $R = 1.07 \times A^{1/3}$ fm and $a = 0.545$ fm. The variation of the nuclear charge distribution density $\rho(r)$ with the distance r is shown in Figure 1.17. At the center of the nucleus, the saturation density $\rho_0 \approx 0.14$ nucleons · fm $^{-3}$ (see application 1.10).

As shown in Figure 1.17, the nuclear charge distribution density is approximately constant inside the nucleus, regardless of the nucleus considered. This is due to the saturation property of short-range nuclear forces (see section 1.6.5).

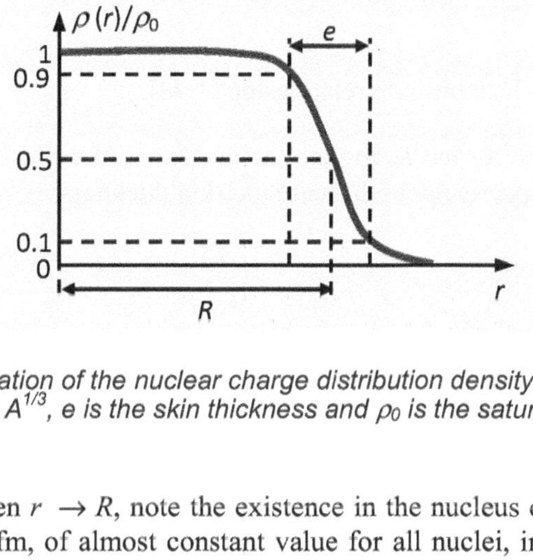

Figure 1.17. *Variation of the nuclear charge distribution density ρ (r); the nucleus radius R = r₀ A$^{1/3}$, e is the skin thickness and ρ₀ is the saturation density*

In addition, when $r \rightarrow R$, note the existence in the nucleus of a diffuse zone of approximately 2.2 fm, of almost constant value for all nuclei, in which the density gradually decreases. When $r > R$ the nuclear charge distribution density $\rho(r)$ gradually decreases at the surface and does not fall abruptly to zero. To translate this progressive decay, let us introduce the parameter, e, called *skin thickness* This parameter is the length for which the density $\rho(r)$ decreases from 90% to 10% of the saturation density ρ_0 (Figure 1.17). The skin thickness is linked to the diffusivity parameter, a, by the relationship (see application 1.11):

$$e = 4a\ln 3 \qquad [1.48]$$

For $a = 0.545$ fm, using [1.48] we find: $e = 2.4$ fm. If we make the approximation $a \approx 0.5$ fm, then we find e $\approx$ 2.2 fm. This result shows that the length of the diffuse zone in the nucleus of approximately 2.2 fm corresponds to the skin thickness.

APPLICATION 1.10.– Show that the saturation density is worth approximately 0.14 nucleon · fm $^{-3}$.

Let $r_0 = 1.2$ fm.

ANSWER.– The saturation density corresponds to the nucleon density per unit volume. For a spherical core of volume V and mass number A we have:

$$\rho_0 = \frac{A}{V} = \frac{3}{4\pi r_0^3} \quad [1.49]$$

NOTE.–

$$\rho_0 = \frac{3}{4\pi \times (1.2)^3} = 0.13815 \approx 0.14 \text{ nucleon} \cdot \text{fm}^{-3}$$

APPLICATION 1.11.– Demonstrate relationship [1.48].

ANSWER.– Let us use R_1 and R_9 to note the nuclear radii corresponding to densities $0.1\rho_0$ and $0.9\,\rho_0$, respectively. In this case the skin thickness:

$$e = R_1 - R_9 \quad [1.50]$$

Using [1.47], we obtain:

$$\begin{cases} 0.1\rho_0 = \dfrac{\rho_0}{1+\exp\left(\dfrac{R_1 - R}{a}\right)} \\ 0.9\rho_0 = \dfrac{\rho_0}{1+\exp\left(\dfrac{R_9 - R}{a}\right)} \end{cases} \Rightarrow \begin{cases} R_1 - R = +a\ln 9 \\ R_9 - R = -a\ln 9 \end{cases}$$

Thus, according to [1.50]: $e = R_1 - R_9 = 2a\ln 9 = 2a\ln(3^2) \Rightarrow e = 4a\ln 3$.

David S. Saxon was an American physicist. He is famous for having proposed, in 1954 along with R. D. Woods (biography not mentioned in the available literature), what is known as the Woods–Saxon potential, thus describing the scattering of protons on heavy nuclei, such as platinum or nickel.

Box 1.9. *Saxon (1920–2005)*

1.5. Nomenclature of nuclides

1.5.1. *Isotopes, isobars, isotones*

In 1886, Crookes gave a speech to the British Association in which he referred to the concept of isotopes. He declared:

> When we say, for example, that the atomic mass of calcium is 40, we are actually saying that the majority of the calcium atoms have a real atomic mass of 40, that there are a small number of atoms which have a mass 39 or 41, others much smaller are 38 or 42, and so forth. [EVA 61]

Today, we know that calcium is a mixture of isotopes of mass numbers 40, 42, 43, 44, 46, and 48. Crookes' predictions were therefore correct; the mass numbers 38 and 39 corresponding to the isotopes of the element potassium. From an experimental perspective, Frederick Soddy [SOD 11, SOD 14] was the first to prove the existence of isotopes in radioactive substances. In addition, he highlighted in his work the existence of substances having the same mass number but different chemical properties, because of the difference in their nuclear charges. These substances were called isobars.

The nomenclature of nuclear species distinguishes between groups of nuclei that form three large families: isotopes, isobars and isotones.

Generally, a nucleus is characterized by its proton number, Z, and its nucleon number, A [SAK 16] is referred to as a nuclide. A nuclide of a chemical element, X, is symbolized by the notation ${}^{A}_{Z}X$, which we recall here. The neutron number, $N = A - Z$.

– *Isotopes* is the name given to nuclei with the same proton number but different neutron numbers, N. For example, ${}^{13}_{6}C$ $(N = 7)$ and ${}^{14}_{6}C$ $(N = 8)$;

– *isobars* refer to nuclei with the same mass number but different proton numbers. For example, ${}^{14}_{6}C$ and ${}^{14}_{7}N$;

– *isotones* refer to nuclei with the same neutron number but different proton numbers. For example ${}^{12}_{6}C$ $(N = 6)$ and ${}^{13}_{7}N$ $(N = 6)$.

Generally, for a family of given nuclei, the nuclide content is different. The most stable nuclei have the highest content, while unstable nuclei exist in the form of traces. Table 1.2 presents the contents of some isotopes. Today, 90 natural elements and 19 artificial elements are known, corresponding to approximately 1,500 different nuclides, approximately 350 of which are natural. Of the 1,500 nuclides identified, 256 are stable, the others spontaneously convert to more stable nuclides. Let us specify some of the isotopes of the first ten elements of the periodic table in addition to those of iron and uranium.

– Hydrogen (H) has three isotopes: ^{1}H, ^{2}H, and ^{3}H. The remaining isotopes from ^{4}H to ^{7}H have been synthesized in the laboratory. They are particularly unstable and have never been observed in nature.

– Helium (He) has eight known isotopes (^{2}He, ^{3}He, ^{4}He, ^{5}He, ^{6}He, ^{7}He and ^{8}He). However, only the two isotopes ^{3}He and ^{4}He are stable. All other radioisotopes have a short lifetime. The longest lifetime is that of the isotope ^{6}He with a half-life of 806.7 ms; the isotope ^{2}He is a hypothetical isotope that would be composed of two protons without any neutron, therefore called the "diproton".

– Lithium (Li) consists of two stable isotopes, ^{6}Li and ^{7}Li, the majority being the latter.

– Beryllium (Be) has twelve known isotopes, with mass numbers ranging from 5 to 16. Only the isotope ^{9}Be is stable. For this reason, beryllium is called a *monoisotopic element*. A monoisotopic element is an element with only one stable isotope. 26 monoisotopic elements can currently be enumerated. Fluorine-19, sodium-23, aluminum-27 and phosphorus-31 can be cited in addition to beryllium-9.

– Boron (B) has fourteen isotopes, of which only the two isotopes, ^{10}B and ^{11}B, are stable. All have a short lifetime. The isotope ^{8}B, with a half-life of 770 ms, has the longest lifetime.

– Carbon (C) has 15 isotopes, with mass numbers ranging from 8 to 22. Of these isotopes, only ^{12}C and ^{13}C are stable. The radioisotope with the longest lifetime is carbon 14, with a half-life of 5,730 years. This isotope is widely used for the dating of dead plant or animal species (see Chapter 1, Volume 2), and is the only radioisotope present in nature. It is formed by the reaction: $^{14}N + {}^{1}n \rightarrow {}^{14}C + {}^{1}H$. All other radioisotopes have a short lifetime. The isotope ^{11}C, with a half-life of 20.334 min, has the longest lifetime.

– Nitrogen (N) has 16 isotopes with mass numbers between 10 and 25, and a nuclear isomer, ^{11m}N, with a half-life of $6.90(80) \times 10^{-22}$ s (see the note at the end of this section for the definition of a nuclear isomer). Only the two isotopes ^{14}N and ^{15}N, which are present in nature, are stable.

– Oxygen (O) has 17 isotopes with mass numbers between 12 and 28, three of which, ^{16}O, ^{17}O, and ^{18}O, are stable. The radioisotope with the longest lifetime is ^{15}O, with a half-life of 122.24 s.

– Fluorine (F) is a monoisotopic element like beryllium. It consists of the only stable isotope, ^{19}F.

– Neon (Ne) has 19 isotopes with mass numbers ranging from 16 to 34, three of which, ^{20}Ne, ^{21}Ne and ^{22}Ne, are stable. The radioisotope with the longest lifetime is ^{24}Ne, with a half-life of 3.38 min.

Among the most significant heavy isotopes are those of iron and uranium. Indeed, of all known nuclides, the most stable is that of iron isotope 56. In addition, the uranium-235 isotope is widely used in nuclear power plants for the production of electrical energy through the fission reaction. The stability data for iron and uranium isotopes are as follows.

– Iron (Fe) has 28 isotopes with mass numbers ranging from 45 to 72, as well as six nuclear isomers [^{52m}Fe: 45.9 (6) s, ^{53m}Fe: 2.526 (24) min, ^{54m}Fe: 364 (7) ns, ^{61m}Fe: 250 (10) ns, ^{65m}Fe: 430 (130) ns and ^{67m}Fe: 64 (17) μs]; the values given in square brackets correspond to the half-lives of the nuclear isomers. Of these isotopes, only ^{54}Fe, ^{56}Fe, ^{57}Fe and ^{58}Fe are stable. The lightest isotope, ^{45}Fe, decays mainly (70%) by two-proton emission into ^{43}Cr. The isotope ^{56}Fe is the most stable of all nuclei (see section 1.6.4).

– Uranium (U) has 26 isotopes, with mass numbers ranging from 217 to 242. It also has seven nuclear isomers [^{234m}U: 33.5 (20) ms; ^{235m}U: ~ 26 min; ^{238m}U: 280 (6) ns; ^{239m1}U: >250 ns; ^{239m2}U: 780 (40) ns; ^{236m1}U: 100 (4) ns; ^{236m2}U: 120 (2) ns]. Uranium is a natural radioactive element and therefore has no stable isotope. The two isotopes ^{235}U and ^{238}U, present in appreciable quantities in the Earth's crust with their decay products, have long half-lives. ^{235}U is the only fissile isotope.

The molar percentages of the most significant isotopes of the first ten elements of the periodic table and those of iron and uranium are presented in Table 1.2.

NOTE.– generally, an excited nucleus normally returns to the lowest energy ground state after a fraction of a second. However, the nuclear transition to ground level can, in exceptional cases, be inhibited and significantly slowed down. The nucleus then remains in an abnormally long excited state. Such excited states are called *metastable states*. *Nuclear isomerism* is introduced to account for the existence of nuclei in different energy states. The term *nuclear isomers* describes the isomers of a nucleus that are all in a metastable state. The nuclear isomers are noted by attaching the letter "m" for "metastable" to the ^{A}X isotope being considered. Thus, a nuclear isotope is noted ^{Am}X. This is the case with the nucleus ^{11m}N of the isotope ^{11}N of nitrogen. A particularly significant example of a nuclear isomer in nuclear medicine is technetium^{99m}Tc (half-life 6.0058 hours) that remains several hours before returning to the normal state of technetium-99. Technetium-99m is obtained in hospitals from a precursor radioactive nucleus, molybdenum ^{99}Mo, generally produced in a reactor and is very useful in nuclear medicine (see Chapter 3, Volume 2).

Isotope	Percentage		Isotope	Percentage
${}_{1}^{1}H$	99.9885%		${}_{8}^{16}O$	99.757%
${}_{1}^{2}H$	0.00115%		${}_{8}^{17}O$	0.038%
${}_{1}^{3}H$	Traces		${}_{8}^{18}O$	0.205%
${}_{2}^{3}He$	0.000134%		${}_{9}^{19}F$	100%*
${}_{2}^{4}He$	99.999866%		${}_{10}^{20}Ne$	90.48%
${}_{3}^{6}Li$	7.59%		${}_{10}^{21}Ne$	0.27%
${}_{3}^{7}Li$	92.41%		${}_{10}^{22}Ne$	9.25%
${}_{4}^{9}Be$	100%*		${}_{26}^{57}Fe$	5.845%
${}_{5}^{10}B$	19.9%		${}_{26}^{56}Fe$	91.754%
${}_{5}^{11}B$	80.1%		${}_{26}^{57}Fe$	2.119%
${}_{6}^{12}C$	98.93%		${}_{26}^{58}Fe$	0.282%
${}_{6}^{13}C$	1.07%		${}_{26}^{60}Fe$	traces
${}_{6}^{14}C$	traces		${}_{92}^{234}U$	0.0050 to 0.0059%
${}_{7}^{14}N$	99.636%		${}_{92}^{235}U$	0.7198 to 0.7202%
${}_{7}^{15}N$	0.364%		${}_{92}^{238}U$	99.2739 to 99.2752%

Table 1.2. *Molar percentages of the most significant isotopes of the first ten elements of the periodic table, and those of iron and uranium. The asterisk (*) refers to the monoisotopic elements being the only stable elements*

With the discovery of isotopes, *mass spectrometry* saw the light of day for isotopic analysis. Thus, instruments called *mass spectrographs* were manufactured, which allowed the *separation of isotopes* of a given substance according to their masses [EVA 61, OLS 91, SAK 16]. The principle of a mass spectrograph involves subjecting the isotope ions of a given chemical to the action of electric and magnetic fields. The electric field makes it possible to accelerate the ions while the magnetic field enables their trajectories to be bent. In the magnetic field, the lower the ion's velocity, the more its trajectory is curved. In addition, the radius of curvature of the trajectory depends on the mass-to-charge ratio of the ion. The ions are then received on a photographic plate. Measuring the distance between the impact points of two ions on the plate enables them to be identified.

Frederick Soddy was a British radiochemist. In 1902 he established, together with Rutherford, the exponential law of radioactive decay. Soddy was awarded the Nobel Prize in Chemistry 1921 for his contributions on radioactive substances and for his research on the nature of isotopes.

Box 1.10. *Soddy (1877-1956)*

1.5.2. *Mirror nuclei, Magic nuclei*

Nuclei known as *mirror nuclei* are used to study neutron/proton symmetry. These are pairs of nuclei with the same mass number, A, and where the proton number, Z, in one equals the neutron number, N, in the other. Thus, for mirror nuclei, the relationship $Z - N = \pm 1$ is confirmed. For example, ${}^{7}_{3}Li$ $(Z - N = -1)$ and ${}^{7}_{4}Be$ $(Z - N = +1)$.

In addition, there exist particularly stable nuclei called "magic nuclei". For a magic nucleus, the number of protons, Z, or neutrons, N, corresponds to complete *nucleon shells* (see Chapter 2, section 2.3). For such a nucleus, Z or N is thus a magic number. The magic numbers are: 2, 8, 20, 28, 50, 82 and 126. For example, ${}^{4}_{2}He$ $(Z = 2; N = 2)$, ${}^{40}_{20}Ca$ $(Z = 20; N = 20)$, ${}^{208}_{82}Pb$ $(Z = 82; N = 126)$. Note that these nuclei are also said to be *doubly-magic nuclei* since Z and N are all magic numbers.

APPLICATION 1.12.– Determine the isospin of mirror nuclei. Conclude.

ANSWER.– For mirror nuclei, $Z - N = \pm 1 \Rightarrow N - Z = \pm 1$. Using [1.29], we obtain: $T_z = \pm 1/2$. The isospin of mirror nuclei is therefore equal to $T = 1/2$.

CONCLUSION.– Mirror nuclei have the same isospin.

1.6. Nucleus stability

1.6.1. *Atomic mass unit*

The *mass-energy equivalence* relationship reflects the fact that mass is a form of energy. It is expressed by the relationship $E = mc^2$ with m the relativistic mass of the particle, c is velocity of light in a vacuum; $E_0 = m_0c^2$ is called the *mass energy*, or the *resting energy*, of the particle of *resting mass*, m_0.

As shown in Table 1.1, the mass of a nucleon is of the order of 10^{-27} kg. This mass, expressed in kg, is therefore very low. To have an appreciable mass, let us introduce a unit (more convenient than kg for calculations), called the atomic mass unit, noted *u*. By definition, the *atomic mass unit* is a twelfth of the mass of the carbon isotope ^{12}C, i.e. [SAK 16]:

$$1\text{u} = \frac{1}{12} m\,(^{12}\text{C}) \qquad [1.51]$$

Yet, $m\,(^{12}\text{C}) = M\,(\text{C})/N_A = 12/N_A$ with N_A Avogadro's number. Hence [1.51] is written:

$$1\text{u} = \frac{1}{N_A}\ (\text{g}) \qquad [1.52]$$

Knowing that $N_A = 6.02214179 \times 10^{23}$ mol $^{-1}$, then numerically:

$$1\text{u} = 1.660538783 \times 10^{-27}\ \text{kg} \qquad [1.53]$$

As result [1.53] indicates, the value of the atomic unit expressed in kg is of the order of the mass of a nucleon. Another much more convenient unit of mass is then introduced: MeV/c^2.

Applying the mass-energy equivalence relationship gives:

$$E = mc^2 = \text{u}c^2 \qquad [1.54]$$

Given that $c = 299792458$ m · s $^{-1}$ and 1 eV = $1.6021794 \times 10^{-19}$ J, using [1.54] and value [1.53], we obtain:

$$E = \frac{1.660538783 \times 10^{-27} \times (299792458)^2}{1.6021794 \times 10^{-19}} = 931.4923\ \text{MeV} \qquad [1.55]$$

Hence:

$$1\text{u} = 931.5\ \text{MeV}/c^2 \qquad [1.56]$$

Table 1.3 summarizes the nucleon and electron masses.

Particles	Mass in atomic unit (u)	Mass in MeV/c^2
Electron	$5.485\ 799\ 094\ 3 \times 10^{-4}$	0.510999910
Neutron	1.00866491597	939.565346
Proton	1.00727646677	938.272013

Table 1.3. *Electron and nucleon masses in u and MeV/c^2*

1.6.2. *Segrè diagram, nuclear energy surface*

To study the stability of the nuclei, it is convenient to represent each isotope with a point on a diagram (Z, N), with abscissa Z and ordinate N. The resulting graph (Figure 1.18) is called the *Segrè diagram*.

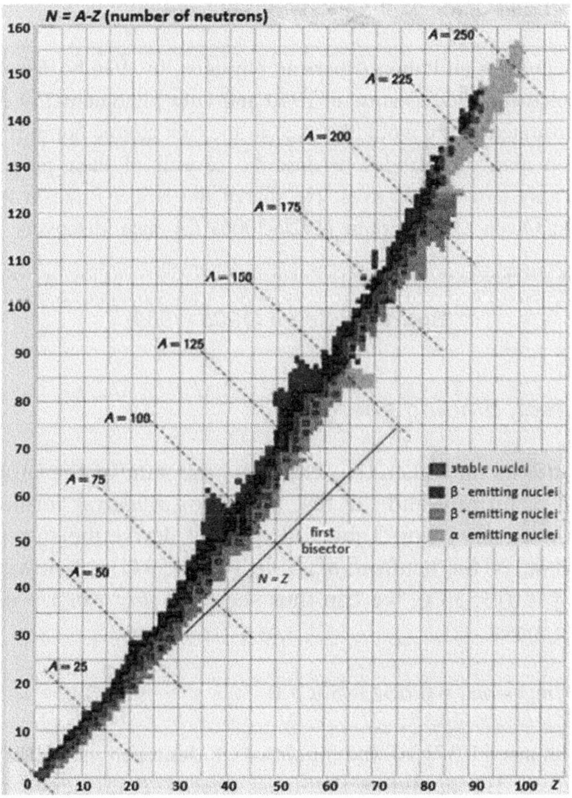

Figure 1.18. *Segrè diagram, the nuclear energy surface indicated in red, grouping together the stable nuclei for which Z = N*

This diagram reveals the existence of four nuclide distribution zones according to the values of Z and N:

– a red area, called the *nuclear energy surface*, or *valley of stability,* where the most stable nuclei are distributed; for $Z < 30$, note that the stable nuclei are located near the first $N = Z$ bisector;

– an orange area occupied by unstable heavy nuclei that decay by α radioactivity;

– a blue area occupied by nuclei presenting excess neutrons compared to stable nuclei with the same mass number, A; they decay by β^- radioactivity;

– a green area occupied by nuclei presenting excess protons compared to stable nuclei with the same mass number, A; they decay by β^+ radioactivity.

The modes of α and β decay are discussed in Chapter 3.

Emilio Gino Segrè was an Italian-American physicist. In 1936 he discovered technetium, the first artificial element, then astatine in 1940 and later plutonium-239, whose fissionable character he demonstrated (^{239}Pu would be used in the first atomic bomb [FER 64]). In 1945, Segrè established a map of nuclides to describe several of their properties graphically, marking them on a system of N/Z axes. He shared the Nobel Prize in Physics 1959 with American physicist Owen Chamberlain (1920-2006) for their discovery of the antiproton (negative proton).

Box 1.11. *Segrè (1905–1989)*

1.6.3. *Mass defect, binding energy*

Let us compare, as an illustrative example, the sum of the proton and neutron masses separated at infinity and the rest to that of *deuteron* formed by the bound system {neutron-proton}. Using Table 1.3, we can derive the sum of the mass, m_p, of the proton and m_n of the neutron: $m_p + m_n = 2.015941$ u. Knowing that the mass of deuteron, $m_d = 2.013553$ u, we can thus see that: $(m_p + m_n) > m_d$. The deuteron nucleus lacks one mass:

$$m = (m_p + m_n) - m_d = 0.002385 \text{ u} > 0 \qquad [1.57]$$

The missing mass [1.57] of the nucleus of deuteron is called a *mass defect*, counted as positive. By applying the mass-energy equivalence relationship, the missing mass is not lost. It is converted into *binding energy* thus ensuring the cohesion of the deuteron.

Generally, the mass defect of a nucleus noted Δm is the difference between the mass of nucleons (protons and neutrons), separated at rest (infinity), and the mass of the nucleus formed at rest (Figure 1.19).

By definition, for a nucleus of mass $M(A, Z)$, mass number A, proton number Z and neutron number $N = (A - Z)$:

$$\Delta m = [Zm_p + (A - Z)m_n - M(A, Z)] \qquad [1.58]$$

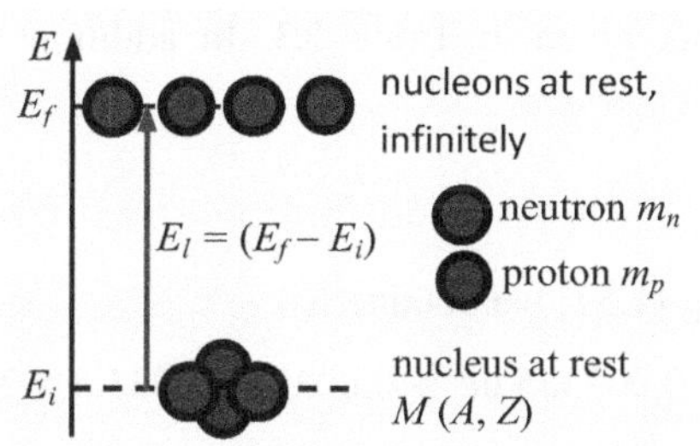

Figure 1.19. *Energy E_l necessary to separate, to infinity, the nucleons of a nucleus (example here: helium-4 nucleus)*

In addition, the binding energy, E_l, is the energy that must be provided to separate, at rest, to infinity, the nucleons of a nucleus as shown in Figure 1.18. Let us express the *principle of energy conservation* when separating nucleons:

– initial state (nucleus at rest): $E_i = M(A, Z)\, c^2$; [1.59]

– final state (nucleons at rest): $E_f = ZM_pc^2 + (A - Z)\, m_nc^2$.

Using [1.59], the energy variation is then written:

$$\Delta E = E_f - E_i = [Zm_pc^2 + (A - Z)\, m_nc^2] - M(A, Z)\, c^2$$

This relationship enables us to derive the expression for the binding energy, $E_l = \Delta E$, i.e.:

$$E_l = \{[Zm_pc^2 + (A - Z)\, m_nc^2] - M(A, Z)\}\, c^2 \qquad [1.60]$$

Taking into account the mass defect [1.58] gives:

$$E_l = \Delta mc^2 \qquad [1.61]$$

Relationship [1.61] reflects the mass-energy equivalence in the case of a nucleus. As the mass defect, Δm, is positive, then according to [1.61], the binding energy, E_l, is also positive. This sign is characteristic of a bound system such as the nucleus. Thus, for an unbound system, $\Delta m < 0 \Rightarrow E_l < 0$.

APPLICATION 1.13.– Determine the binding energy of deuteron, the α particle, the iron isotope ^{56}Fe (Z = 26; M_{Fe} = 55.9206 u), and the uranium-238 isotope, ^{238}U (Z = 92; M_{U} = 238.0003 u). Conclude.

We will use the data given in Table 1.3. In addition: m_α = 4.0015 u; 1u = 931.5 MeV/c^2.

ANSWER.–

– For deuteron, using [1.57] we obtain:

$$E_l\,(^2\text{H}) = \Delta mc^2 = 0.002385\ \text{u}c^2 = 0.002385 \times 931.5 = 2.22\ \text{MeV}$$

– For the α particle (helium nucleus):

$$E_l\,(^4\text{He}) = 2 \times [m_p + m_n]\,c^2 - m_\alpha c^2$$

NOTE.–

$$E_l\,(^4\text{He}) = (2 \times 2.015941 - 4.0015) \times 931.5 = 28.30\ \text{MeV}$$

– For the iron-56 nucleus and uranium, we obtain, as above:

$$E_l\,(^{56}\text{Fe}) = (26 \times 1.00727646677 + 30 \times 1.00866491597 - 55.9206) \times 931.5 = 492.331\ \text{MeV}$$

$$E_l\,(^{238}\text{U}) = (92 \times 1.00727646677 + 146 \times 1.00866491597 - 238.0003) \times 931.5 = 1801.719\ \text{MeV}$$

In summary:

$$E_l\,(^2\text{H}) = 2.22\ \text{MeV}; \qquad E_l\,(^4\text{He}) = 28.30\ \text{MeV} \qquad [1.62a]$$

$$E_l\,(^{56}\text{Fe}) = 492.331\ \text{MeV};\ E_l\,(^{238}\text{U}) = 1801.719\ \text{MeV} \qquad [1.62b]$$

CONCLUSION.– the more nucleons there are in the nucleus, the greater the binding energy.

1.6.4. *Binding energy per nucleon, Aston curve*

As we found in application 1.13, the binding energy of a nucleus rises with the increase in the number of nucleons. However, experience shows that a nucleus is not more stable the higher its binding energy is. For this reason, binding energy is not a good indicator for comparing nucleus stability. To achieve this, let us define the *binding energy per nucleon*. We thus show that a nucleus is not more stable the higher its binding energy (E/A) is. The binding energy per nucleon is of the order of 8 MeV.

APPLICATION 1.14.– Verify by calculating which of the nuclei of iron-56 and uranium-238 is most stable.

ANSWER.– Using [1.62b], we obtain:

$$E_l(^{56}\text{Fe})/A = 492.331/56 = 8.79 \text{ MeV/nucleon} \quad [1.63a]$$

$$E_l(^{238}\text{U})/A = 1801.719/238 = 7.57 \text{ MeV/nucleon} \quad [1.63b]$$

$E_l\,(^{56}\text{Fe})/A > E_l\,(^{238}\text{U})/A$: the nucleus of iron-56 is more stable than that of uranium-238.

In the general case, let us look at the variation of the curve – $E_l/\text{A} = f\,(A)$. A valley-shape curve is obtained whose bottom is occupied by the nucleus of iron-56 with binding energy per nucleon equaling 8.8 MeV (Figure 1.20) in accordance with result [1.63a]. This graph, called the *Aston curve* shows that the nucleon of iron-56 is the most stable of all nuclei.

The shape of the Aston curve calls for four main observations:

– light nuclei such as deuteron ^{2}H and triton ^{3}H can fuse to give heavier stable nuclei located in the valley;

– the curve presents something of a plateau for $50 < A < 100$ with a minimum for $A = 56$. These nuclides correspond to the most stable nuclei;

– for $A > 100$, the curve increases slightly; the corresponding heavy nuclei are less stable;

– very heavy nuclei for $A > 200$ may undergo fission reactions to give lighter, more stable nuclei.

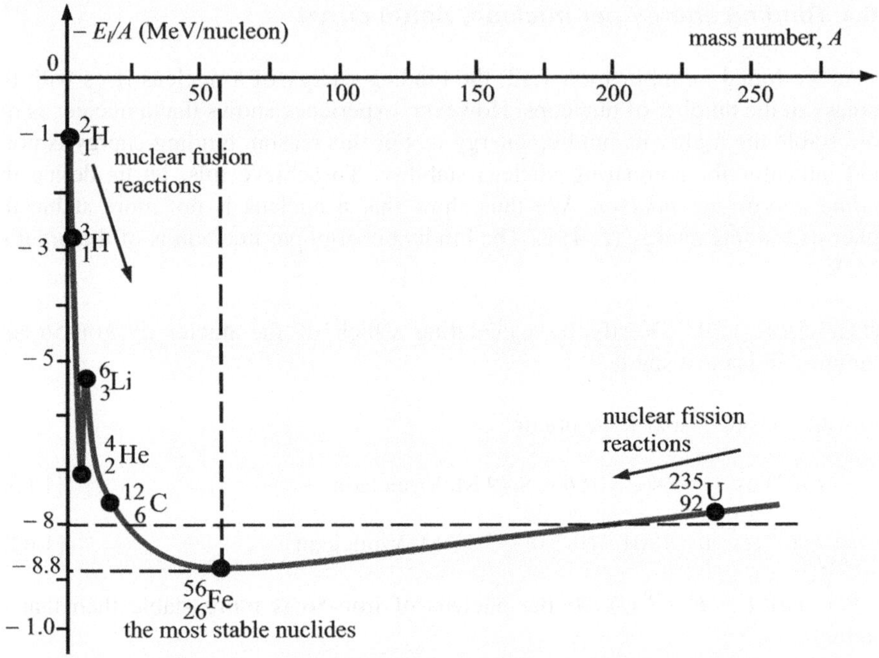

Figure 1.20. *Aston curve*

Thermonuclear fusion and nuclear fission reactions are nuclear reactions that are induced, i.e., not spontaneous. These reactions are therefore not studied in this book. However, as an illustration, let us give an example of a fusion reaction and another example of a fission reaction to account for the fact that the products of these induced reactions are more stable than the starting nuclides.

– *Fusion of deuteron* ^{2}H and triton ^{3}H:

$$ {}^2_1H + {}^3_1H \rightarrow {}^4_2He + {}^1_0n \tag{1.64} $$

– *Fission of uranium-235* under the impact of a neutron, *n*:

$$ {}^1_0n + {}^{235}_{92}U \rightarrow {}^{94}_{38}Sr + {}^{140}_{54}Xe + 2\,{}^1_0n + \gamma \tag{1.65} $$

The nucleus of helium ^{4}He obtained after the fusion reaction [1.64] is much more stable than deuteron ^{2}H and triton ^{3}H, in accordance with their positions on the Aston curve (Figure 1.20). Likewise, products of fission ^{94}Sr (strontium-94) and ^{140}Xe (xenon-140) are more stable than the original ^{235}U uranium nucleus (see application 1.15).

APPLICATION 1.15.– Compare the stability of the nuclei of strontium-54, uranium-235 and xenon-140 using a horizontal axis. We will classify the nuclides in order of increasing stability. In addition, we will use the data given in Table 1.3.

Additional data: M (^{235}U) = 235.04392 u; M (^{94}Sr) = 93.91536 u; M (140 Xe) = 139.91879 u.

ANSWER.– Simply compare the binding energy per nucleon of each nuclide. Using [1.60], we obtain:

– E_l (^{94}Sr) = (38 × 1.00727646677 + 56 × 1.00866491597 – 93.91536) × 931.5 = 788.404 MeV;

– E_l (^{140}Xe) = (54 × 1.00727646677 + 86 × 1.00866491597 – 139.91536) × 931.5 = 1138.993 MeV;

– E_l (^{235}U) = (92 × 1.00727646677 + 143 × 1.00866491597 – 235.04392) × 931.5 = 1736.873 MeV.

The binding energies per nucleon are thus equal to:

$$E_l\,(^{94}\text{Sr})/A = 788.404/94 = 8.4 \text{ MeV/nucleon} \quad [1.66a]$$

$$E_l\,(^{140}\text{Xe})/A = 1138.993/140 = 8.1 \text{ MeV/nucleon} \quad [1.66b]$$

$$E_l\,(^{235}\text{U})/A = 1736.873/235 = 7.4 \text{ MeV/nucleon} \quad [1.66c]$$

Considering results [1.66], the nuclides considered are classified in order of increasing binding energy per nucleon (Figure 1.21).

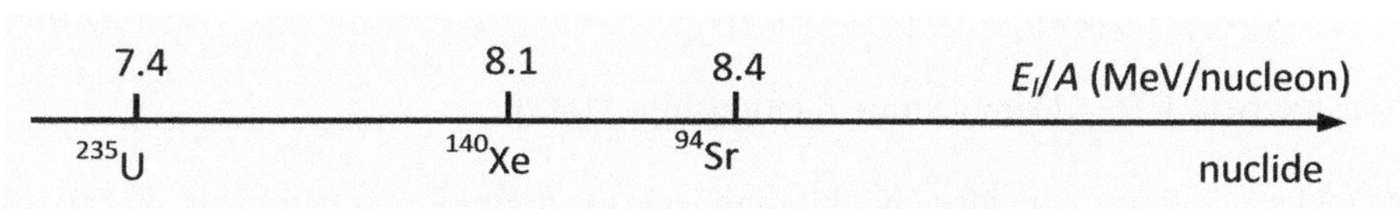

Figure 1.21. *Comparison of the stability of the nuclei of uranium-238, strontium-54 and xenon-140. The products of the fission reaction [1.65] are much more stable than the starting uranium-238*

Francis William Aston was a British chemist and physicist. He is famous for having developed the Aston spectrograph in 1919. The curve for locating the most stable nuclides in the valley of stability is named in his honor. He was awarded the Nobel Prize in Chemistry 1922 for his discovery, using his mass spectrometer, of isotopes of a large number of non-radioactive elements and for his formulation of the whole-number rule.

Box 1.12. *Aston (1877-1945)*

1.6.5. *Separation energy of a nucleon*

The experimental study of the variation in the separation energy of a nucleon from a nucleus provides an idea of its cohesion. By definition, the *separation energy of a nucleon* is the energy required to remove a nucleon from the nucleus. Nucleon separation energies are analogous to the first-ionization potential of neutral atoms (extraction of an electron).

For a nucleus of mass $M(Z, N)$, an isotope of mass $M(Z, N-1)$ is obtained if a neutron is removed, and an isotone of mass $M(Z-1, N)$ if a proton is removed. By convention, $S_n(Z, N)$ is used to designate the *separation energy of a neutron* and $S_p(Z, N)$ designates the *separation energy of a proton*. These energies are given by the following respective expressions:

$$S_n(Z, N) = [M(Z, N-1) + m_n - M(Z, N)]c^2 \qquad [1.67a]$$

$$S_p(Z, N) = [M(Z-1, N) + m_p - M(Z, N)]c^2 \qquad [1.67b]$$

According to the binding energies, E_l, relationships [1.67] are written:

$$S_n(Z, N) = E_l(Z, N) - E_l(Z, N-1) \qquad [1.68a]$$

$$S_p(Z, N) = E_l(Z, N) - E_l(Z-1, N) \qquad [1.68b]$$

APPLICATION 1.16.– Demonstrate relationships [1.48].

ANSWER.– Let us consider the binding energy definition relationship [1.60] by noting the mass $M(A, Z)$ of the nucleus with $M(Z, N)$. Knowing that $N = A - Z$, we obtain:

$$E_l(Z, N) = \{[ZM_p + Nm_n] - \mathrm{M}(Z, N)\}c^2 \qquad [1.69]$$

For the isotope of mass M (Z, $N-1$) corresponding to the removal of a neutron, the binding energy is calculated from relationship [1.69]. Thus:

$$E_l\,(Z, N-1) = \{[Zm_p + (N-1)\,m_n] - M\,(Z, N-1)\}c^2 \qquad [1.70a]$$

For the isotone of mass M ($Z-1$, N) corresponding to the removal of a proton, we obtain:

$$E_l(Z-1, N) = \{[(Z-1)\,m_p + Nm_n] - M\,(Z-1, N)\}c^2 \qquad [1.70b]$$

By subtracting [1.69] from [1.70], we find, respectively:

$$E_l\,(Z, N) - E_l\,(Z, N-1) = \{M\,(Z, N-1) + m_n - M\,(Z, N)\}c^2 \qquad [1.71a]$$

$$E_l\,(Z, N) - E_l\,(Z-1, N) = \{M\,(Z-1, N) + m_p - M\,(Z, N)\}c^2 \qquad [1.71b]$$

By comparing relationships [1.67] and [1.71], relationships [1.68] are found.

As shown in equations [1.68], the separation energy of a neutron represents the increase in the binding energy, E_l (Z, $N-1$) when a neutron is added to the smallest isotope (Z, $N-1$), giving the isotope (Z, N) (equation [1.68a]). Similarly, the separation energy of a proton represents the increase in the binding energy, E_l ($Z-1$, N) when a proton is added to the smallest isotone ($Z-1$, N), giving the isotone (Z, N) (equation [1.68b]). For these reasons, the separation energies of a neutron, S_n (Z, N) or a proton, S_p (Z, N) are also referred to, respectively, as the *last neutron separation energy* (i.e. the last neutron added) or *last proton separation energy* (the last proton added).

APPLICATION 1.17.– Determine the separation energy of the last neutron for the ${}^{7}_{3}Li$ nucleus and the separation energy of the last proton for the ${}^{16}_{8}O$ nucleus. We will start by writing the nucleon extraction equations for the two cases considered.

Given data: m_n = 1.0087 u; m_p = 1.0073 u; M (3.4) = 7.0143 u; M (3.3) = 6.0135 u; M (8.8) = 15.9905 u;

M (7.8) = 14.9963 u; 1u = 931.5 MeV/c^2.

ANSWER.– For the extraction equations of a lithium-7 neutron and an oxygen-16 proton, we obtain:

– extraction of a lithium-7 neutron, the lithium-6 isotope is obtained:

$${}^{7}_{3}Li \xrightarrow{extraction} {}^{6}_{3}Li + {}^{1}_{0}n \qquad [1.72a]$$

– extraction of an oxygen-16 proton, the nitrogen-15 isotone is obtained:

$$^{16}_{8}O \xrightarrow{extraction} {}^{15}_{7}N + {}^{1}_{1}H \qquad [1.72b]$$

Using [1.67], we obtain:

$$S_n(3.4) = [M(3.3) + m_n - M(3.4)]c^2 \qquad [1.73a]$$

$$S_p(8.8) = [M(7.8) + m_p - M(8.8)]c^2 \qquad [1.73b]$$

Numerically, we thus obtain:

$$S_n(3.4) = [6.0135 + 1.0087 - 7.0143] \times 931.5 = 7.359 \text{ MeV} \approx 7.4 \text{ MeV}$$

$$S_p(3.4) = [14.9963 + 1.0073 - 15.9905] \times 931.5 = 2.203 \text{ MeV} \approx 12.2 \text{ MeV}$$

1.6.6. *Nuclear forces*

In principle, nuclei should be unstable due to Coulomb repulsion between protons. *Nuclear forces* are manifested at a very short range equaling around 2×10^{-13} cm). At this distance, they are attractive and thus mask the forces of Coulomb repulsion between protons of the same nucleus. This explains the stability of the nuclei. Nevertheless, for distances under than 5×10^{-14} cm, nuclear forces are repellent as if there is an impenetrable hard core of nucleons within the nucleus. As we saw in section 1.3.4, the introduction of isospin is based on the fundamental assumption that *pp*, *pn* or *nn* nuclear interaction is the same owing to the charge independence of nuclear forces. The cohesion of the nuclei is explained by the existence of so-called nuclear forces. These forces are attractive and independent of the electrical charge

1.7. Exercises

EXERCISE 1.1.– Lifetime of the Rutherford planetary atom

Let us account for the instability of the planetary atomic model according to Rutherford. For this purpose, let us consider the electron of the hydrogen atom in motion around the proton.

a) Show that the electron is subjected to a central acceleration.

b) According to the conceptions of classical electrodynamics, any charged particle animated by accelerated motion emits electromagnetic waves. The *loss of energy per unit time* (radiated power) is expressed by the relationship:

$$-\frac{dE}{dt}=\frac{2}{3}\frac{e^2}{4\pi\varepsilon_0 c^3}a^2 \qquad [1.74]$$

In [1.74], a designates the acceleration of the particle.

i) Is Rutherford's atomic model then stable according to the conceptions of classical electrodynamics? Substantiate the answer.

ii) Show that the radius, r, of the electron orbit confirms the differential equation:

$$\frac{d}{dr}\left(\frac{1}{r}\right)=\frac{A}{r^4} \qquad [1.75]$$

with A a constant that will be explained below.

iii) Considering the fall of the electron from its initial position, $r = a_0$, express the duration, Δt, of the falling motion of the electron on the proton. Deduce therefrom the lifetime of the hydrogen atom.

Given datum: *Bohr radius*, a_0= 52.9 pm.

EXERCISE 1.2.– Discovery of isotopes

In 1886, Goldstein discovered positively charged rays that he named "*canal rays*". Sir J.J. Thomson showed in 1912, using the device presented in Figure 1.22, that these "canal rays" were composed of ions of two neon isotopes.

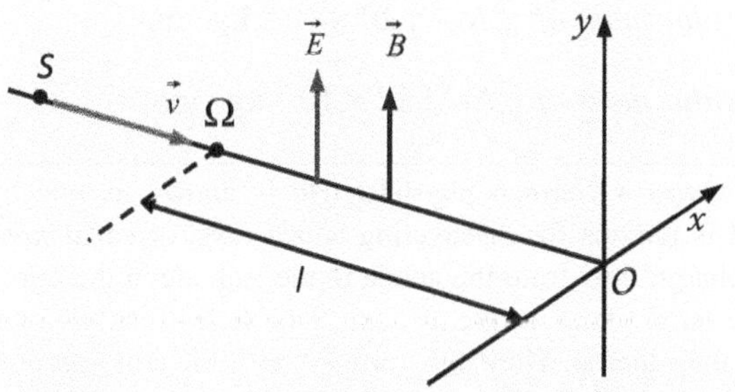

Figure 1.22. *Simplified Thomson device for identifying the canal rays discovered by Goldstein*

A beam consisting of positive ions moving along the *SO* axis, and different velocity vectors, $\vec{v}$, is subjected to the simultaneous actions of an electric field, $\vec{E}$, and a magnetic field, $\vec{B}$, which are parallel and facing the same way (in the *Oy* axis direction). Let $l = \Omega O$ be the width of the portion of space on which both fields are active.

a) Express the vertical deviation, y, of an ion of charge q, velocity v and mass m, subjected to the action of the electric field, $\vec{E}$, alone.

b) Similarly, express the horizontal deviation, x, of the same ion, subjected to the action of the magnetic field, $\vec{B}$, alone. We will assume that the deviation angle is small.

c) Under the conditions of the experiment, the x and y coordinates of the particle when the $\vec{E}$ and $\vec{B}$ fields are started up simultaneously are virtually equal to those obtained when they act alone. Establish the Cartesian equation of the trace formed on the screen by the impacts of ions of the same charge and mass but of different velocities. What shape is this trace?

d) What happens if the beam consists of ions of different mass charges, q/m?

e) In the case where the ion source is neon, two arcs of parabola can be observed on the screen, the extensions of which are tangent at O to the Ox axis.

i) Why are the parabolas not drawn up to point O?

ii) On the first parabola, note the Cartesian coordinates of a point, M_1: x_1 = 12.0 mm; y_1 = 4.5 mm. On the second parabola, the coordinates of a point M_2 are noted: x_2 = 10.9 mm; y_2 = 4.1 mm. Show that these are the traces of the two isotopes, ${}^{20}_{10}Ne^{+}$ and ${}^{22}_{10}Ne^{+}$, of neon.

Given data:

– *atomic molecular mass of* ${}^{20}_{10}Ne^{+}$: 20×10^{-3} kg · mol $^{-1}$;

– *atomic molecular mass of* ${}^{22}_{10}Ne^{+}$: 22×10^{-3} kg · mol $^{-1}$.

> Eugen Goldstein was a German physicist. He is known as one of the first to study discharge tubes and is famous for discovering anode rays, or canal rays. The positive ions produced in the discharge flow from the anode to the cathode in the opposite direction to the cathode rays. These ions, which arrive at one of the holes (canals) of the cathode, cannot reach it, because of their inertia. They thus form a "ray", the cross-section of which is that of the hole. In 1886, Goldstein called these positive rays "canal rays" because they appeared to pass through a canal.

Box 1.13. *Goldstein (1850–1930)*

EXERCISE 1.3.– Mass spectrograph

An ionization chamber (*IC*) produces $^{79}Br^-$ and $^{81}Br^-$ isotope ions of charge q and respective masses, m_1 and m_2. These ions penetrate with negligible velocity at O into a vacuum chamber (A) where they are accelerated by a constant voltage, U_0.

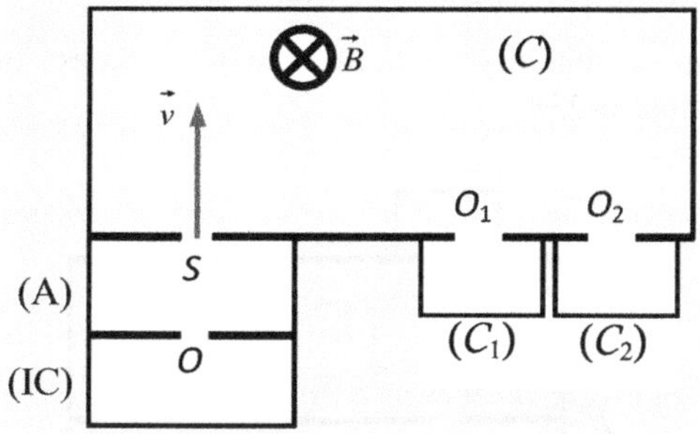

Figure 1.23. *Simplified diagram of a mass spectrograph*

They then exit at S with velocity $\vec{v}$ to penetrate into a magnetic deflector (C) where there is a uniform magnetic field $\vec{B}$ perpendicular to the plane of the figure. The two types of ions are then collected in the collectors (C_1) and (C_2), where they penetrate from orifices O_1 and O_2, respectively (Figure 1.23). All orifices are assumed to be point-like.

a) Express the input velocity, v, of an ion of mass m and charge q in the magnetic deflector according to m, q and U_0.

b) Show that the trajectory of an ion in the magnetic deflector is planar and circular. What, then, is the nature of the motion?

c) Identify by name the ion collected in the collector (C_1). Now plot the trajectories of the ions up to the two collectors.

d) Express the distance $D = O_1O_2$. Perform the numerical application.

e) Within one minute, the collectors (C_1) and (C_2) receive, respectively, the following amounts of electricity: $Q_1 = -\ 6.60 \times 10^{-8}$ C and $Q_2 = -\ 1.95 \times 10^{-8}$ C. Determine the composition of the isotope ion mixture.

Given numerical data:

– *absolute value of the ion charge*: $|q| = 1.6 \times 10^{-19}$ C;

– *applied voltage*: $|U_0| = 4$ kV;

– *ion masses*: $m_1 = 1.310\,4 \cdot 10^{-25}$ kg; $m_2 = 1.343\,6 \times 10^{-25}$ kg;

– *magnetic field value*: $B = 0.10$ T.

EXERCISE 1.4.– Separation of the uranium isotopes, ^{235}U and ^{238}U

Natural uranium essentially contains two isotopes of different masses: uranium-235 and uranium-238.

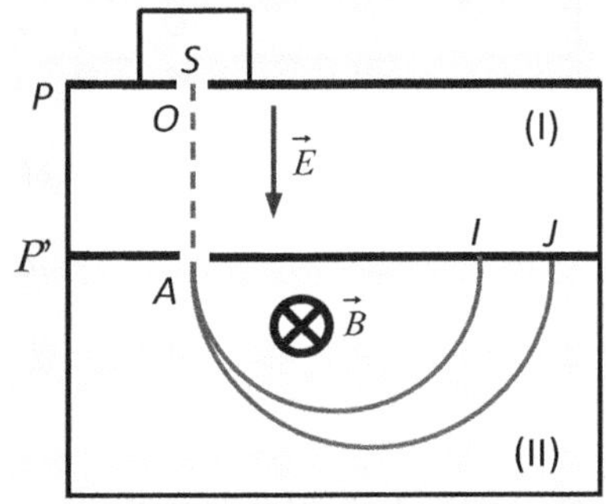

Figure 1.24. *Uranium isotope trajectories in a magnetic deflector*

One of the historical processes for separating these isotopes is based on the difference between the radii of curvature of the trajectories of ionized atoms in motion in a uniform magnetic field using a mass spectrograph. First, the neutral atoms are injected into an ion source where they each lose one electron. Thus, the ions formed in this way each carry the same electrical charge, $q = +e$. They exit the ion source with negligible velocity, then they are accelerated from O to A by a constant electrical voltage, $U_0 = V_P - V_{P'}$, applied between the two metal plates, P and P' (Figure 1.24). The ions then penetrate at A with a velocity of direction perpendicular to the plates in a region where there is a uniform magnetic field, $\vec{B}$, perpendicular to the direction of the electric field, $\vec{E}$, created between the two plates.

a) The ions of an isotope i ($i = 1, 2$) penetrate into the magnetic deflector with a velocity of v_I. Express values v_1 and v_2 of the respective velocities of the uranium 235 and 238 ions according to the charge, the mass of the corresponding ion and the accelerating voltage. Calculate v_1 and v_2.

b) Beyond point A, the ions travel with circular trajectories located in the plane of Figure 1.23. R_1 designates the radius of trajectory AI and R_2 the radius of trajectory *AJ*. Calculate R_1 and R_2 and express the distance $l = IJ$ between the two traces *I* and *J* of the isotope ions. Calculate *l*.

c) The ion current from the source corresponds to an intensity of 10 μA. Knowing that natural uranium contains, in number of atoms, 0.7% light isotope (the only fissile), what is the mass, in μg, of this isotope collected in 24 hrs?

Given data:

– $B = 0.10$ T; $U_0 = 4$ kV; $e = 1.6 \times 10^{-19}$ C;

– *atomic mass unit* (u): 1 u = 1.66×10^{-27} kg;

– *Avogadro's number*: $N_A = 6.02 \times 10^{23}$ mol $^{-1}$;

– *mass of the uranium-235 ion*: $m_1 = 235$ u;

– *mass of the uranium-238 ion*: $m_2 = 238$ u.

1.8. Solutions to exercises

SOLUTION 1.1.– Lifetime of the Rutherford planetary atom

a) *Nature of electron acceleration*

In the Frenet frame $(M, \vec{\tau}, \vec{n})$, a particle in circular motion with the velocity $\vec{v}$ is subjected to the acceleration given by the relationship:

$$\vec{a} = \frac{d\vec{v}}{dt}\vec{\tau} + \frac{v^2}{\rho}\vec{n} \qquad [1.76]$$

In relation [1.76], ρ is the radius of curvature of the trajectory.

The motion of the electron of the hydrogen atom is uniform, therefore v = const. By writing $\rho = R$, we obtain, according to [1.76]:

$$\vec{a} = \vec{a}_n = \frac{v^2}{R}\vec{n} \qquad [1.77]$$

b) *Planetary atomic model*

i) Stability

According to the conceptions of classical electrodynamics, the electron of the hydrogen atom subjected to central or centripetal acceleration [1.77] radiates energy.

This loss of energy through radiation induces a falling motion in the electron that eventually falls on the nucleus. Yet this does not happen because the hydrogen atom exists as a stable bound system: the Rutherford atomic model is therefore unstable.

ii) Demonstration

The total mechanical energy of the bound system {proton – electron} constituting the hydrogen atom is given by the relationship:

$$E(r) = -k\frac{e^2}{2r} \qquad [1.78]$$

Furthermore, the electric force between the proton and the electron being a central force in $1/r^2$, the kinetic energy $E_c(r) = -E(r)$, thus, by using:

$$E_c(r) = \frac{1}{2}mv^2 = k\frac{e^2}{2r} \qquad [1.79]$$

By deriving v from [1.79], we obtain the relationship:

$$v^4 = k^2\frac{e^4}{m^2r^2} \qquad [1.80]$$

Given that acceleration $a = a_n = v^2/r$, the radiated power [1.74] is written, taking into account [1.80]:

$$-\frac{dE(r)}{dt} = \frac{2}{3}\times\frac{ke^2}{c^3}\frac{v^4}{r^2} = \frac{2}{3}\times\frac{k^3e^6}{m^2c^3}\times\frac{1}{r^4} \qquad [1.81]$$

If we then derive [1.78] with respect to time and then take into account [1.81], we obtain:

$$\frac{dE(r)}{dt} = -\frac{ke^2}{2}\frac{d}{dt}\left(\frac{1}{r}\right) = \frac{2}{3}\times\frac{k^3e^6}{m^2c^3}\times\frac{1}{r^4}$$

Which then gives:

$$\frac{d}{dt}\left(\frac{1}{r}\right) = \frac{A}{r^4}, \quad A = \frac{4}{3}\frac{k^2e^4}{m^2c^3} \qquad [1.82]$$

iii) Lifetime of the hydrogen atom

At instant $t_0 = 0$, the electron is at the distance $r = a_0$ from the proton. When it falls on the proton at instant t_c, then $r = 0$. By integrating [1.82] between the limits t_0, t_c and a_0 and 0, we obtain, using the expression of A, according to [1.82]:

$$\Delta t = A(t_c - t_0) = \frac{a_0^3}{3} \Rightarrow \Delta t = \frac{m^2 c^3 a_0^3}{4k^2 e^4} \quad [1.83]$$

NOTE.–

$$\Delta t = \frac{(9.1\times10^{-31})^2 \times (3\times10^8)^3 \times (5.29\times10^{-11})^3}{4\times(9\times10^9)^2 \times (1.6\times10^{-19})^4} = 1.559\times10^{-11} \approx 1.6\times10^{-11} \text{ s.}$$

SOLUTION 1.2.– Discovery of isotopes

a) *Expression for vertical deviation*

The electric field acts alone. Drawing on the *theorem of the center of inertia*, we obtain:

$$\vec{F} = m\vec{a} = q\vec{E} \Rightarrow \begin{cases} \dfrac{d^2x}{dt^2} = 0 \\ \dfrac{d^2y}{dt^2} = \dfrac{qE}{m} \end{cases} \Rightarrow \begin{cases} x = vt \\ y = \dfrac{1}{2}\dfrac{qE}{m}t^2 \end{cases} \quad [1.84]$$

On the screen, $x = l \Rightarrow t = l/v$. We obtain, according to [1.84]:

$$y = \frac{1}{2}\frac{qEl^2}{mv^2} \quad [1.85]$$

b) Expression for horizontal deviation

The magnetic field acts alone. Likewise we obtain:

$$\vec{F} = m\vec{a} = q\vec{v} \wedge \vec{B} \Rightarrow \begin{cases} \dfrac{d^2y}{dt^2} = 0 \\ \dfrac{d^2x}{dt^2} = \dfrac{qvB}{m} \end{cases} \Rightarrow \begin{cases} y = vt \\ x = \dfrac{1}{2}\dfrac{qvB}{m}t^2 \end{cases} \quad [1.86]$$

On the screen , $y = l \Rightarrow t = l/v$. Using [1.86], we obtain:

$$y = \frac{1}{2}\frac{qEl^2}{mv^2} \qquad [1.87]$$

c) Cartesian equation, the nature of the trace formed on the screen

Using [1.85] and [1.87] we obtain:

$$y = 2\frac{m}{q}\frac{E}{mB^2l^2}x^2 \qquad [1.88]$$

The trace is a parabola of vertex O.

d) Case of an ion beam of different mass charges

For ions of different mass/charge ratios, q/m, separate parabolas are obtained on the screen: the isotopes are then separated.

e) Observation on the screen

i) Explanation

Since the velocities of the ions are low, they are deflected before arriving at O. The impact points of ions of the same charge and mass but of different velocities are then distributed over two arcs of parabola whose extensions are tangent at O to the Ox axis.

ii) Isotope identification

For an ion (i) we obtain, according to [1.88]:

$$\frac{q}{m_i} = 2\frac{E}{B^2l^2}\frac{x_i^2}{y_i} \qquad [1.89]$$

Using [1.89], we then obtain:

$$\frac{m_2}{m_1} = \frac{x_1^2}{x_2^2}\frac{y_2}{y_1} \qquad [1.90]$$

Experimental results: in M_1: x_1 = 12.0 mm; y_1 = 4.5 mm; in M_2: x_2 = 10.9 mm; y_2 = 4.1 mm. We then find, numerically, according to [1.90]:

$$\begin{cases} \dfrac{m_2}{m_1} = \dfrac{12^2}{10.9^2}\dfrac{4.1}{4.5} = 1.104 \\ \dfrac{m_2}{m_1} = \dfrac{A_2 u}{A_1 u} = \dfrac{22}{20} = 1.100 \end{cases} \qquad [1.91]$$

The results [1.91] show that the isotopes are those of neon, $^{20}Ne^+$ and $^{22}Ne^+$.

SOLUTION 1.3.– Mass spectrograph

a) Expression for input velocity

Let us apply the *theorem of kinetic energy*. We obtain:

$$\frac{1}{2}mv^2 = |q|U_0 \Rightarrow v = \sqrt{\frac{2|q|U_0}{m}} \qquad [1.92]$$

b) Demonstration

– Nature of the trajectory

Let us consider a portion of the trajectory of an ion of charge q in the Frenet frame (Figure 1.25).

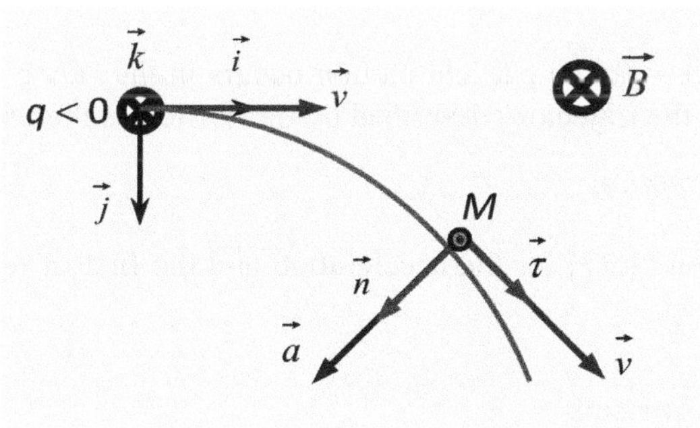

Figure 1.25. *Portion of the trajectory of a particle of charge q in motion in a magnetic deflector*

Let us determine the power developed by the *Lorentz force* acting on the particle. Taking into account the *properties of the vector product*, we obtain:

$$P=\vec{F}\cdot\vec{v}=(q\vec{v}\wedge\vec{B})\cdot\vec{v}=0 \qquad [1.93]$$

Using [1.93], the kinetic energy theorem gives:

$$\Delta E_c = W(\vec{F}) = P\Delta t = 0 \Rightarrow v = \text{constant} \qquad [1.94]$$

The motion of an ion in the magnetic deflector is therefore uniform according to [1.94]. Let us show that the trajectory described by the ion is planar.

Let us determine the scalar product $\vec{a}\cdot\vec{k}$ in the Frenet frame. Knowing that $\vec{B} = B\cdot\vec{k}$ (see Figure 1.25), we obtain, by deriving the acceleration from the first of relationships [1.86]:

$$\vec{a}\cdot\vec{k} = \left(\frac{q}{m}\vec{v}\wedge\vec{B}\right)\cdot\vec{k} = 0 \qquad [1.95]$$

Knowing that $a_z = 0$ according to [1.95] and that at instant $t = 0$, the particle is at the origin point, O ($z_0 = 0$, $v_{0Z} = 0$), then the time equation of the motion along Oz is written:

$$z(t) = \frac{1}{2}a_z t^2 + v_{0z}t + z_0 = 0 \qquad [1.96]$$

The z coordinate $= 0$, the particle motion occurs in the xOy plane, orthogonal to the magnetic field; the trajectory described by the particle is therefore planar.

– *Nature of the motion*

Using expression [1.77] for the acceleration and the first of relationships [1.86], we obtain:

$$R = \frac{mv}{|q|B} = \text{Constant} \qquad [1.97]$$

The velocity [1.94] and the radius of curvature [1.97] are constant: the motion of an ion in the magnetic deflector is uniform circular motion.

c) Identification of the ion collected in the collector (C_1)

According to [1.97], the ion collected in the collector (C_1) has a trajectory with the smallest radius and therefore with the smallest mass, $m = A$u: it is $^{79}Br^-$ and $^{81}Br^-$. The shape of the trajectories of the two ions is shown in Figure 1.26.

d) Expression of the distance, D, numerical application

The distance $D = O_1O_2 = 2(R_2 - R_1)$. Thus, using [1.97] and [1.92]:

$$D = \frac{2}{eB}(m_2v_2 - m_1v_1) = \frac{2}{B}\sqrt{\frac{2|U_0|}{e}}\left(\sqrt{m_2} - \sqrt{m_1}\right) \quad [1.98]$$

NOTE.–

$$D = \frac{2}{0.1}\sqrt{\frac{2\times 4\times 10^3}{1.6\times 10^{-19}}}\left(\sqrt{1.3436\times 10^{-25}} - \sqrt{1.3104\times 10^{-25}}\right) = 0.0204 \text{ m}$$

Thus: $D = 20$ mm

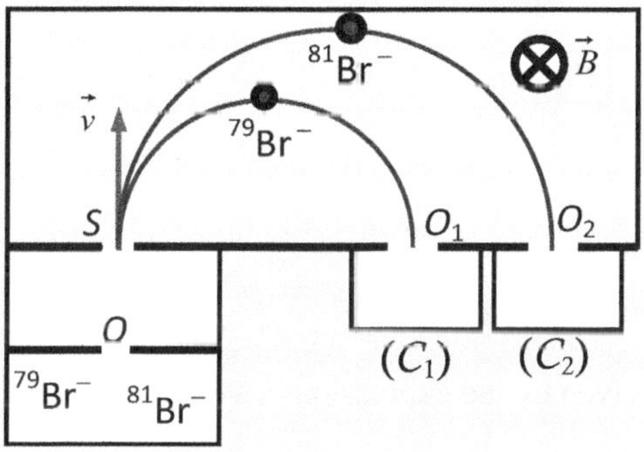

Figure 1.26. *Circular trajectories of isotope ions in the magnetic deflector*

e) Composition of the isotope ion mixture

In one minute, the collectors (C_1) and (C_2) receive the amounts of electricity, Q_1 and Q_2, respectively. The total charge, $Q = |Q_1| + |Q_2|$. This then gives:

$Q_1 = -6.60 \times 10^{-8}$ C; $Q_2 = -1.95 \times 10^{-8}$ C $\Rightarrow Q = 8.55 \times 10^{-8}$ C.

$^{79}Br^-$: $x_1 = |Q_1|/Q = 77.19\%$; $^{80}Br^-$: $x_2 = |Q_2|/Q = 22.81\%$

SOLUTION 1.4.– Separation of the uranium isotopes, ^{235}U and ^{238}U.

a) Velocity expressions and values

Drawing on the theorem of kinetic energy, we obtain:

$$m_2 v_2^2 = m_2 v_2^2 = 2qU_0 \Rightarrow v_i = \sqrt{\frac{2qU_0}{m_i}} = \sqrt{\frac{2eU_0}{A_i u}} \qquad [1.99]$$

NOTE.– ^{235}U: $v_1 = 5.73 \times 10^4$ m · s^{-1}; ^{238}U: $v_2 = 5.69 \times 10^4$ m · s^{-1}

b) Radius values: expression and value of the length between traces

– *Radius value*

The radius of curvature of the trajectory of an ion is given by [1.97]. Taking account of [1.99], for an ion, i, we obtain:

$$R = \frac{m_i v_i}{|q| B} = \frac{1}{B}\sqrt{\frac{2U_0 A_i u}{e}} \qquad [1.100]$$

NOTE.– $^{235}U^+$: $R_1 = 1.397$ m; $^{238}U^+$: $R_1 = 1.406$ m.

– *Value of the length between traces*

The value of l is given by the expression [1.98], which we write in the form:

$$l = \frac{2}{B}\sqrt{\frac{2U_0 u}{e}}\left(\sqrt{A_2} - \sqrt{A_1}\right) \qquad [1.101]$$

NOTE.– $l \approx 18$ mm.

c) Mass of uranium-235 collected

Let Q be the amount of electricity that has circulated. Using N to designate the number of particles (here uranium-235 isotope ions), n for the number of moles of ions and N_A for Avogadro's number, the following relationships are obtained:

$$\begin{cases} Q = Ne = It \\ n = \dfrac{N}{N_A} = \dfrac{m}{A} \Rightarrow N = \dfrac{mN_A}{A} = \dfrac{It}{e} \end{cases} \quad [1.102]$$

Knowing that the ^{235}U isotope contributes 0.7% to the ion current, using [1.102], we obtain a mass:

$$m_1 = 0.7\,\% \frac{ItA_1}{eN_A} \quad [1.103]$$

NOTE.– $m_1 \approx 15$ µg.

2

Nuclear Deexcitations

Overall objective	
To describe the processes of nuclear deexcitation	
Specific objectives	
To compare the internal conversion probabilities of a *K*-electron, *L*-electron, and *M*-electron	To determine a γ-deexcitation probability based on Weisskopf estimates
To compare the shell structures derived from a harmonic potential and a Woods–Saxon potential	To differentiate between bound and virtual levels of nuclei
To know the expression of the quantified energy of a nucleon subjected to a harmonic potential	To distinguish between the different internal conversion γ-deexcitation processes
To know the order of the quantum states of a nucleon within the framework of the shell model	To distinguish the parity of an electric multipole from that of a magnetic multipole
To know the quantum numbers characterizing the individual state of a nucleon	To distinguish between electric multipole transitions and magnetic multipole transitions
To know the conditions of nuclear deexcitation by nucleon emission	To deduce the expression of binding energy from the Bethe–Weizsäcker formula
To know the spectroscopic notation of a nucleonic state	To establish the relationship between the half-life of an excited nuclear level and the internal conversion coefficient

For a color version of all of the figures in this chapter, see www.iste.co.uk/sakho/nuclear1.zip.

To know the classification of the γ-transitions according to the multipole order	To explain the mechanism of nuclear deexcitation by delayed-neutron emission
To know the principle for determining coefficients in the Bethe–Weizsäcker formula	To explain the shell structure derived from a Woods–Saxon potential, with spin-orbit coupling
To describe the Hamiltonian of a nucleon subjected to the Woods–Saxon potential with spin-orbit coupling	To express the total half-life of an excited state as a function of partial half-lives of deexcitation by γ-photon emission and conversion electron emission
To describe the nuclear shell model	To express the total conversion coefficient according to the partial conversion coefficients
To describe the liquid-drop model	To perform the energy balance of nuclear deexcitation by nucleon emission
To define the multipole order of γ-radiation	To correlate virtual nuclear levels with deexcitation processes by emission of particles (n, p, α, etc.)
To define the internal conversion coefficient	To make the link between X-photon emission and internal conversion and electronic capture phenomena, thus creating an electronic gap
To define the total probability per unit time of nuclear deexcitation	To interpret the shape of the harmonic potential of depth V_0
To determine the J^{π} of the ground state of a given nucleus	To interpret the shape of the Woods–Saxon potential
To determine the J^{π} of the excited states of a given nucleus	To interpret the variation in the α_K conversion coefficient with the atomic number
To determine the K-shell internal conversion coefficient for electric multipole transitions and magnetic multipole transitions	To interpret the variation in the K–shell internal conversion coefficient with the energy of nuclear transition and of the multipole order
To determine the degree of degeneracy of a nucleonic state according to the shell model	To interpret the terms in the Bethe–Weizsäcker formula
To determine the distribution of nucleons of a light nucleus according to the shell model	To justify the preference for the Woods–Saxon potential over the Yukawa potential
To determine an atomic mass from the Bethe–Weizsäcker formula	To use the nucleonic level filling rules
To determine the angular momentum and parity of a nucleonic state	To use the principles of angular momentum and parity conservation

Prerequisites	
Spin-orbit coupling in hydrogen-like systems	Laws of restricted relativity
Stationary Schrödinger equation	Energy and momentum conservation laws
Hamiltonian of hydrogen-like systems	Atomic model of the electron shells
Hamiltonian of hydrogen-like systems in the weakly relativistic domain	Classical harmonic oscillator
Orbital angular momentum operator properties	Three-dimensional quantum harmonic oscillator
Potential wells in quantum mechanics	

2.1. Nuclear shell model

2.1.1. *Overview of nuclear models*

Nuclear models enable the internal structure and properties of nuclei to be studied. None of the models proposed offer a precise explanation of all of the experimental observations related to the structure and properties of nuclei. Each of these models is limited to a certain field of validity, according to the basic hypotheses formulated to develop them. In general, depending on the type of weak or strong interaction adopted, nuclear models can be classified into two categories: the model of independent particles, in which nucleon interactions are weak (such as the nucleon shell model) and the model of strongly correlated particles, in which nucleon interactions are very strong (such as the liquid-drop model, the collective model, etc.).

The shell model is similar to the electronic shell model of atomic systems and allows the interpretation of physical phenomena related to nuclear angular momenta and magnetic moments, parity and dipole moment of ground states and weakly excited states of nuclei. This model also helps to explain the particular stability of the magic nuclei.

The liquid-drop model helps explain nuclear phenomena that it is not possible to describe within the framework of the shell model. Many phenomena related to the binding energy of ground levels of nuclei, the energy of α and β decay, nuclear fission, etc., are successfully interpreted within the framework of the liquid-drop model. In this book, we will focus the study on the case of the shell model and the liquid-drop model for the reasons set out above.

NOTE.– *Within the framework of the collective model, nucleons in strong interaction are described as evolving in vibrational and rotational motions. In this case, the nucleus can be likened to a molecule, and can therefore undergo vibrational and rotational excitations.*

2.1.2. *Individual state of a nucleon*

By analogy with atomic electrons, the state of a nucleon in the nucleus is described by the following quantum numbers [EVA 61, BLA 99, BRÖ 01, STÖ 07, MAY 17]:

1) the *principal quantum number*, n, characterizing each nucleon shell; n, a positive integer;

2) the *orbital quantum number*, ℓ, characterizing the motion of each nucleon in the nucleus; it takes the values: 0, 1, 2, etc. $n - 1$; this quantum number is associated with the *orbital magnetic quantum number*, m_ℓ, characterizing the projection of the orbital angular momentum, $\vec{l}$, along a preferred direction; it takes the possible values: $-\ell \leq m_\ell \leq +\ell$, that is, in total, (2+1) values;

3) the *spin*, $s = 1/2$; the spin is associated with the *magnetic quantum number of spin*, m_s, characterizing the projection of the orbital angular momentum, $\vec{s}$, along a preferred direction; it takes two possible values: $m_s = \pm\ 1/2$, that is, in total, $(2s + 1)$ values;

4) the *total quantum number*, j, taking the spin-orbit coupling into account; this quantum number is associated with the *total magnetic quantum number*, m_j, characterizing the projection of the orbital angular momentum, $\vec{j}$, along a preferred direction: it takes the possible values: $-j \leq m_j \leq +j$, that is, in total, $(2j + 1)$ values.

According to the vector addition rule, the *total angular momentum*, $\vec{j}$, reflecting the *spin-orbit coupling* is given by the relationship:

$$\vec{j} = \vec{l} + \vec{s} \tag{2.1}$$

Since the spin of a nucleon can only have two possible orientations (up and down), j takes the possible values:

$$j = \ell \pm \frac{1}{2} \tag{2.2}$$

When $\ell = 0$, the "+" sign is considered in relationship [2.2], since $j > 0$. Moreover, j is always a half-integer according to [2.2].

Unlike the electronic shells attracted by a nucleus placed in the center of the atom, there is no center of matter in a nucleus that would exert an attractive force on the A nucleons. To make the analogy with central-nucleus electron shells, it is assumed that each nucleon is subjected to a central attractive force from the $(A - 1)$ nucleons. Thus, the shell model describes the motion of the nucleons as independent particles, each of which is subjected to an *average potential*, $V(r)$, generated by the nucleons themselves.

If r designates the distance between a nucleon and the center of the nucleus and a the diffusivity parameter, then two main types of potential are used to describe the interaction occurring between a nucleon and the $(A - 1)$ other nucleons:

– The *potential well* of depth V_0, such that:

$$V(r) = -V_0 \text{ for } r \leq a;\ V(r) = 0 \text{ for } r > a \quad [2.3]$$

– The *harmonic potential.*

These two types of potential are studied below.

2.1.3. *Form of the harmonic potential*

Within the framework of the constant nucleon density model, the radius of the nucleus assumed to be spherical is $R = r_0 A^{1/3}$. In the case of the harmonic potential (potential energy), the potential function, $V(r)$, is written in the form:

$$V(r) = -V_0\left(1 - \frac{r^2}{R^2}\right) \quad [2.4]$$

It is known that for a classical oscillator of mass m with a dimension x, subjected to the restoring force $F = -kx$, the potential energy, $V(x)$, is given by the relationship:

$$V(x) = \frac{1}{2}kx^2 + V_0 \quad [2.5]$$

In relationship [2.5], V_0 represents the value of the potential energy at the origin of the coordinates at $x = 0$.

Moreover, for a classical oscillator, the pulsatance, ω, of the oscillations is given by the relationship:

$$\omega = \sqrt{\frac{k}{m}} \tag{2.6}$$

Using [2.6], the elastic potential energy [2.5] takes the form:

$$V(x) = \frac{1}{2}m\omega^2 x^2 + V_0 \tag{2.7}$$

Let us then write [2.4] in a form analogous to [2.7], by introducing the pulsatance, ω, of the oscillations of a nucleon into the nucleus.

Let us thus expand [2.4] as follows:

$$V(r) = -V_0 + V_0 \frac{r^2}{R^2} \tag{2.8}$$

Let m be the mass of a nucleon. The last term of the right-hand member of [2.8] can be transformed as follows:

$$V(r) = -V_0 + \frac{1}{2}m\omega^2 r^2 \tag{2.9}$$

In relationship [2.9], the pulsatance, ω, is given by the relationship:

$$\omega = \sqrt{\frac{2V_0}{mR^2}} \tag{2.10}$$

Relationship [2.9] shows that at the center of the nucleus ($r = 0$), assumed attractive, $V(r) = -V_0$. In addition, for a nucleon located at the surface of the nucleus ($r = R$), $V(r) = 0$. This result means that beyond the nucleus surface, strong nucleon interactions are no longer felt (this is justified by the fact that nuclear forces are short-range). The shape of potential [2.9] is shown in Figure 2.1.

APPLICATION 2.1.– Demonstrate relationship [2.5].

ANSWER.– The restoring force curl is null:

$$\vec{\nabla} \wedge \vec{F} = \vec{0} \tag{2.11}$$

The force $\vec{F}$ therefore derives from the gradient of the scalar potential, $V(x)$, i.e.:

$$\vec{F} = -\vec{\nabla} V(x) \Rightarrow Fdx = -\,dV \qquad [2.12]$$

By applying $F = -\,kx$ in [2.12], we obtain [2.5] after integration.

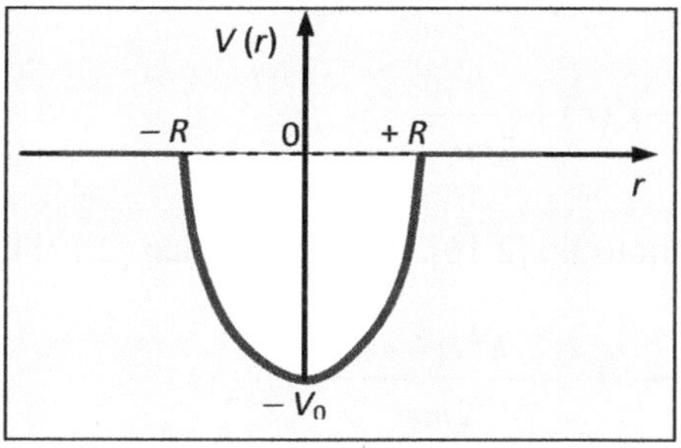

Figure 2.1. *Profile of a harmonic potential well of depth V_0*

2.1.4. *Shell structure derived from a harmonic potential*

The problem comes down to determining the energy, E, of a nucleon in the harmonic potential well within the framework of quantum mechanics. By noting with $\Psi\ (r,\theta,\varphi)$ the function describing the state of a Hamiltonian H nucleon, the Schrödinger equation is written:

$$H\Psi = E\,\Psi \qquad [2.13]$$

In equation [2.16], the Hamiltonian

$$H = H_r + H_\ell \qquad [2.14]$$

In equation [2.14], H_r is the radial part of the Hamiltonian associated with the kinetic energy of the particle (here the nucleon) and with the potential energy $V(r)$ and H_ℓ is the operator associated with the rotational kinetic energy of the particle (here the nucleon) around the attractive center of the nucleus. These operators are analogous to those defined in the case of hydrogen-like systems [SAK 12]:

$$H_r = -\frac{\hbar^2}{2m}\nabla_r^2 + V(r)\,;\, H_\ell = \frac{l^2}{2mr^2} \qquad [2.15]$$

In quantum mechanics, only the square l^2 and one of the l_z projections of the orbital angular momentum operator, $\vec{l}$, have determined values. Thus:

$$l^2 = \hbar^2 \ell(\ell+1) \; ; \; l_z = m_\ell \hbar \qquad [2.16]$$

Using [1.59], the Hamiltonian H is then written:

$$H = -\frac{\hbar^2}{2m}\nabla_r^2 + V(r) + \frac{l^2}{2mr^2} \qquad [2.17]$$

Using the first relationship [2.16], the Hamiltonian [2.17] takes the form:

$$H = -\frac{\hbar^2}{2m}\nabla_r^2 + V(r) + \frac{\hbar^2 \ell(\ell+1)}{2mr^2} \qquad [2.18a]$$

In spherical coordinates, the radial part of the Laplacian is written:

$$\Delta = \nabla_r^2 = \frac{1}{r^2}\frac{d}{dr}\left(r^2 \frac{d}{dr}\right) = \frac{d^2}{dr^2} + \frac{2}{r}\frac{1}{dr} \qquad [2.18b]$$

Given that the *observables* $\mathrm{H_r}$, l^2 and l_z constitute a CSCO, they therefore have the same *eigenfunction*, $\Psi(r,\theta,\varphi) = R_{n,\ell}(r)\mathrm{Y}_\ell^m(\theta,\varphi)$. The *Schrödinger radial equation* is then written [SAK 12]:

$$\frac{d^2 R_{n,\ell}(r)}{dr^2} + \frac{2}{r}\frac{dR_{n,\ell}(r)}{dr} + \frac{2m}{\hbar^2}\left(E - V(r) - \frac{\hbar^2 \ell(\ell+1)}{2mr^2}\right)R_{n,\ell}(r) = 0 \qquad [2.18c]$$

In the case of a *three-dimensional quantum harmonic oscillator*, the resolution of equation [2.18c] shows that the energy, E, is quantified and given by the relationship (see Appendix 1):

$$E = h\omega\left(n + \frac{3}{2}\right) \qquad [2.19]$$

In relationship [2.19], the integer n is the principal quantum number with $n \geq 0$. In the case of a nucleon subjected to potential [2.12], a solution analogous to [2.19] is obtained, i.e.:

$$E_N = h\omega\left(N + \frac{3}{2}\right) \qquad [2.20]$$

For the *principal quantum number*, $N \geq 0$, each value of N determines the number of a nucleon shell. For $N = 0$, the nucleons distributed over shell I are obtained, for $N = 1$, the nucleons distributed over shell II are obtained, for $N = 2$, the nucleons distributed over shell III are obtained, and so on. For the first three shells, the energy of a nucleon is equal to:

$$E_0 = \frac{3}{2} h\omega \,;\, E_2 = \frac{5}{2} h\omega \,;\, E_3 = \frac{7}{2} h\omega$$

Moreover, within the framework of the harmonic potential, each nucleon is characterized by a group of four quantum numbers: k, ℓ , m_ℓ and m_s, with k the radial quantum number (often noted n_r) equal to the number of nodes of the wave function describing the state of the nucleon, $k > 0$. The quantum numbers ℓ, m_ℓ and m_s are already defined in section 2.3.2. The principal quantum number, N, verifies the relationship:

$$N = 2(k-1) + \ell \qquad [2.21]$$

As with electrons, two identical nucleons cannot have the same spin orientation under *Pauli's exclusion principle*.

For a given value of N, the subshells are determined by the pairs of values (k, ℓ). In addition, for a shell N, the magnetic quantum number of spin, m_s, takes a total of ($2s + 1$) values, while the orbital magnetic quantum number takes ($2\ell + 1$) possible values. The *degree of degeneracy* or the number of quantum states characterized by the same energy value, E_N, is given by the relationship ($s = 1/2$ for a given nucleon):

$$g_N = (2s+1) \times (2\ell + 1) = 2(2\ell + 1) \qquad [2.22]$$

Using relationships [2.21] and [2.22], we determine the number of identical nucleons per shell. We can thus find some of the magic numbers characterizing particularly stable nuclei.

1) Shell I: $N = 0 \Rightarrow 0 = 2(k - 1) + \ell$. We thus obtain:

– for $k = 1$, $\ell = 0$: subshell 1s containing 2 ($2 \times 0 + 1$) = 2 nucleons;

– for $k = 2$, $0 = 2 + \ell$: there is no value of ℓ.

CONCLUSION.– The shell $N = 0$ therefore contains 2 identical nucleons in the 1s state.

2) Shell II: $N = 1 \Rightarrow 1 = 2(k - 1) + \ell$. We obtain:

– for $k = 1$, $\ell = 1$: subshell 1p containing 2 ($2 \times 1 + 1$) = 6 nucleons;

– for $k = 2$, $1 = 2 + \ell$: there is no value of ℓ.

CONCLUSION.– The shell $N = 1$ therefore contains 6 identical nucleons in the 1p state.

3) Shell II: $N = 2 \Rightarrow 2 = 2(k - 1) + \ell$. We obtain:

– for $k = 1$, $\ell = 2$: subshell 1d containing $2\,(2 \times 2 + 1) = 10$ nucleons;

– for $k = 2$, $\ell = 0$: subshell 2s containing $2\,(2 \times 0 + 1) = 2$ nucleons;

– for $k = 3$, $2 = 4 + \ell$: there is no value of ℓ.

CONCLUSION.– Shell $N = 2$ thus contains 12 identical nucleons in the 1d and 2s states (the second state, "s", the first state, "s", corresponding to $N = 0$).

4) Shell III: $N = 3 \Rightarrow 3 = 2(k - 1) + \ell$. We obtain:

– for $k = 1$, $\ell = 3$: subshell 1f containing $2\,(2 \times 3 + 1) = 2$ nucleons;

– for $k = 1$, $\ell = 1$: subshell 2p containing $2\,(2 \times 1 + 1) = 6$ nucleons;

– for $k = 3$, $3 = 4 + \ell$: there is no value of ℓ.

CONCLUSION.– Shell $N = 3$ thus contains 20 identical nucleons in the 1f and 2p states (the second "p" state, the first "p" state corresponding to $N = 1$).

5) Shell III: $N = 4 \Rightarrow 4 = 2(k - 1) + \ell$. We obtain:

– for $k = 1$, $\ell = 4$: subshell 1g containing $2\,(2 \times 4 + 1) = 18$ nucleons;

– for $k = 1$, $\ell = 2$: subshell 2d containing $2\,(2 \times 2 + 1) = 10$ nucleons;

– for $k = 2$, $\ell = 0$: subshell 3s containing $2\,(2 \times 0 + 1) = 2$ nucleons;

– for $k = 4$, $4 = 6 + \ell$: there is no value of ℓ.

CONCLUSION.– Shell $N = 4$ thus contains 30 identical nucleons in the 1g, 2d and 3s states.

Table 2.1 shows the order of the quantum states, as well as the number of identical nucleons within the framework of the harmonic potential.

Using the previous results, the essential properties of a nucleon subjected to a harmonic potential are summarized in Table 2.2.

Orbital quantum number ℓ	0	1	2	0	3	1	4
Order of states	1s	1p	1d	2s	1f	2p	1g
Maximum number of nucleons	2	6	10	2	14	6	18

Table 2.1. *Order of the quantum states of a nucleon within the framework of the shell model. The maximum number of nucleons of the same nature is also indicated*

With the magic numbers being 2, 8, 20, 28, 50, 82 and 126, Table 2.2 shows that only the numbers 2, 8 (2 + 6) and 20 (8 + 12) are found. From $N = 3$ onwards, the magic numbers are no longer reproduced. This justifies the limits of the harmonic potential that do not take spin-orbit coupling into account.

Shell no.	N	E_N	k	ℓ	State	Maximum number of nucleons	Number of nucleons on a shell
I	0	$\frac{3}{2}\hbar\omega$	1	0	1s	2	2
II	1	$\frac{5}{2}\hbar\omega$	1	1	1p	6	6
III	2	$\frac{7}{2}\hbar\omega$	1	2	1d	10	12
			2	0	2s	2	
IV	3	$\frac{9}{2}\hbar\omega$	1	3	1f	14	20
			2	1	2p	60	
V	4	$\frac{11}{2}\hbar\omega$	1	4	1g	18	30
			2	2	2d	10	
			3	0	2s	2	
VI	5	$\frac{13}{2}\hbar\omega$	1	5	1h	22	42
			2	3	2f	14	
			3	1	3p	6	

Table 2.2. *Properties of a nucleon subjected to a harmonic potential*

APPLICATION 2.2.– Determine the number of nucleons for $N = 5$ and $N = 6$. We will specify the subshells into which the nucleons are distributed.

ANSWER.– For $N = 5$, we find 42 identical nucleons distributed in the 1h, 2f, and 3p states. For $N = 6$, we find 56 identical nucleons distributed in the 1i, 2g, 3d, and 4s states.

As an example, using a diagram, let us give the distribution of nucleons according to the shell model for the nuclei ${}^{4}_{2}He$ and ${}^{7}_{3}Li$:

– the helium-4 nucleus contains 2 protons and 2 neutrons distributed in the 1s state of each nucleon; the *s* state is saturated at 2 identical nucleons;

– the lithium-7 nucleus contains 3 protons and 4 neutrons; 2 protons in the 1s state and 1 proton in the 1p state; 2 neutrons in the 1s state and 2 neutrons in the 1p state; the *p* state is saturated at 6 identical nucleons.

The distribution shown in Figure 2.2 is then obtained. As shown in this figure, the 1s state of the helium-4 nucleus contains 4 nucleons. It is then understood that saturation is relative to nucleons of the same nature and not to the number *A* of nucleons of the nucleus. This state thus contains a maximum of 2 protons and 2 neutrons. The same applies to the lithium-7 nucleus.

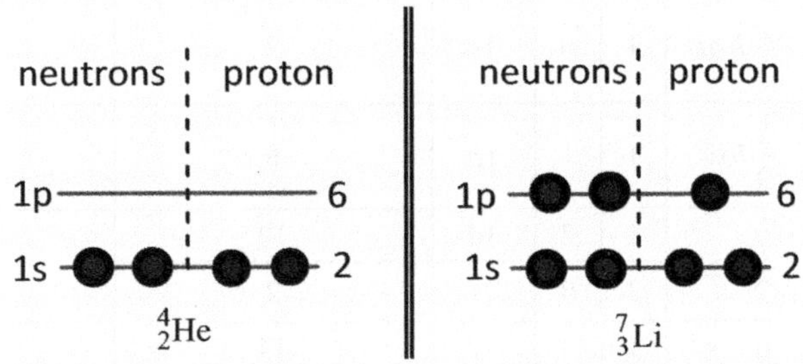

Figure 2.2. *Nucleon distribution according to the shell model for helium-4 and lithium-7 nuclei*

APPLICATION 2.3.– Determine, using a diagram, the nucleon distribution according to the shell model for the ${}^{18}_{10}Ne$ nucleus.

ANSWER.– The neon-18 nucleus contains 10 protons and 8 neutrons distributed as follows:

– 2 protons in the 1s state, 6 in the 1p state and 2 in the 1d state;

– 2 neutrons in the 1s state, 6 in the 1p state and 0 in the 1d state.

The distribution shown in Figure 2.3 is then obtained.

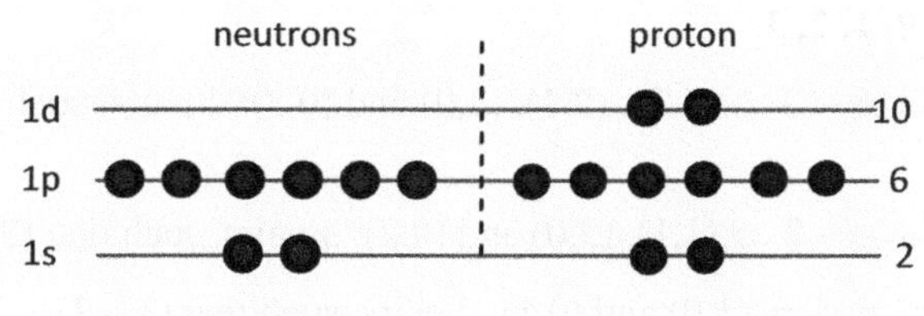

Figure 2.3. *Nucleon distribution according to the shell model for the neon-18 nucleus*

NOTE.– *Another expression of the degree of degeneracy,* g_N *[2.22], can be established as a function of the principal quantum number, N [MAY 17, BES 17]. For this purpose, a quantum number,* $n_i \geq 0$ $(i = x, y, z)$ *is assigned to each degree of freedom of the nucleon, n. The principal quantum number is then given by the relationship:* $N = n_x + n_y + n_z$. *The degree of degeneracy,* g_N, *then corresponds to triplets* (n_x, n_y, n_z), *giving the same energy value,* E_N. *To express* g_N, *we fix* n_x *(varying from 0 to N) and look for the number* $C_{ny,nz}$ *of pairs* (n_y, n_z), *such that:*

$$n_y + n_z = N - n_x \qquad [2.22a]$$

Taking into account the fact that the magnetic quantum number of spin takes $(2s + 1)$ values, the degree of degeneracy is then given by the relationship:

$$g_N = (2s+1)\sum_{n_x=0}^{N} C_{nx,ny} \qquad [2.22b]$$

It now remains to express $C_{ny,nz}$ as a function of n_x, taking into account [2.22a]

Let us reason in terms of the particular cases where $N = 0$, 1, 2 and 3.

– $N = \mathbf{0}$, $n_x = \mathbf{0}$; $n_y + n_z = 0 \Rightarrow (0,0)$: 1 pair, such that $(\mathbf{0} - \mathbf{0}) + 1 = 1$;

– $N = \mathbf{1}$, $n_x = \mathbf{0}, \mathbf{1}$;

- $n_x = \mathbf{0}$; $n_y + n_z = \Rightarrow (1,0)$ and $(0,1)$: 2 pairs, such that $(\mathbf{1} - \mathbf{0}) + 1 = 2$;

- $n_x = \mathbf{1}$; $n_y + n_z = 1 \Rightarrow (0,0)$: 1 pair, such that $(\mathbf{1} - \mathbf{1}) + 1 = 1$.

– $N = \mathbf{2}$, $n_x = \mathbf{0}, \mathbf{1}, \mathbf{2}$;

- $n_x = \mathbf{0}$; $n_y + n_z = 2 \Rightarrow (1,1)$, $(2,0)$ and $(0,2)$: 3 pairs, such that $(\mathbf{2} - \mathbf{0}) + 1 = 3$;

- $n_x = \mathbf{1}$; $n_y + n_z = \Rightarrow (1,0)$ and $(0,1)$: 2 pairs, such that $(\mathbf{2} - \mathbf{1}) + 1 = 1$.

- $n_x = \mathbf{2}$; $n_y + n_z = 0 \Rightarrow (0,0)$: 1 pair, such that $(\mathbf{2} - \mathbf{2}) + 1 = 1$.

– $N = \mathbf{3}$, $n_x = \mathbf{0, 1, 2, 3}$;

- $n_x = \mathbf{0}$; $n_y + n_z = 3 \Rightarrow$ (1,2), (2,1), (3,0) and (0,3): 4 pairs, such that $(\mathbf{3} - \mathbf{0}) + 1 = 4$;

- $n_x = \mathbf{1}$; $n_y + n_z = 2 \Rightarrow$ (1,1), (2,0) and (0,2): 3 pairs, such that $(\mathbf{3} - \mathbf{1}) + 1 = 3$.

- $n_x = \mathbf{2}$; $n_y + n_z = \Rightarrow$ (1,0) and (0,1): 2 pairs, such that $(\mathbf{3} - \mathbf{2}) + 1 = 2$.

- $n_x = \mathbf{3}$; $n_y + n_z = 0 \Rightarrow$ (0,0): 1 pair, such that $(\mathbf{3} - \mathbf{3}) + 1 = 1$.

It can therefore be admitted that in the general case, for given values of N and n_x, the number of pairs $C_{ny,nz} = (\boldsymbol{N} - \boldsymbol{n_x}) + 1$. Relationship [2.22b] is then written:

$$g_N = (2s+1) \sum_{n_x=0}^{N} \left[(N - n_x) + 1\right] \tag{2.23}$$

By expanding [2.22c], we obtain:

$$g_N = (2s+1)\left\{(N+1)\sum_{n_x=0}^{N} 1 - \sum_{n_x=0}^{N} n_x\right\} = (2s+1)\left\{(N+1)(N+1) - \frac{N(N+1)}{2}\right\}$$

Note that the sum on n_x is the sum of an arithmetic sequence of first term $U_0 = 0$ and reason $r = 1$ ($U_{nx} = n_x$). This sum is therefore equal to $N(N+1)/2$. This then gives:

$$g_N = (2s+1)\left\{(N+1)(N+1) - \frac{N(N+1)}{2}\right\}$$

That is, after arrangement, knowing that for the nucleon $s = 1/2$:

$$g_N = (N+1)(N+2) \tag{2.24}$$

Using [2.24], we obtain, for $N = 0, 1, 2, 3, 4, 5$, etc., respectively: 2, 6, 12, 20, 30, 42, etc. The maximum number of nucleons per shell is found, as shown in Table 2.3. Therefore, g_N can be determined using [2.22] or the simpler relationship [2.24].

2.1.5. *Shell structure derived from a Woods–Saxon potential*

Within the framework of the shell model based on the potential well of depth V_0, the *Woods–Saxon potential* is generally used. To account for the set of magic numbers, this potential is corrected by adding a term that takes spin-orbit coupling into account.

First case: the spin-orbit coupling is unknown

Let us express the nuclear charge density [1.47] as a function of the skin thickness, e. Knowing that (see application 1.4) $e = 4a\ln 3$, then the diffusivity parameter value is:

$$a = \ln\frac{e}{4\ln 3} = 0.2276\,e \approx 0.228\,e \qquad [2.25]$$

The nuclear charge density can then be written as:

$$\rho(r) = \frac{\rho_0}{1+\exp\left(\dfrac{r-R}{0.228\,e}\right)} \qquad [2.26]$$

In the event that the spin-orbit coupling is unknown, we use the Woods–Saxon potential taking the form of the charge density [2.26], i.e.:

$$V(r) = \frac{-V_0}{1+\exp\left(\dfrac{r-R}{0.228\,e}\right)} \qquad [2.27]$$

In relationship [2.27], $R = r_0\,A^{1/3}$, $r_0 = 1.2$ fm. The shape of potential [2.27] is shown in Figure 2.4.

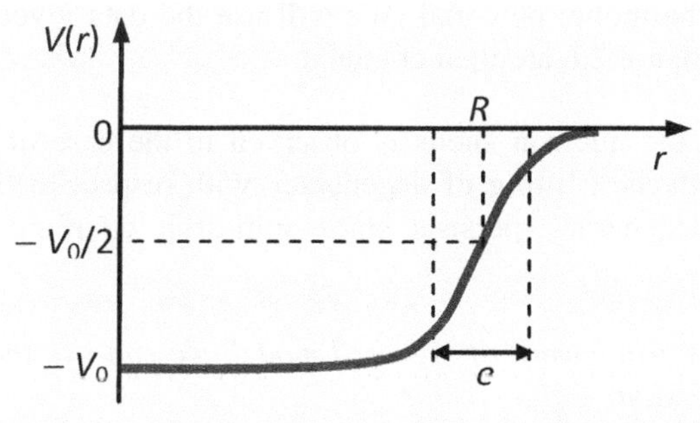

Figure 2.4. *Shape of the Woods–Saxon potential*

The value of V_0 varies approximately between 50 MeV [CAR 79, MOU 11] and 53 MeV [BES 17], as shown in Figure 2.5 for nuclei of mass number 16, 40, 120, and 129 [BES 17].

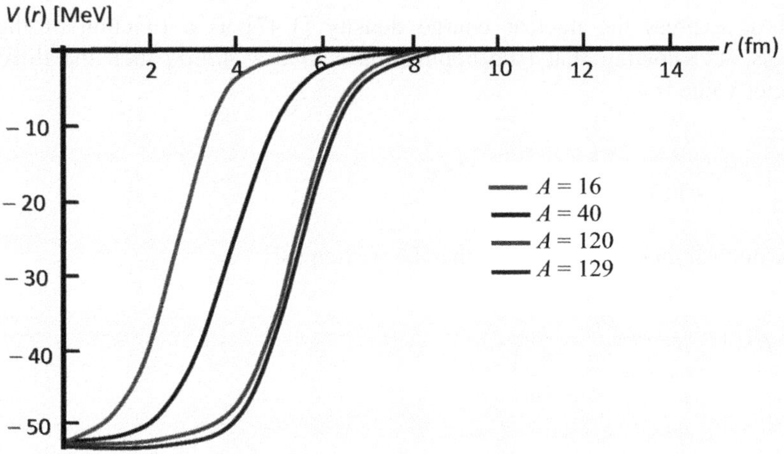

Figure 2.5. *Shape of the Woods–Saxon potential for nuclei of mass number 16, 40, 120, and 129, $-V_0 = -53$ MeV*

Because of the shape of the potential [2.27] the analytical resolution of the Schrödinger equation [2.20] is impossible except for the 1s state for which $\ell = 0$. For $\ell \neq 0$, different approximations are used to determine the energy, E, of a nucleon, see, for example, [MOU 11]. As the calculations are highly complex, we will limit ourselves to analyzing the results obtained and comparing them with those derived from the harmonic potential. We will use the data given in Table 2.2. The results shown in Figure 2.6 are then obtained.

A splitting of the nucleon shells is observed in the case of the Woods–Saxon potential. This reflects a lifting of degeneracy with respect to the orbital quantum number. Spin degeneracy persists since spin-orbit coupling is not taken into account.

NOTE.– *Within the framework of the shell model, we can use the Yukawa potential given by the expression*:

$$V(r) = \frac{-r_0 V_0}{r \exp\left(r / r_0\right)} \qquad [2.28]$$

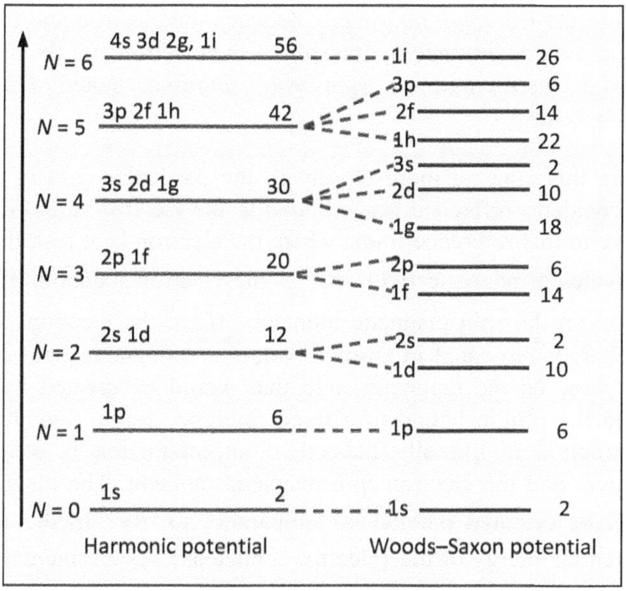

Figure 2.6. *Comparison of shell structures derived from a harmonic potential and a Woods–Saxon potential*

In this relationship , $r_0 = 1.2$ fm and $V_0 = 50$ MeV.

Nevertheless, the Woods–Saxon potential is preferable since it is closer in form to that of the nuclear charge density [2.26].

Second case: taking the coupling into consideration

Decisive progress was made in the development of the shell model, by introducing a strong coupling between the spin angular momentum, $\vec{s}$, and the orbital angular momentum, $\vec{l}$, of the nucleon, called spin-orbit interaction, or spin-orbit coupling.

Indeed, inspired by the fine structure of spectral lines, in 1949 Maria Goeppert-Mayer and Hans Jensen independently demonstrated that all of the magic numbers observed experimentally can be reproduced by adding a spin-orbit coupling term of the type $\propto \vec{l} \cdot \vec{s}$ to the Woods–Saxon potential.

Let us briefly recall the concept of spin-orbit coupling studied in atomic physics by considering hydrogen-like systems [SAK 20].

In addition to the *Coulomb interaction* between the electron and the nucleus, an additional interaction determined by the spin of the electron and the nuclear charge, called *spin-orbit interaction*, or *spin-orbit coupling*, noted *LS*, appears in hydrogen-like systems.

To illustrate this concept more closely in the particular case of the hydrogen atom, let us consider a reference frame bound to the electron in motion around the proton. Relative to this reference frame where the electron is at rest, the proton is in motion and creates a magnetic field, $\vec{B}_0$, at the electron location. This field then exerts its action on the spin magnetic moment, $\vec{M}_s$ of the electron. As the proton and electron charges are equal in absolute value, the magnetic field created by the proton is the same as the magnetic field that would be created by the electron rotating around the proton bound to a fixed reference frame. For this reason, the spin-orbit interaction is formally likened to an interaction between the orbital magnetic moment and the electron spin magnetic moment. The magnetic moment, $\vec{M}_s$, can then be oriented parallel or antiparallel to $\vec{B}_0$. In the first case, the interaction potential energy of the {electron – nucleus} system increases and, in the second case, decreases. As a result, each energy level of the atom splits into two sub-levels under the effect of the spin-orbit interaction.

Although there is no positive center within the nucleus around which a nucleon would gravitate, the previous interpretation of spin-orbit coupling is drawn from to correct the Woods–Saxon potential. We then find a corrective term whose validity is justified by the simple fact that its introduction into the Woods–Saxon potential makes it possible to correctly reproduce all of the magic numbers observed experimentally. In the following, we will draw from the correction relative to spin-orbit coupling applied to the Hamiltonian for hydrogen-like systems in the weakly relativistic domain.

In the weakly relativistic domain, the term of the fine-structure Hamiltonian relating to spin-orbit interaction (*LS* coupling) is given by the expression [SAK 20]:

$$W_{SO} = \frac{1}{2m^2c^2}\frac{1}{R}\frac{dV(\vec{R})}{dR}\vec{L}\cdot\vec{S} \qquad [2.29]$$

Using lowercase letters, the term [2.29] is written:

$$W_{so} = \frac{1}{2m^2c^2}\frac{1}{R}\frac{dV(\vec{R})}{dR}\vec{l}\cdot\vec{s} \qquad [2.30]$$

Drawing from [2.30], we then add to the Woods–Saxon potential [2.27] a corrective term proportional to $\vec{l} \cdot \vec{s}$, such that:

$$W_{so} = -C\vec{l} \cdot \vec{s} \qquad [2.31]$$

In corrective term [2.31], C is a positive constant.

Taking into account [2.27] and [2.31], the corrected potential, $V_{cor}(r)$, is written as:

$$V_{cor}(r) = \frac{-V_0}{1+\exp\left(\dfrac{r-R}{0.228e}\right)} - C\vec{l} \cdot \vec{s} \qquad [2.32]$$

Let us now determine the solution to the Schrödinger equation where the scalar potential is of type [2.32]. Let $H_{ws} = H$, the Hamiltonian [2.19a] of a nucleon subjected to the Woods–Saxon potential. Taking into account the spin-orbit coupling, the total Hamiltonian is written:

$$H_{cor} = -\frac{\hbar^2}{2m}\nabla_r^2 + V(r) + \frac{\hbar^2 \ell(\ell+1)}{2mr^2} - C\vec{l} \cdot \vec{s} = H_{WS} + W_{so} \qquad [2.33]$$

In Hamiltonian [2.33], $V(r)$ is the Woods–Saxon potential, given by [2.27].

The problem then amounts to solving the Schrödinger equation:

$$H_{cor}\Psi = E_{cor}\Psi \qquad [2.34]$$

Using the last equality in [2.33], Hamiltonian [2.34] is written:

$$(H_{WS} + W_{so})\Psi = (E_{WS} + E_{so})\Psi \Rightarrow E_{cor} = E_{WS} + E_{so} \qquad [2.35]$$

As [2.35] indicates, the resolution of equation [2.34] comes down to determining E_{so}, since E_{WS} is already known. For this, let us use the base $|k,\ell,j,m_j\rangle$ in the space of the states of a nucleon. By definition:

$$E_{so} = \langle W_{so} \rangle = \langle k,\ell,j,m_j | W_{so} | k,\ell,j,m_j \rangle \qquad [2.36]$$

This gives, using [2.31]:

$$E_{so} = -C\langle k,\ell,j,m_j|\vec{l}\cdot\vec{s}|k,\ell,j,m_j\rangle \quad [2.37]$$

Let us now express the scalar product in the right-hand member of [2.37]. Using expression [2.10] of the total angular momentum, we obtain:

$$\vec{j}^2 = (\vec{l}+\vec{s})^2 = \vec{l}^2 + \vec{s}^2 + 2\vec{l}\cdot\vec{s}$$

That is:

$$\vec{l}\cdot\vec{s} = \frac{1}{2}\left(\vec{j}^2 - \vec{l}^2 - \vec{s}^2\right) \quad [2.38]$$

By applying [2.38] in [2.37], we obtain:

$$E_{so} = -\frac{C}{2}\langle k,\ell,j,m_j|(\vec{j}^2 - \vec{l}^2 - \vec{s}^2)|k,\ell,j,m_j\rangle \quad [2.39]$$

In quantum mechanics, the eigenvalue of the square f^2 of an angular momentum, $\vec{f}$, is equal to $\hbar^2 f(f+1)$, i.e.:

$$\vec{f}^2|k,\ell,j,m_j\rangle = \hbar^2 f(f+1)|k,\ell,j,m_j\rangle \quad [2.40]$$

Taking into account [2.40], the average value [2.39] is then written:

$$E_{so} = -\frac{C\hbar^2}{2}\left[j(j+1) - \ell(\ell+1) - s(s+1)\right] \quad [2.41]$$

The corrected energy [2.35] taking spin-orbit coupling into account is ultimately written:

$$E_{cor} = E_{WS} - \frac{C\hbar^2}{2}\left[j(j+1) - \ell(\ell+1) - s(s+1)\right] \quad [2.42]$$

Given that s = 1/2 and j = $\ell \pm$ 1/2 according to [2.11], two possible cases are obtained:

$$E_{cor} = E_{WS} - \frac{C\hbar^2}{2}\left[\left(\ell \pm \frac{1}{2}\right)\left(\ell \pm \frac{1}{2} + 1\right) - \ell(\ell+1) - \frac{3}{4}\right] \quad [2.43]$$

For $j = \ell+1/2$, we obtain, by expanding the term in square brackets of [2.43]:

$$\left(\ell+\frac{1}{2}\right)\left(\ell+\frac{1}{2}+1\right)-\ell(\ell+1)-\frac{3}{4}$$
$$=\ell^2+\frac{\ell}{2}+\ell+\frac{\ell}{2}+\frac{1}{4}+\frac{1}{2}-\ell^2-\ell-\frac{3}{4}=\ell$$

The same is obtained for $j = \ell-1/2$:

$$\left(\ell-\frac{1}{2}\right)\left(\ell-\frac{1}{2}+1\right)-\ell(\ell+1)-\frac{3}{4}$$
$$=\ell^2-\frac{\ell}{2}+\ell-\frac{\ell}{2}+\frac{1}{4}-\frac{1}{2}-\ell^2-\ell-\frac{3}{4}=-(2\ell+1)$$

Taking into account the previous results, we ultimately obtain:

$$\begin{cases} E_{cor} = E_{WS} - \dfrac{C\hbar^2}{2}\ell\,;\ j=\ell+\dfrac{1}{2} \\ E_{cor} = E_{WS} + \dfrac{C\hbar^2}{2}(2\ell+1);\ j=\ell-\dfrac{1}{2} \end{cases} \quad [2.44]$$

APPLICATION 2.4.– Determine the expression of E_{cor} in a state determined by $\ell = 0$.

ANSWER.– As the total quantum number, j, is strictly positive, for $\ell = 0$, the expression satisfied for $j=\ell+1/2=1/2$ is considered. This then gives $E_{cor} = E_{WS}$ ($E_{so} = 0$) according to [2.44]. This result was predictable since the spin-orbit coupling is absent in all quantum states determined by $\ell = 0$ (states k, 0).

To account for the set of magic numbers in accordance with the experimental observations, let us first introduce the spectroscopic notation of the nucleonic states using atomic physics as inspiration.

In the case of hydrogen-like systems [SAK 20], the definition of the total quantum number, j, allows the *spectroscopic notation* of quantum states to be written using the label: $n\ ^{2s+1}L_j$. In this notation, n is the principal quantum number, $(2s+1)$ designates the *level multiplicity*, and L denotes the quantum state considered, determined by the value of the orbital quantum number, ℓ. By analogy, a nucleonic state is designated by the label $n\ l_j$ (L is replaced with l). However, in this notation, n is not the principal quantum number. It designates a number indicating the number of occurrences of the orbital quantum number, ℓ. For example (see Table 2.2), for the 1s, 1p and 1d states, $n = 1$, for the 2s, 2d and 2f states, $n = 2$, and so on. Note that for a given value of j, the total magnetic quantum number, m_j, takes $(2j + 1)$ possible values. The degree of degeneracy of a $n\ l_j$ state is:

$$g_j = (2j + 1) \qquad [2.45]$$

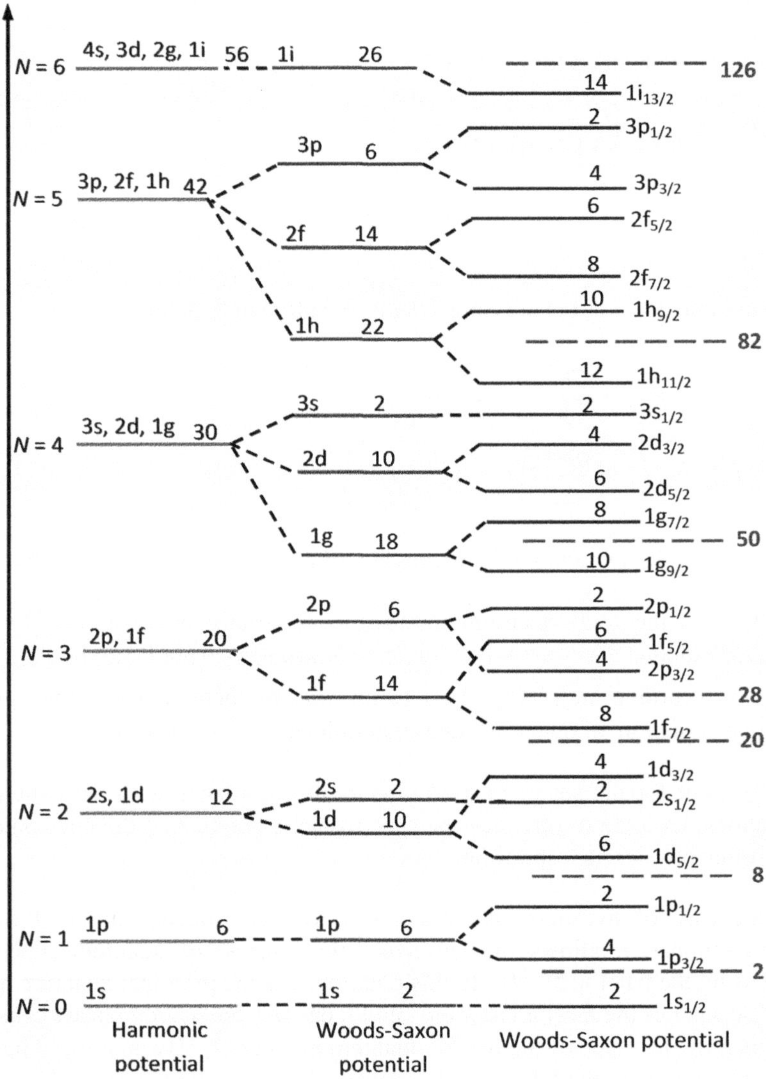

Figure 2.7. *Comparison of shell structures derived from a harmonic potential and from a Woods–Saxon potential, with or without a spin-orbit coupling*

Each $n\ l_j$ state then contains a maximum of $(2j + 1)$ nucleons.

In addition, the parity, π, of a monoelectronic atomic system is defined by the simple relationship: $\pi = (-1)^{\ell}$. This notation is linked to the properties of hydrogen-like wave functions (see the note at the end of this section). Taking the parity into account makes it possible to note a nucleonic state using the label nl_j^{π}.

Let us now represent the nucleonic states compared to the energy levels derived from the Woods–Saxon potential.

To account for all magic numbers, 2, 8, 20, 28, 50, 82 and 126, let us consider all of the values of $0 \leq \ell \leq 5$. The results obtained are shown in Figure 2.7. For greater clarity, the parity of states is omitted.

As relationships [2.44] indicate, for the same value of the orbital quantum number, ℓ, the state characterized by $j = \ell + 1/2$ is deeper than the state characterized by $j = \ell - 1/2$. In addition, the difference in energy between the $j = \ell + 1/2$ and $j = \ell - 1/2$ states increases with ℓ. These two observations enable magic numbers 28, 50, 82 and 126 to be taken into account.

– The separation of the $1f_{5/2}$ and $1f_{7/2}$ levels induces lowering of the $1f_{7/2}$ level, which thus forms a single shell allowing magic number 28 to be reproduced.

– Similarly, the separation of the $1g_{7/2}$ and $1g_{9/2}$, $1h_{9/2}$ and $1h_{11/2}$ and $1i_{11/2}$ (this level is not shown) and $1i_{13/2}$ levels induces the lowering of the $1g_{9/2}$, $1h_{11/2}$, and $1i_{13/2}$ levels, allowing magic numbers 50, 82 and 126, respectively, to be reproduced.

Beyond the 7th magic number, 126, the shell structure is written in order (with the number of nucleons saturating the subshell indicated in parentheses) $2g_{9/2}$ (10) $\rightarrow$ $5d_{5/2}$ (6) $\rightarrow$ $1i_{11/2}$ (12) $\rightarrow$ $2g_{7/2}$ (8) $\rightarrow$ $4s_{1/2}$ (2) $\rightarrow$ $3d_{3/2}$ (4) $\rightarrow$ $1j_{15/2}$ (16). This gives 58 nucleons. By adding 126 we find the 8th magic number, 184.

In summary, contrary to the harmonic potential and the Woods–Saxon potential, taking corrective term [2.31] into account enables all of the magic numbers 2, 8, 20, 28, 50, 82, 126 and 184 to be correctly reproduced.

NOTE.– *Semi-magic numbers.*

Noted in the literature are what are known as *semi-magic numbers*, corresponding to the saturation of a nuclear subshell, obtained by adding together the magic number of the saturated shell and the number of nucleons occupying the subshell located just above the line (here, a dotted red line), indicating the magic number considered. For example, these are the numbers 6 (magic number 2 + 4 nucleons of the $1p_{3/2}$ subshell); the sum of the semi-magic number 6 and the 2 nucleons of the $1p_{1/2}$ subshell gives the 2nd magic number, 8; 14 (magic number 8 + 6 nucleons of the $1d_{5/2}$ subshell); 16 (semi-magic number 14 + 2 nucleons of the $2s_{1/2}$ subshell). By adopting the same approach, we find the semi-magic numbers: 32, 38, 40, 58, 64, 68, 70, 92, 100, 106, 110, 112 (= magic number 126 – 14 nucleons of the last $1i_{13/2}$ shown in Figure 2.9). It should be kept in mind that magic numbers are highlighted quite clearly by the experiment. By contrast, none of the semi-magic numbers have been observed experimentally. Therefore, only magic numbers have a precise physical meaning within the framework of the shell model, since they correspond to particularly stable nuclei.

NOTE.– *Concept of parity [SAK 20].*

Within the framework of *LS* coupling, a *spectral term*, or, simply, a *term*, is noted by the symbol ${}^{2S+1}L_J$. To determine the complete writing of a term, the parity of the system already introduced in the case of hydrogen-like systems should be taken into account. For these systems with a well-determined orbital angular momentum, l , the parity of the system or the quantum state, $\pi = (-1)^{\ell}$. This result stems from the action of a mathematical being called the *parity operator*, noted Π.

The parity operation on a wave function of the type $f(r, \theta, \varphi)$ consists of a symmetry operation with respect to the origin of the reference axes, such that:

$$r \Rightarrow r;\ \theta \Rightarrow \pi - \theta;\ \varphi \Rightarrow \pi + \varphi \qquad [2.46]$$

Inversion operations [2.46] show that if the radial part, $R_{n,\ell}(r)$, of the wave function $\psi_{n,\ell,m}(r,\theta,\varphi) = R_{n,\ell}(r)\mathrm{Y}_\ell^m(\theta,\varphi)$ remains unchanged after inverting the coordinates, the angular part, $\mathrm{Y}_\ell^m(\theta,\varphi)$, however, determined by spherical harmonics, changes sign:

$$\mathrm{Y}_\ell^m(\pi - \theta, \varphi + \pi) = (-1)^\ell\, \mathrm{Y}_\ell^m(\pi - \theta, \varphi + \pi) \qquad [2.47]$$

By definition, the parity operator is an operator whose action on a wave function involves inverting the coordinates of spaces with respect to the center of reference. Thus, by taking [2.47] into account:

$$\Pi\psi_{n,\ell,m}(r,\theta,\varphi) = R_{n,\ell}(r)\Pi\mathrm{Y}_\ell^m(\theta,\varphi) = (-1)^\ell\,\psi_{n,\ell,m}(r,\theta,\varphi) \qquad [2.48]$$

According to [2.48] the eigenvalues of the parity operator $\pi = (-1)^{\ell}$ are equal + 1 (even parity for even ℓ) and – 1 (odd parity for odd ℓ).

Hideki Yukawa was a Japanese physicist. He is known for having established, around 1930, the nuclear potential bearing his name, often used alongside the Woods–Saxon potential. He was awarded the Nobel Prize in Physics 1949 for his prediction of the existence of mesons (particles having an intermediate mass between that of the electron and that of the proton) based on theoretical work on nuclear forces. Yukawa was the first Japanese Nobel Prizewinner.

Maria Gertrud Käte Goeppert, a.k.a. Maria Goeppert-Mayer was a German-American physicist. She is famous for theoretically demonstrating the existence of two-photon absorption (TPA) by atoms in 1929, with TPA only highlighted experimentally for the first time 30 years later, with the invention of lasers. In addition, she conducted research on nuclear decay processes, including double β decay (see Chapter 3), and nuclear structures. She was jointly awarded half of the Nobel Prize in Physics 1963, which she shares with Daniel Jensen for their discoveries relating to the shell structure of the atomic nucleus.

Johannes Hans Daniel Jensen was a German physicist. He is famous for his work on the shell structure of the atomic nucleus. He was joint winner of half the Nobel Prize in Physics 1963 (the other half being awarded to the Hungarian and naturalized-American physicist, Eugene Paul Wigner (1902-1995)).

Box 2.1. *Yukawa (1905–1981); Goeppert-Mayer (1906–1970); Jensen (1907–1973)*

2.2. Angular momentum and parity

2.2.1. *Angular momentum and parity of ground state*

In atomic physics, when spin-orbit coupling is predominant with respect to the Coulomb repulsion between electrons, the so-called (*j*, *j*) *bond* encountered in the case of heavy atoms ($Z \geq 30$) [SAK 20] is obtained. Within the framework of *jj* coupling, the individual orbital ($\vec{l}_i$) and spin ($\vec{s}_i$) momenta of each electron are added together to give the total momentum, $\vec{j}_i$:

$$\vec{j}_i = \vec{l}_i + \vec{s}_i \quad . \tag{2.49}$$

The total angular momentum, $\vec{J}$, is thus written, for N electrons:

$$\vec{J} = \sum_{i=1}^{N} \vec{j}_i = \vec{j}_1 + \vec{j}_2 + ... + \vec{j}_N \qquad [2.50]$$

In nuclear physics, the strong coupling of nucleons is treated within the framework of jj coupling. The individual orbital ($\vec{l}_i$) and spin ($\vec{s}_i$) momenta of each nucleon are added together to give the total momentum, $\vec{j}_i$, according to the vector addition rule [2.49]. For a nucleus containing A nucleons, the total angular momentum, $\vec{J}$, is written according to [2.50]:

$$\vec{J} = \sum_{i=1}^{A} \vec{j}_i = \vec{j}_1 + \vec{j}_2 + ... + \vec{j}_A \qquad [2.51]$$

In addition, the total parity of a nuclear state with A nucleons is given by the product of the parities of the individual nucleonic states, i.e.:

$$\pi = \prod_{i=1}^{A} (-1)^{\ell_i} = (-1)^{\ell_1} \times (-1)^{\ell_2} \times ....(-1)^{\ell_A} \qquad [2.52]$$

The term of the state of a nucleus is then written using the label, J^π. This notation is then used to determine the nature of J^π of the ground state of a nucleus with A nucleons.

Let us first set out the *nucleonic level filling rules* [MAY 17]:

Identical nucleons on the same energy level tend to form a $J^\pi = 0^+$ *pair.*

Let us therefore identify two cases, depending on whether A is even or odd, to state the rules for determining the J^π of the ground state of a nucleus.

– *In their ground state, the even-even nuclei (even Z – even N) are in the* $J^\pi = 0^+$ *state.*

– *In their ground state, the* J^π *of nuclei of odd A (even Z – odd N or odd Z – even N) is determined by the unpaired nucleon.*

– *In their ground state, the* J^π *of odd-odd nuclei (odd Z – odd N) is determined by the coupling of the two unpaired nucleons.*

To apply the preceding rules, we start by mapping out the distribution of nucleons according to the shell model, in order to determine the individual states, nl_j, of each nucleon within the framework of the spin-orbit coupling. Next, [2.52] is used to determine π parity, then one of the previous rules is applied to determine the J^π.

Let us illustrate the three preceding rules using simple examples by considering the ${}^4_2\text{He}$ and ${}^6_3\text{Li}$ nuclei.

Nucleon distribution according to the shell model for helium-4 and lithium-7 nuclei was presented in Figure 2.4. Taking into account the spin-orbit coupling (Figure 2.9), the 1s level is a single level, $1s_{1/2}$. Each of levels 1p and 1d, however, split into two components: $1p_{1/2}$ and $1p_{3/2}$, $1d_{3/2}$, and $1d_{5/2}$.

– **Case of the helium-4 nucleus**

The ${}^4_2\text{He}$ nucleus contains 2 protons distributed on the $1s_{1/2}$ level and 2 neutrons distributed over the $1s_{1/2}$ level. The $1s_{1/2}$ level contains a pair of identical nucleons: $J = 0 \Rightarrow J^\pi = 0^\pi$. For the 1s state, $\ell = 0$, hence $\pi = +1$. For the ground state, $J^\pi = 0^+$. The nucleon distribution shown in Figure 2.8 is then obtained for the ground state of the ${}^4_2\text{He}$ nucleus.

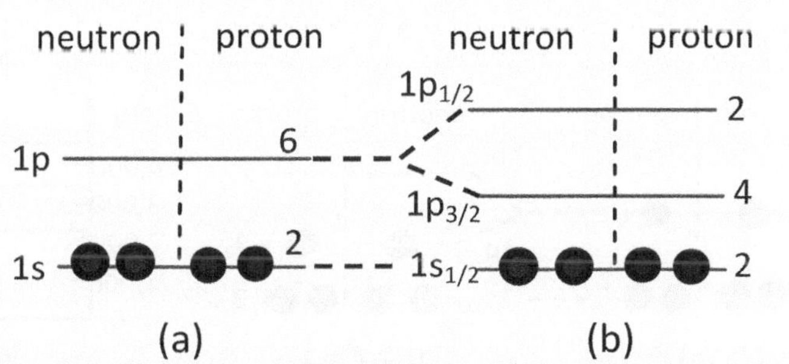

Figure 2.8. *Ground state of the helium-4 nucleus, $J^\pi = 0^+$; (a) distribution derived from a harmonic potential; (b) distribution derived from the Woods–Saxon potential with spin-orbit coupling*

– **Case of the lithium-6 nucleus**

The ${}^6_3\text{Li}$ nucleus contains 3 protons, 2 of which are distributed over the $1s_{1/2}$ level and 1 over the $1p_{3/2}$ level (this level is filled before the $1p_{1/2}$ higher level, see Figure 2.7), and 3 neutrons (2 over the $1s_{1/2}$ level and 1 over the $1p_{3/2}$ level). Under the third rule, the J^π of the ground state is determined by the coupling of the $1p_{3/2}$ unpaired proton and the $1p_{3/2}$ unpaired neutron. For this state, the total quantum number, $j = 3/2$. According to the vector addition rule [2.51], the total angular momentum:

$$\vec{J} = \sum_{i=1}^{2} \vec{j}_i = \vec{j}_1 + \vec{j}_2 \qquad [2.53]$$

As with two-electron atomic systems, the angular momentum, J, takes all of the values comprised between $j_1 + j_2$ and $|j_1 - j_2|$. Thus, the values:

$$j_1 + j_2, j_1 + j_2 - 1, j_1 + j_2 - 2, \text{etc.}, |j_1 - j_2| \qquad [2.54]$$

Given that $j_1 = j_2 = 3/2$, J takes the values 3, 2, 1 and 0 according to [2.54]. This corresponds to $J^\pi = 0^\pi$, 1^π, 2^π or 3^π.

Using [2.52], we determine the parity, π. Knowing that $\ell = 1$ for the $1p_{3/2}$ level, we find: $\pi = (-1)\times(-1) = +1$. This then gives the states $J^\pi = 0^+$, 1^+ or 3^+. Thus, the shell model predicts four possible states for the ground level of the $^6_3$Li nucleus. Experimentally, the ground level of the $^6_3$Li nucleus corresponds to $J^\pi = 1^+$. The other three values in the order 3^+, 0^+, 2^+, correspond to excited states (see Figure 2.9c).

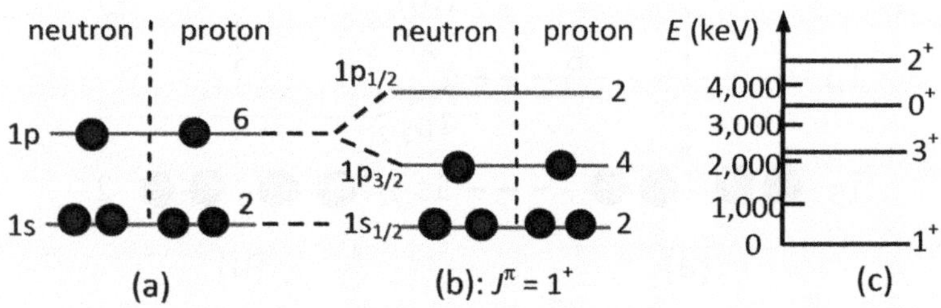

Figure 2.9. *Ground state of the lithium-6 nucleus, $J^\pi = 1^+$; and three excited states revealed by the experiment. (a) distribution derived from a harmonic potential; (b) distribution derived from the Woods–Saxon potential with spin-orbit coupling; (c) experimental observations*

APPLICATION 2.5.– Determine the J^π of the ground state of the helium-3 nucleus.

ANSWER.– The $^3_2$He nucleus contains 2 protons distributed over the $1s_{1/2}$ level and 1 neutron over the $1s_{1/2}$ level. Under the third rule, the J^π of the ground state is determined by the unpaired neutron; i.e. $J = j = 1/2$ and $\pi = +1$. This then gives $J^\pi = 1/2^+$. This prediction is confirmed by the experiment.

2.2.2. *Angular momentum and parity of an excited state*

Experimentally, all nuclei with an even number of nucleons have a ground state characterized by $J^\pi = 0^+$. The excitation process then consists of breaking a pair of identical nucleons and carrying one of the nucleons to the higher level. This separation would correspond to an energy cost due to the strong neutron-neutron coupling. For a nucleus containing an unpaired nucleon, excitation requires less energy. The situation is similar in atomic physics: it is easier to excite an unpaired electron than to excite a paired electron. We can thus identify two rules for determining the J^π of an excited nucleus.

– *The J^π of odd-A nuclei is determined by the unpaired nucleon.*

– *The J^π of the even-A nuclei is determined by the possible values of J given by the relationship:*

$$J = 2j - 1, 2j - 3, 2j - 5, 2j - 7, \text{etc.}, 0 \qquad [2.55]$$

In relationship [2.55], j represents the quantum number of the initial state of excitation of one of the identical nucleons located on the same level, nl_j, of the outer shell.

Let us illustrate the two previous rules in the case of lithium-7 and neon-18.

– Case of the lithium-7 nucleus

The $^7_3$Li nucleus contains 3 protons, 2 of which are distributed over the $1s_{1/2}$ level and 1 over the $1p_{3/2}$ level) and 4 neutrons (2 over the $1s_{1/2}$ level and 2 over the $1p_{3/2}$ level). Under the second rule verified for odd A, the J^π of the ground state is determined by the unpaired nucleon here the proton located on the $1p_{3/2}$ level; i.e., $J = j = 3/2$ and $\pi = -1$. This then gives $J^\pi = 3/2^-$. By applying the proton on the $1p_{1/2}$ level, an excited state of lithium-7 corresponding to $J^\pi = 1/2^-$ is obtained. The theoretical predictions for the ground level ($J^\pi = 3/2^-$) and the first excited level ($J^\pi = 1/2^-$) are corroborated by the experiment (Figure 2.10).

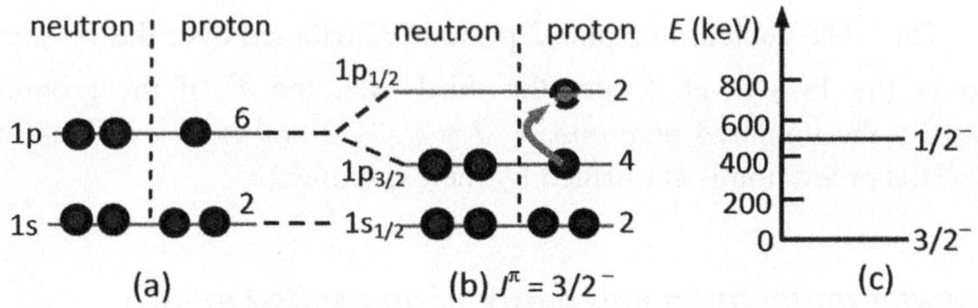

Figure 2.10. *Ground state of the nucleus of lithium-6, $J^\pi = 3/2^-$; and the first excited state revealed by the experiment. (a) distribution derived from a harmonic potential; (b) distribution derived from the Woods–Saxon potential with spin-orbit coupling, the arrow indicates an excitation process; (c) experimental observations*

– Case of the neon-18 nucleus

The $^{18}_{10}Ne$ nucleus contains 10 protons (2 distributed over the $1s_{1/2}$ level; 4 over the $1p_{3/2}$ level; 2 over the $1p_{1/2}$ level and 2 over the $1d_{5/2}$ level) and 8 neutrons (2 distributed over the $1s_{1/2}$ level; 4 over the $1p_{3/2}$ level; 2 over the $1p_{1/2}$ level). The study of the neon-18 excitation process can be envisaged as follows:

– either we envisage breaking the proton pair from the $1d_{5/2}$ level and carrying one of the protons to the $1d_{5/2}$ level (or to the nearest $2s_{1/2}$ level);

– or we envisage breaking the pair of neutrons from the $1p_{1/2}$ level and carrying one of the neutrons to the $1d_{5/2}$ level.

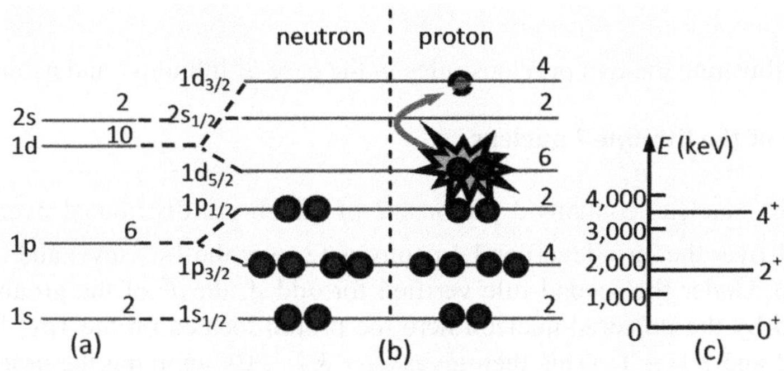

Figure 2.11. *Ground state, $J^\pi = 0^+$; and excited states ($J^\pi = 2^+$, $J^\pi = 4^+$) of the neon-18 nucleus, revealed by the experiment. (a) shell structure derived from a harmonic potential; (b) distribution derived from the Woods–Saxon potential with spin-orbit coupling, the arrow indicates a proton excitation process; (c) experimental observations*

The most probable process is when exciting the protons of the $1d_{5/2}$ level, since the neutrons in magic number (8) form particularly stable shells. For the $1d_{5/2}$ level, $j = 5/2$. The possible values of J are deduced using relationship [2.55]. This then gives $J = 4$, 2 and 0. For the $1d_{5/2}$ level, $\ell = 2$; hence the parity $\pi = +1$. This then gives $J^\pi = 4^+$, 2^+ or 0^+. The level $J^\pi = 0^+$ corresponds to the ground state and the levels $J^\pi = 4^+$ and 2^+ correspond to the excited states. These predictions are consistent with the experimental observations (Figure 2.11).

APPLICATION 2.6.– Experimentally, two excited levels of energy 585 keV and 974 keV are found for the magnesium-25 nucleus, corresponding to $J^\pi = 1/2^+$ and $J^\pi = 3/2^+$, respectively. Determine the J^π of the ground state and then theoretically find the J^π values of the excited states. Create a diagram showing the ground state, the two excited states of the magnesium-25 nucleus revealed by the experiment; (a) the shell structure derived from a harmonic potential; (b) the distribution of its nucleons derived from the Woods–Saxon potential with spin-orbit coupling, two arrows indicating the possible excitation processes of one of the nucleons to be specified; (c) experimental observations. Energy range: 0, 1,000 and 2,000 keV.

ANSWER.– The $^{25}_{12}Mg$ nucleus contains 12 protons (2 distributed over the $1s_{1/2}$ level, 4 over the $1p_{3/2}$ level, 2 over the $1p_{1/2}$ higher level and 4 over the $1d_{5/2}$ level) and 13 protons (2 distributed over the $1s_{1/2}$ level, 4 over the $1p_{3/2}$ level, 2 over the $1p_{1/2}$ higher level and 5 over the $1d_{5/2}$ level). Under the second rule verified for odd A, the J^π of the ground state is determined by the single nucleon, here the unpaired neutron located on the $1d_{5/2}$ level; i.e., therefore, $J = j = 5/2$ and $\pi = +1$ since $\ell = 2$. This then gives, for the ground state $J^\pi = 5/2^+$.

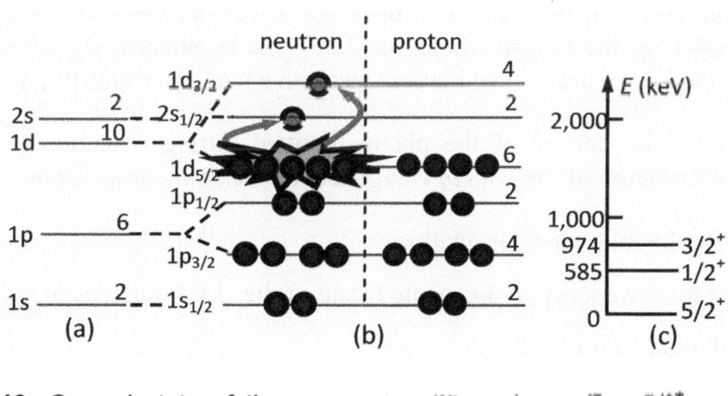

Figure 2.12. *Ground state of the magnesium-25 nucleus, $J^\pi = 5/2^+$; and the two excited states revealed by the experiment. (a) shell structure derived from a harmonic potential; (b) distribution derived from the Woods–Saxon potential with spin-orbit coupling, the arrow indicates an excitation process; (c) experimental observations*

Exciting the uncoupled neutron consists of either carrying it on the $2s_{1/2}$ intermediate level, thus giving $J^{\pi} = 1/2^+$ ($J = j = 1/2$ and $\pi = +1$, $\ell = 0$); or on the $1d_{3/2}$ higher level, giving $J^{\pi} = 3/2^+$ ($J = j = 3/2$ and $\pi = +1$, $\ell = 2$). The J^{π} values of the two excited states of the magnesium-25 nucleus are thus found, in accordance with the experiment (Figure 2.12). Note that the 5 neutrons of the $1d_{3/2}$ level are indiscernible, therefore they have the same probability of being excited at the $2s_{1/2}$ or $1d_{3/2}$ higher level.

2.3. Gamma deexcitation

2.3.1. *Definition, deexcitation energy*

Gamma deexcitation corresponds to the emission of a γ-ray or photon by a low-energy excited nucleus (excited-state energy less than the separation energy of a nucleon). γ-deexcitation often accompanies α- or β-type radioactive decay processes (see Chapter 3). This deexcitation may also result from a radiative capture process, inelastic shock, an induced nuclear reaction such as fission, etc. In the general case, the excited nucleus can go through various intermediate excited states before moving to the final ground level. There then follows a process of successive deexcitations generating a spectrum of γ-photon lines. During the nuclear transition between one state of the excited nucleus noted ${}_{Z}^{A}X^*$ and the final state of the stable nucleus, ${}_{Z}^{A}X$, the γ-deexcitation equation is written:

$$ {}_{Z}^{A}X^* \rightarrow {}_{Z}^{A}X + \gamma \qquad [2.56]$$

Since the emission of a γ-photon does not affect the mass number, A, or the charge number, Z, the two notations in [2.56] can be simplified, considering the symbols X^* and X for the excited nucleus and stable nucleus, respectively.

To express the energy of the photon emitted during transition [2.56], it is sufficient to make use of the *laws of energy and momentum conservation*.

Let us use the following designations:

– E (X^*) for the energy of the excited state of the ${}_{Z}^{A}X^*$ nucleus of resting mass, m (X^*), and momentum $\vec{p}_{X*}$;

– E (X) for the energy of the ground state of the ${}_{Z}^{A}X$ nucleus of resting mass, m (X), and momentum $\vec{p}_X$;

– E_γ for the energy of the γ-photon of momentum $\vec{p}_\gamma$.

During deexcitation, the X nucleus undergoes a recoil. Let $E_{cr}(X)$ be this kinetic energy. The laws of conservation of energy and momentum are written, respectively:

$$E(X^*) = E(X) + E_{cr}(X) + E_\gamma \qquad [2.57]$$

$$\vec{p}_{X^*} = \vec{p}_X + \vec{p}_\gamma \qquad [2.58]$$

Since the initial nucleus, X^*, is at rest, its momentum is null. It follows that according to [2.58], the momentum of the final nucleus, X, is non-null; this allows verification of the law of conservation of momentum. In addition, the resting energy of a particle of resting mass, m_0, is equal to m_0c^2. Equations [2.57] and [2.58] are then written:

$$m(X^*)\,c^2 = m(X)\,c^2 + E_{cr}(X) + E_\gamma \qquad [2.59]$$

$$\vec{0} = \vec{p}_X + \vec{p}_\gamma \Rightarrow \vec{p}_X = -\vec{p}_\gamma \qquad [2.60]$$

Equation [2.60] shows that the momentum of the X nucleus is directly opposite that of the γ-photon; this justifies its recoil during deexcitation of the X^* nucleus.

Using [2.59], the energy of the γ-photon is written:

$$E_\gamma = m(X^*)\,c^2 - m(X)\,c^2 - E_{cr}(X) \qquad [2.61]$$

With a very good approximation, the energy, E_γ can be directly calculated by overlooking the recoil kinetic energy, $E_{cr}(X)$, in [2.61]. We will nevertheless take this into account and establish the general expression of E_γ. For this, let us express the recoil kinetic energy of the X nucleus in function E_γ

The photon energy is linked to its momentum by the relationship $E_\gamma = p_\gamma c$. Using [2.60], we obtain:

$$\begin{cases} p_X = p_\gamma \\ E_{cr}(X) = \dfrac{p_X^2}{2m(X)} \Rightarrow E_{cr}(X) = \dfrac{E_\gamma^2}{2m(X)c^2} \\ p_\gamma = \dfrac{E_\gamma}{c} \end{cases} \qquad [2.62]$$

Using the last relationship, [2.62], equation [2.61] can be put in the form:

$$E_\gamma = m\,(X^*)c^2 - m\,(X)c^2 - \frac{E_\gamma^2}{2m\,(X)c^2} \quad [2.63]$$

Let us write, in [2.63]:

$$\Delta E = m\,(X^*)\,c^2 - m\,(X)\,c^2 \quad [2.64]$$

We thus obtain:

$$\frac{E_\gamma^2}{2m\,(X)c^2} + E_\gamma - \Delta E = 0 \quad [2.65]$$

Equation [2.65] is second degree in E_γ. Its discriminant $\Delta = 1 + 2\Delta E/m\,(X)c^2$. Its physically acceptable solutions are given by the relationship:

$$E_\gamma = m\,(X)c^2\left[-1 \pm \sqrt{1 + \frac{2\Delta E}{m\,(X)c^2}}\right] \quad [2.66]$$

Since $E_\gamma > 0$, the physically acceptable solution is then given by the relationship:

$$E_\gamma = m\,(X)c^2\left[\sqrt{1 + \frac{2\Delta E}{m\,(X)c^2}} - 1\right] \quad [2.67]$$

Consider the case where $m\,(X)\,c^2 >> \Delta E$.

First, let us recall the following binomial expansion:

$$(1+x)^n = 1 + nx + \frac{n(n-1)}{2!}x^2 + \frac{n(n-1)(n-2)}{3!}x^3 + \ldots\ldots \quad [2.68]$$

For $x << 1$, equation [2.68] is written in second-order form:

$$(1+x)^n \approx 1 + nx + \frac{n(n-1)}{2!}x^2 \quad [2.69]$$

By writing $x = 2\Delta E/m\ (X)\ c^2 << 1$, the limited expansion of the term in square brackets on the right side of equation [2.67] is in the shape.

$$E_\gamma \approx m\,(X)c^2\left[1+\frac{1}{2}\frac{2\Delta E}{m\,(X)c^2}-\frac{1}{8}\left(\frac{2\Delta E}{m\,(X)c^2}\right)^2-1\right]$$

i.e. after simplification and arrangement:

$$E_\gamma = \Delta E - \frac{1}{2}\frac{\Delta E^2}{m(X)c^2} \qquad [2.70]$$

As equation [2.70] indicates, if the energy gap, ΔE, is small, we obtain:

$$E_\gamma \approx \Delta E = m(X^*)c^2 - m(X)c^2 \qquad [2.71]$$

This amounts to overlooking in equation [2.61] the recoil kinetic energy, $E_{cr}\ (X)$ of the final nucleus, X. Relationship [2.71] is often verified with a very good approximation, as highlighted above.

To confirm this assertion, let us determine the energy of the gamma photon emitted during deexcitation of the yttrium-89 nucleus according to the equation:

$$^{89}_{39}\mathrm{Y}^* \rightarrow {}^{89}_{39}\mathrm{Y} + \gamma \qquad [2.72]$$

Given data:

$m(\mathrm{Y}^*) = 88.9037013$ u; $m(\mathrm{Y}) = 88.9027212$ u; 1 u 931.5 MeV/c^2.

Using [2.71], we obtain:

$E_\gamma \approx \Delta E = [88.9037013 - 88.9027212]\ \mathrm{u}c^2 = 0.91296315$ MeV

Which then gives:

$$E_\gamma \approx \Delta E \approx 913\ \mathrm{keV} \qquad [2.73]$$

Let us now take into account the recoil kinetic energy of the Y* nucleus. The energy of the γ-photon is deduced from [2.70]. Let:

$$E'_\gamma = \Delta E - \frac{1}{2}\frac{\Delta E^2}{m(\mathrm{Y})c^2} \qquad [2.74]$$

Let us determine the following relative gap:

$$\frac{\Delta E_\gamma}{E_\gamma}=\frac{E_\gamma - E_\gamma^{'}}{E_\gamma}=\frac{\Delta E - E_\gamma^{'}}{E_\gamma} \qquad [2.75]$$

Knowing that $E_\gamma \approx \Delta E$, and using [2.74], we obtain:

$$\frac{\Delta E_\gamma}{E_\gamma}=\frac{1}{2E_\gamma}\frac{\Delta E^2}{m\,(Y)c^2}=\frac{1}{2}\frac{\Delta E}{m\,(Y)c^2} \qquad [2.76]$$

Numerically we find:

$$\frac{\Delta E_\gamma}{E_\gamma}=\frac{1}{2}\times\frac{0.91296315}{88.9027212\times 931.5}=5.5\times 10^{-6} \qquad [2.77]$$

Result [2.77] shows that correction [2.74] applied to the energy of the γ-photon by taking into account the recoil energy of the final nucleus is very low. It will be accepted that in the general case and unless otherwise indicated, approximation [2.71] may be used to determine the γ-deexcitation energy.

2.3.2. *Angular momentum and multipole order of γ-radiation*

As with atomic electrons and nucleons in a nucleus, γ-radiation has an angular momentum, $\vec{l}$, its square, l^2, and one of its projections on an arbitrary axis (*Oz*, for example) having determined values equal, respectively, to:

$$l^2 = \hbar^2 \ell(\hbar + 1)\,;\, l_z = m_\ell \hbar \qquad [2.78]$$

For the photon $\ell > 0$

From a classical perspective, the emission of a γ-photon is perceived as the electromagnetic radiation emitted by a distribution of nuclear charges (protons) contained in the nucleus. This charge distribution can be dipole, quadrupole, etc., or multipole. Thus, γ-radiation is characterized by a *multipole order,* $k_\ell = 2^\ell$.

– For $\ell = 0$, $k_0 = 1$: a *monopole* is obtained. Electromagnetic radiation corresponds to the simultaneous propagation in the space of electric and magnetic fields. While an *electric monopole* (elementary charged particle) exists, no *magnetic monopole* exists in nature, however (manifestation of an elementary magnetic charge). Therefore, the value $\ell = 0$ is not possible. This justifies that for the photon, $\ell > 0$.

– For $\ell = 1$, $k_1 = 2$: a *dipole* is obtained. The corresponding radiation is called *dipole radiation*.

– For $\ell = 2$, $k_2 = 4$: a *quadrupole* corresponding to *quadrupolar radiation* is obtained.

– For $\ell = 3$, $k_3 = 8$: an *octopole* corresponding to *octopolar radiation* is obtained, and so on.

2.3.3. Classification of γ-transitions, parity of γ-radiation

Like any particle, the photon is characterized by its spin, $s = +1$. The value of the magnetic quantum number of spin is thus $m_s = \pm s$, i.e. $m_s = -1$, 0 or +1. That is, $(2s + 1)$ values. In addition, the quantum number, m_s indicates the different propagation directions of the photon. When the projection of the photon spin, $\vec{s}$, on the propagation axis, is equal to + 1 (spin parallel to the propagation direction), the light wave is said to have a right-hand polarization (*right-handed helicity state*). Otherwise ($m_s = -1$: spin antiparallel to the propagation direction) the polarization is said to be left-hand (*left-handed helicity state*). Since electromagnetic waves consist of photon fluxes and only two propagation directions are sufficient to describe the electromagnetic field, it is accepted that the photon has two possible polarizations:

– a *right-hand polarization* ($m_s = +1$) corresponding to the right-handed helicity state, $\left|\vec{k}, m_s = +1\right\rangle$;

– a *left-hand polarization* ($m_s = -1$) corresponding to the left-handed helicity state, $\left|\vec{k}, m_s = -1\right\rangle$, with $\vec{k}$ as the wave vector indicating the photon propagation direction.

Based on current knowledge, the state in which the spin vector projection is equal to $m_s = 0$ does not exist according to quantum electrodynamics [SAK 12, SAK 19].

For each multipole order, there can therefore only be two different waves by the polarization of the photon. Each value of ℓ corresponds to an electric radiation quantum and a magnetic radiation quantum having the same angular momentum, but different parities. The *parity of an electric multipole* is opposite to the *parity of an magnetic multipole*, i.e.:

– for an electric multipole: $\pi = (-1)^{\ell}$;

– for a magnetic multipole: $\pi = -(-1)^{\ell} = (-1)^{\ell+1}$. [2.79]

For the purpose of their classification, the γ-transitions are characterized by their multipole order, k_{ℓ} and their type: E (electrical) or M (magnetic). They are then noted E_{ℓ} (E_1, E_2, E_3, etc.) for *electric multipole transitions* or M_{ℓ} (M_1, M_2, M_3, etc.) for *magnetic multipole transitions*. Table 2.3 summarizes the classification of γ-transitions according to the multipole order. Monopole γ-transitions (with zero quantum number) have never been observed. They are therefore excluded from Table 2.3.

Classification of γ-transitions	Type	Quantum number	Multipole order
Electric dipole	E_1	1	2
Magnetic dipole	M_1	1	2
Electric quadrupole	E_2	2	4
Magnetic quadrupole	M_2	2	4
Electric octopole	E_3	3	8
Magnetic octopole	M_3	3	8
Electric multipole	E_{ℓ}	ℓ	2^{ℓ}
Magnetic multipole	M_{ℓ}	ℓ	2^{ℓ}

Table 2.3. *Classification of the γ-transitions according to the multipole order*

2.3.4. *γ-transition probabilities, Weisskopf estimates*

As mentioned above, the process of successive deexcitations generate a spectrum of γ-photon lines. Each γ-transition then occurs with a certain probability that can be evaluated theoretically using *Weisskopf estimates* [WEI 51, DAU 99, MAY 17]. It should be noted, however, that these estimates do not constitute a theoretical framework with predictions that can be compared with the observations. They are used to understand the phenomenon of γ-deexcitation by comparing the probabilities of the different γ-transitions studied. If we use λ_{γ} to designate the *deexcitation probability*, γ, per unit time, the Weisskopf estimates for electric (E_ℓ) and magnetic (M_ℓ) transitions are written, respectively (see Exercise 2.9 for demonstrations):

$$T_E(\ell) = \lambda_\gamma(E_\ell) = C_\ell(E) A^{2\ell/3} E_\gamma^{2\ell+1} \quad [2.80]$$

$$T_M(\ell) = \lambda_\gamma(M_\ell) = C_\ell(M) A^{(2\ell-2)/3} E_\gamma^{2\ell+1} \quad [2.81]$$

In these relationships, A is the mass number of the nucleus, E_γ is the energy of the γ-photon expressed in MeV, $C_\ell(E)$ and $C_\ell(M)$ are estimated coefficients for each value of ℓ (Table 2.4); the probabilities of multipole transitions $\lambda_\ell(E_\ell)$ and $\lambda_\ell(M_\ell)$ are expressed in s^{-1}.

ℓ	1	2	3	4	5
C_ℓ (E)	1.0×10^{14}	7.4×10^{7}	34.5	1.1×10^{-5}	2.5×10^{-12}
C_ℓ (M)	3.1×10^{13}	2.2×10^{7}	10.3	3.3×10^{-6}	7.4×10^{-13}

Table 2.4. *Values of the $C_\ell(E)$ and $C_\ell(E)$ coefficients for γ-transitions*

Table 2.4 shows that the values of the $C_\ell(E)$ and $C_\ell(M)$ coefficients are all the greater, the smaller ℓ is. Therefore, for a small value of ℓ, the probabilities of γ-transitions [2.80] and [2.81] are all the greater, the higher the energy, E_γ, is. We will find the values gathered in exercise 2.9.

2.3.5. *Conserving angular momentum and parity*

As noted in the previous section, the smallest value of ℓ is predominant for a given γ-transition. In the general case, the γ-deexcitation process satisfies the principle of conserving the total angular momentum.

Let $\vec{J}_i$ designate the angular momentum of the initial transition level of γ and $\vec{J}_f$ the angular momentum of the final level of the same. During this transition, the γ-photon emitted is characterized by the angular same transition, $\vec{l}$. Applying the principle of total angular momentum conservation, we obtain:

$$\vec{J}_i = \vec{J}_f + \vec{l} \Rightarrow \vec{l} = \vec{J}_i - \vec{J}_f \quad [2.82]$$

From the last equality [2.82], the possible values of ℓ are taken, i.e.:

$$\left|J_i - J_f\right| \leq \ell \leq J_i + J_f \qquad [2.83]$$

In addition, during a transition, the *principle of conserving parity* should be verified. As indicated in relationships [2.79], the parities of electric multipole transition states are opposite to the parities of magnetic multipole transition states. Using π_i to designate the parity of the initial state of the transition and π_f for the parity of the final state of the same transition, the principle of parity conservation gives:

– for an electric multipole transition (E_l): $\pi_i = \pi_f(-1)^{\ell}$

– for a magnetic multipole transition (M_l): $\pi_i = \pi_f(-1)^{\ell+1}$ [2.84]

Note that as a general rule, an E-type transition is never in competition with an M-type transition. It follows that relationships [2.84] cannot be simultaneously verified. Either one of these relationships or the other is satisfied during a γ-transition.

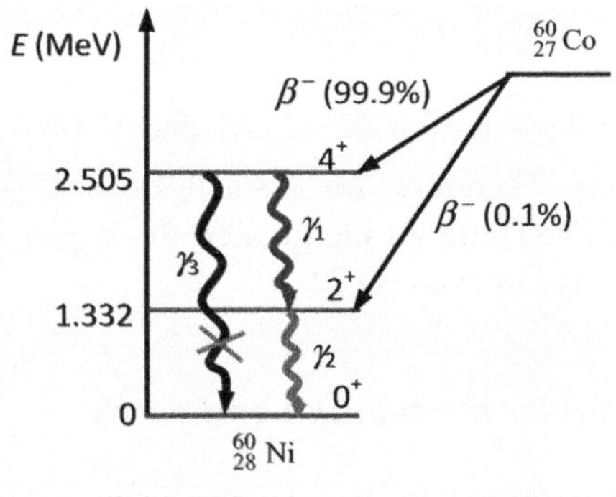

Figure 2.13. *γ-transitions to the ground level of the nickel-60 nucleus*

Let us illustrate by an example, the analysis of the γ-deexcitation spectrum of the nickel-60 nucleus toward the ground level. The β^- decay (see Chapter 3) of cobalt $^{60}_{27}Co$ feeds two excited states of the nickel $^{60}_{28}Ni$ nucleus. Experimentally, these two states are characterized by the state $J^{\pi} = 4^+$ (99.9%) of energy $E^* = 2.505$ MeV and the state $J^{\pi} = 2^+$ (0.1%) of energy $E^* = 1.332$ MeV. Theoretically, γ-transitions to the fundamental level $J^{\pi} = 0^+$ should induce the emission of three photons, γ_1, γ_2 and γ_3 (Figure 2.13).

Experimentally, only two transitions are observed: the $4^+ \rightarrow 2^+$ transition (γ_1) and the $2^+ \rightarrow 0^+$ (γ_2) transition. It follows that the $4^+ \rightarrow 0^+$ transition (γ_3) is absent from the emission spectrum. We propose to verify these experimental observations theoretically by calculating, *inter alia*, the probabilities of transitions based on the Weisskopf estimates.

– Permitted transitions

Let us use the conservations of total angular momentum and parity to determine ℓ and the type of multipole transition, E or M.

– **$4^+ \rightarrow 0^+$ transition**: $\pi_i = \pi_f = +1$; the parity is unchanged: this is an E-type transition according to [2.84]. $J_i = 4$, $J_f = 0$. Since $J_f = 0$, only one value exists, $\ell = 4$ according to [2.83]. It is therefore an E_4 transition;.

– **$4^+ \rightarrow 2^+$ transition**: $\pi_i = \pi_f = +1$; $J_i = 4$, $J_f = 2$. According to [2.83], $2 \leq \ell \leq 6$. The possible values of ℓ are 2, 3, 4, 5 and 6. The most favored transition corresponds to the smallest value of ℓ, i.e. $\ell = 2$. We thus obtain an E_2 transition;

– **$2^+ \rightarrow 0^+$ transition**: $\pi_i = \pi_f = +1$; $J_i = 4$, $J_f = 0$. Therefore, only one value exists, $\ell = 2$. We likewise obtain an E_2 transition;

Ultimately, two dipole transitions are obtained, $4^+ \rightarrow 2^+$ and $2^+ \rightarrow 0^+$, which are widely favored, confirming the experimental observations. The $4^+ \rightarrow 0^+$ transition is therefore unlikely. This result can be verified by calculating the transition probabilities.

– Transition probabilities

For the $4^+ \rightarrow 2^+$ and $2^+ \rightarrow 0^+$ transitions: $\ell = 2$ and the $4^+ \rightarrow 0^+$ transition: $\ell = 4$. Using the Weisskopf formula [2.80], we obtain ($A = 60$ for Ni):

– $4^+ \rightarrow 2^+$ transitions

$$\lambda_{\gamma_1}(E_2) = C_2(E_2) \times A^{4/3} E_{\gamma_1}^5 \qquad [2.85a]$$

– $2^+ \rightarrow 0^+$ transitions

$$\lambda_{\gamma_2}(E_2) = C_2(E_2) \times A^{4/3} E_{\gamma_2}^5 \qquad [2.85b]$$

– $4^+ \rightarrow 0^+$ transitions

$$\lambda_{\gamma3}(E_4) = C_4(E_4) \times A^{8/3} E_{\gamma3}^9 \quad [2.85c]$$

Referring to Figure 2.13 and Table 2.4, we take:

$E_{\gamma1} = (2.505 - 1.332)$ MeV = 1.173 MeV; $C_2(E) = 7.3 \times 10^7$

$E_{\gamma2} = 1.332$ MeV; $E_{\gamma3} = 2.505$ MeV; $C_4(E) = 1.1 \times 10^{-5}$

Numerically, according to [2.85], we obtain:

$$\lambda_{\gamma_1}(E_2) = 7.3\times10^7 \times 60^{4/3} \times (1.173)^5 = 3.8\times10^{10}\ \text{s}^{-1}$$

$$\lambda_{\gamma_2}(E_2) = 7.3\times10^7 \times 60^{4/3} \times (1.332)^5 = 7.2\times10^{10}\ \text{s}^{-1} \quad [2.86]$$

$$\lambda_{\gamma_3}(E_4) = 1.1\times10^{-5} \times 60^{8/3} \times (2.505)^9 = 2.4\times10^3\ \text{s}^{-1}$$

Results [2.86] show that the $4^+ \rightarrow 2^+$ and $2^+ \rightarrow 0^+$ transitions are around 10 million times more probable than the $4^+ \rightarrow 0^+$ transition. Note that the probability of the $4^+ \rightarrow 0^+$ transition is far from negligible. Nevertheless, these are results obtained using the Weisskopf estimates that are not as we stated above, a theoretical framework that gives predictions that may be corroborated by experimental observations. We therefore envisage obtaining two emission lines corresponding to γ_1 and γ_2 photons.

APPLICATION 2.7.– Figure 2.14 shows the excited level fed by the β^- decay of cesium-137. Determine the type of multipole transition, E_ℓ or M_ℓ, corresponding to the observed deexcitation. Estimate the corresponding transition probability.

ANSWER.– For the $11/2^- \rightarrow 3/2^+$ transition, $\pi_i = + = - \pi_f = +1$: there is a change in parity. This is therefore a magnetic multipole transition (M_ℓ). $J_i = 11/2$, $J_f = 3/2$. According to [2.83], $4 \leq \ell \leq 7$. The possible values of ℓ are 4, 5, 6 and 7. The most favored transition corresponds to the smallest value of ℓ, i.e. $\ell = 4$. It is therefore a magnetic 2^4-pole transition, or put simply, a M_4-type transition.

To estimate the transition probability, let us use the Weisskopf estimate formula [2.81]. We obtain:

$$\lambda_\gamma(M_4) = C_4(M_4) \times A^2 E_\gamma^5 \quad [2.87]$$

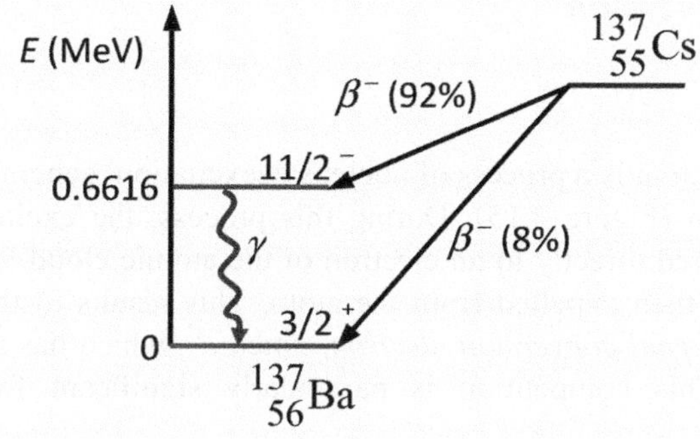

Figure 2.14. *γ-transitions to the ground level of the barium-137 nucleus*

From Table 2.4, we take $C_4 = 3.3 \times 10^{-6}$. According to Figure 2.14, $E_\gamma = 0.6616$ MeV. Numerically, using [2.87], we then find:

$$\lambda_\gamma(M_4) = 4.5\times10^{-6}\times137^2\times(0.6616)^5 = 0.008 \quad [2.88]$$

Mathematically, the magnetic 2^4–pole transition $11/2^- \rightarrow 3/2^+$ of virtually zero probability is unlikely. Yet it is observed experimentally. This contradiction is entirely justified by the fact that the Weisskopf estimates are not based on a theory able to predict results that may be corroborated by the experiment.

Victor Frederik Weisskopf was an American-naturalized Austrian theoretical physicist. In 1937, he formulated the general theory of the statistical model to describe the evaporation model of nucleons of a highly-excited nucleus. In addition, in 1951, he established equations for estimating γ-transition probabilities, these equations bearing the name 'Weisskopf estimates' in his honor.

Box 2.2. *Weisskopf (1908–2002)*

NOTE.– *On the evaporation model*:

Within the framework of the *nucleon evaporation model*, a large number of close and very excited nuclear levels are considered. The emission of a nucleon from a highly-excited nuclear level is analogous to the process of extracting a molecule from an evaporating liquid drop by heating. In the nucleon evaporation model, the usual laws of statistical thermodynamics are applied.

2.4. Internal conversion

2.4.1. *Definition*

Internal conversion is a process of nucleus deexcitation, generally in competition with γ-deexcitation (Figure 2.15). During this process, the excitation energy of a nucleus is transferred directly to an electron of the atomic cloud. The electron of W binding energy is then expelled from the atom. This results in the emission of an electron called *internal conversion electron,* noted e^{C}, which has a well-determined kinetic energy. This competition is particularly significant for low excitation energies.

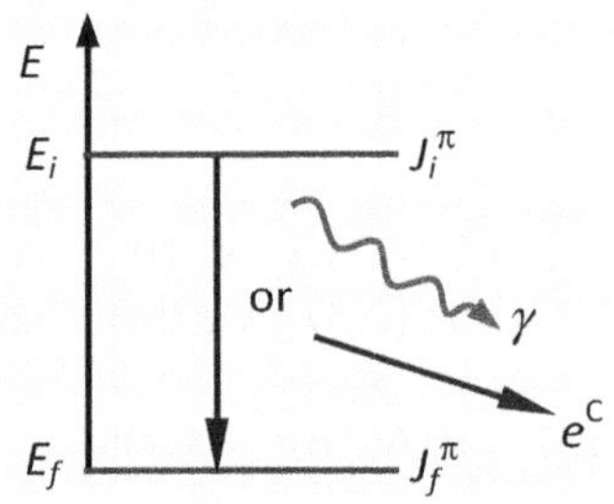

Figure 2.15. *Competition of γ-deexcitation and the internal conversion process*

Let E_I and E_f be the energies of the initial and final levels, respectively, of the excited nucleus. Let $E = (E_I - E_f)$ be the difference in nuclear energy. The kinetic energy of the internal conversion electron is deduced from the principle of energy conservation, i.e.:

$$E_c = E - W \tag{2.89}$$

In the case of the emission of an electron by *photoelectric effect,* a photon of energy $E = h\nu$ is absorbed by a metal of work function W_0. Equation [2.89] then gives the *Einstein equation* of the photoelectric effect:

$$E_c = h\nu - W_0 \tag{2.90}$$

The similarity of equations [2.89] and [2.90] created confusion as to the correct interpretation of the phenomenon of internal conversion, which was likened to a photoelectric effect.

Indeed, in 1922, Ellis [ELL 21, ELL 22] and Lisa Meitner [EVA 61] showed, independently of one another, that equation [2.89] was valid for what was called "the *spectrum of β ray lines*" and is nowadays referred to as "conversion electrons". In interpreting their experimental observations, Ellis and Meitner attributed the spectrum of electron lines produced by nuclear deexcitation to a photoelectric effect. Their model initially assumed the emission of a photon during a process of deexcitation of the nucleus. Subsequently, the photon is absorbed by the inner shells of the atom without ever escaping. This results in the expulsion of an electron by photoelectric effect.

Amongst the best evidence that an internal conversion cannot be likened to a photoelectric effect is the observation of the electric monopole transition, E_0, resulting from a deexcitation of the 0^+ excited level of the zirconium-90 nucleus to the ground level, 0^+ (Figure 2.16).

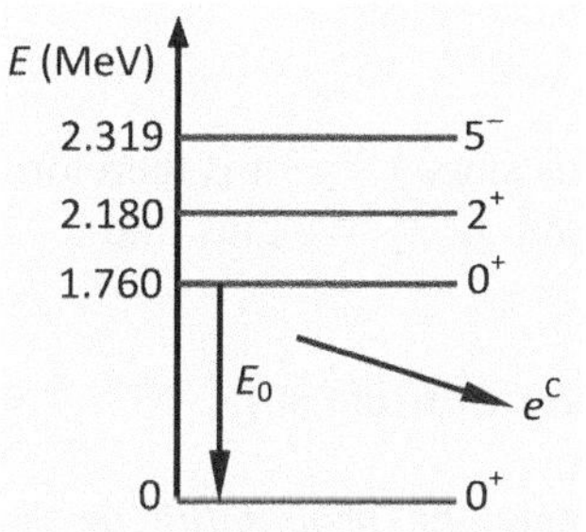

Figure 2.16. *Electric monopole transition, E_0, resulting from a process of initial-level deexcitation of the zirconium-90 nucleus by internal conversion*

The $0^+ \rightarrow 0^+$ transition is forbidden in the case of photon emission, since it corresponds to = 0. Deexcitation from level 0^+, with a very short half-life equaling 61.3 ns, to the ground level, is then accompanied by the emission of an electron by internal conversion. The conversion electron is ejected from the *K*-shell with a kinetic energy equaling 1.742 MeV.

APPLICATION 2.8.– Determine the *K*-shell binding energy of the zirconium atom. We will use the data given in Figure 2.16. Verify the classical or relativistic nature of the internal conversion electron. Let $m_0c^2 = 0.511$ MeV for the electron.

ANSWER.– Using [2.89], the *K*-shell binding energy of the zirconium atom is given by the relationship:

$$W_K = E - E_c \qquad [2.91a]$$

We have $E_c = 1.742$ MeV. Considering the $0^+ \rightarrow 0^+$ transition (Figure 2.16), the difference in nuclear energy, $E = (E_i - E_f) = 1.760$ MeV is determined. According to [2.91], the K-shell binding energy is $W_K = (1.760 - 1.742)$ MeV = 18 MeV.

To verify the classical or relativistic nature of the internal conversion electron, we can use a demonstration by reducing to absurd. Let us assume that it is a classical particle. Its kinetic energy is thus given by the relationship:

$$E_c = \frac{1}{2} m v^2 = \frac{1}{2} m_0 c^2 \frac{v^2}{c^2} \Rightarrow v = \sqrt{\frac{2E_c}{m_0 c^2}}\, c \qquad [2.91b]$$

NOTE.– $E_c = 1.742$ MeV, $m_0c^2 = 0.511$ MeV. We obtain:

$$v = \sqrt{\frac{2 \times 1.742}{0.511}}\, c = 2.61\, c \qquad [2.91c]$$

Result [2.91c] is erroneous since $v < c$: it is therefore a relativistic electron whose velocity is given by relationship [2.2]. We thus find $v = 0.97\ c$.

2.4.2. *Internal conversion coefficients*

Let us use λ_γ to designate the *probability per unit time of deexcitation* by γ-photon emission and λ_C the probability per unit time of deexcitation by conversion electron emission. Taking into account the two processes in competition at low energy, a quantity is introduced called the *internal conversion coefficient,* noted α, which is given by the relationship:

$$\alpha = \frac{\lambda_C}{\lambda_\gamma} = \frac{N_C}{N_\gamma} \qquad [2.92]$$

In the definition relationship [2.92], N_C and N_γ represent, respectively, the numbers of conversion electrons and γ-photons emitted, in the same time interval, by the same sample of identical nuclei characterized by the difference in nuclear energy, W, during the transition being considered. The internal conversion coefficient depends on the atomic number, Z, of the nucleus under consideration, the multipole order, 2^ℓ, and the difference in energy transition, W. The *total probability per unit time of nuclear deexcitation,* λ, is then given by the relationship:

$$\lambda = \lambda_\gamma + \lambda_C \qquad [2.93]$$

Taking into account [2.92], relationship [2.93] can be put in the form:

$$\lambda = (1 + \alpha)\lambda_\gamma \qquad [2.94]$$

Relationship [2.94] allows us to define the half-life, T, of an excited nuclear level, i.e.:

$$T = \frac{\ln 2}{\lambda} = \frac{\ln 2}{(1+\alpha)\lambda_\gamma} \qquad [2.95]$$

Knowledge of the internal conversion coefficient, α, and the internal conversion probability, λ_γ, enables the determination of the half-life, T, of an excited nuclear state.

Using T_γ to designate the *partial half-life of deexcitation* by γ-photon emission and T_C for the partial half-life of deexcitation by conversion electron emission, the *total half-life, T, of an excited state* is also defined by the relationship:

$$\frac{1}{T} = \frac{1}{T_\gamma} + \frac{1}{T_C} \qquad [2.96]$$

APPLICATION 2.9.– The half-life of the 0^+ excited state of the zirconium-90 nucleus (Figure 2.16) is equal to 61.3 ns. Specify the value of half-life T_γ, then determine the probability per unit time of the deexcitation of ^{90}Zr by internal conversion.

ANSWER.– This is an internal conversion. Mathematically, according to [2.96], we obtain: $T = T_C \Rightarrow T_\gamma = \infty$: physically, the γ-emission is unobservable. Using [2.95], we obtain: $\lambda = \lambda_C = \ln 2/T = 1.13 \times 10^7\ s^{-1}$.

2.4.3. *Partial conversion coefficients*

Historically, *electronic shells* were designated by the letters K, L, M, etc., used in X-ray spectroscopy. The *electronic configuration* of an atom can then be written as K^a, L^b, M^c, etc., the shells K, L, M, etc., are marked by the numbers 1, 2, 3, 4, etc., respectively. On the basis of the planetary model, the electrons are distributed in the shells of the electronic cloud in order, starting from the first shell, K. For a given configuration of the type K^a, L^b , M^c, etc., the atomic number $Z = a + b + c +$. Depending on the number of electrons, the K, L, M, etc. shells are saturated at 2, 8, 18, etc. electrons, respectively [SAK 19].

According to the above, the internal conversion electron can originate from one of the shells K, L, M, etc. Thus, a *partial conversion coefficient* is introduced for each electronic shell. The *total conversion coefficient*, α_{tot}, is then equal to the sum of the *partial conversion coefficients*, α_K, α_L, α_M, etc., i.e.:

$$\alpha_{tot} = \alpha_K + \alpha_L + \alpha_M +, \text{etc.} \qquad [2.97]$$

In relationship [2.97], each partial conversion coefficient is relative to the set of electrons distributed on the shell. Thus, α_K is relative to the two K-electrons, α_L to the eight L-electrons, α_M to the 18-M electrons, and so on. Given that the K-shell is closer to the nucleus, the *internal conversion probability of a K-electron* is greater than the *internal conversion probability of an L-electron*, which is itself greater than the *internal conversion probability of an M-electron*, and so on. Theoretically, we should expect to see: $\alpha_K > \alpha_L > \alpha_M >$, etc. In the following, we will limit our study to the case of the K-shell.

2.4.4. *K-shell conversion*

Theoretically, the values of the partial conversion coefficients, α_K, α_L, α_M, etc., are determined by approximated relativistic formulas.

When *Born's rule* $[Z\alpha(v/c)] << 1$ is satisfied, two formulas of interest relating to multipole transitions can be used as examples for the study of K-shell internal conversion. Note that in Born's rule (named after the German physicist, Max Born (1882-1971)), α designates the fine-structure constant: $\alpha = 1/137.036$. To avoid confusion with the internal conversion coefficient α, defined by the relationship [2.92], the approximation $\alpha \approx 1/137$ is applied and Born's rule is written as $[Z/137\,(v/c)] << 1$, with v as the velocity of the conversion electron.

Let us note with $E_K = W_K$, the binding energy of a K-electron, and with $E = W = (E_i - E_f)$, the difference in nuclear energy. If, in addition to Born's rule, the double inequality $E_K << W << m_0c^2$ is verified, the partial conversion coefficient, α_K, can be determined from the approximated formulas [EVA 61]:

– For electric multipole transitions:

$$(\alpha_K)_{el} \approx \frac{\ell}{\ell+1} Z^3 \left(\frac{1}{137}\right)^4 \left(\frac{2m_0c^2}{W}\right)^{\ell+5/2} \qquad [2.98]$$

– For magnetic multipole transitions:

$$(\alpha_K)_{mag} \approx Z^3 \left(\frac{1}{137}\right)^4 \left(\frac{2m_0c^2}{W}\right)^{\ell+3/2} \qquad [2.99]$$

As indicated in formulas [2.98] and [2.99], for a given value of the transition energy, W, the K-shell internal conversion coefficient increases with the atomic number, Z, and multipole order, 2^{ℓ}. Likewise, for fixed Z, α_K is all the greater, the lower the transition energy is and the higher the multipole order is.

APPLICATION 2.10.– Estimate the internal conversion coefficient, α_K, for zirconium-90 (Figure 2.18). Let $m_0c^2 = 0.511$ MeV. Given datum: $Z = 40$ for zirconium.

ANSWER.– For the monopole transition, E_0, $\ell = 0$. Therefore, formula [2.99] is used. Given that $W = (E_I - E_f) = 1.760$ MeV, we obtain:

$$(\alpha_K)_{mag} \approx 40^3\left(\frac{1}{137}\right)^4\left(\frac{2\times 0.511}{1.760}\right)^{3/2} = 8.3\times 10^3$$

Using numerical calculations, the exact theoretical values of the K-shell internal conversion coefficients listed in tables can be determined. Figure 2.17 shows the variation in the conversion coefficient, α_K, with atomic number Z.

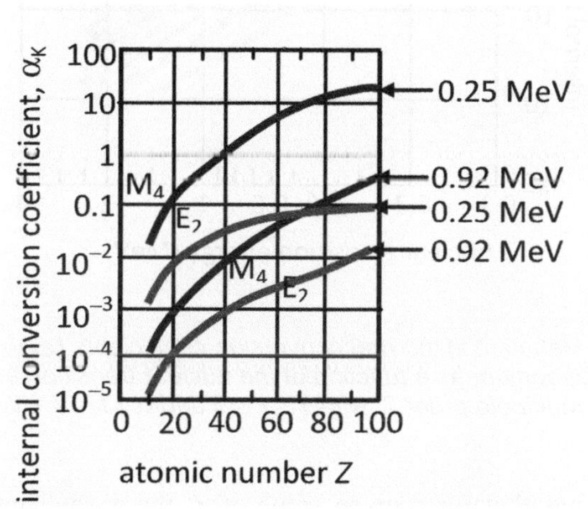

Figure 2.17. *Variation in the internal conversion coefficient, α_K with atomic number Z for two multipoles, E_2 at 0.92 MeV and M_4 at 0.25 MeV*

It is easily verified that α_K is all the greater, the lower the transition energy is (here $W = 0.25$ MeV) and the higher the multipole order is (here magnetic 2^4-pole order, M_4). In particular, it is noted that α_K is all the greater, the lower the transition energy is and the higher ℓ is. This proves that deexcitation by γ-photon emission is disadvantaged for a low value of W and a high value of ℓ. When W increases, the preferred mode of deexcitation is deexcitation by γ-photon emission.

Similarly, in Figure 2.18 we have reproduced the results obtained for zirconium under the conditions $1 \leq \ell \leq 5$ and $0.3\ m_0c^2 \leq W \leq 5\ m_0c^2$ [EVA 61].

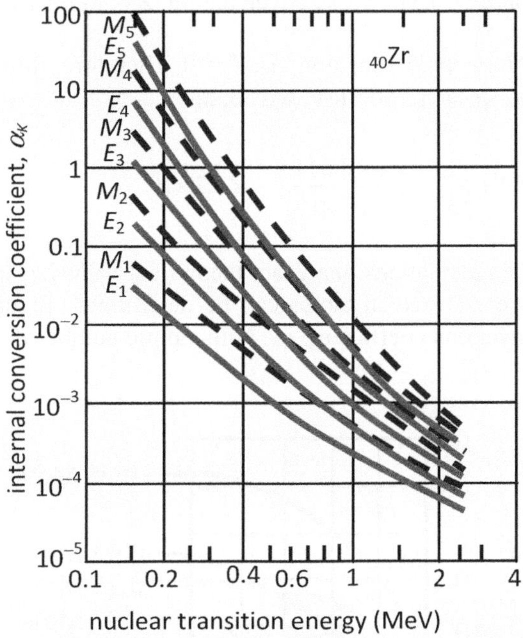

Figure 2.18. *Variation in internal conversion coefficients $(\alpha_K)_{el}$ and $(\alpha_K)_{mag}$ relative to zirconium as a function of the nuclear transition energy, W, and of the multipole order 2^ℓ with $1 \leq \ell \leq 5$ and 0.1 MeV < W < 4 MeV*

This figure shows that for a given value of Z, the K-shell internal conversion coefficient is strongly dependent on nuclear transition energy, W, and the multipole order, 2^ℓ.

According to the above, low-energy nuclear deexcitation is accompanied by the emission of γ-photons in competition with the emission of conversion electrons. The K, L, M, etc. shell conversion process leaves a hole or a vacant orbital in the inner shell. This hole can then be occupied by an electron from an upper shell with emission of a characteristic X-ray, or with emission of an electron called an Auger electron. These processes of *atomic deexcitation* result from a redistribution of energy within the atom, following a rearrangement of the electronic cloud. Although X-ray and Auger electron emissions are derived from atomic phenomena, we will incorporate their study into this book (see section 4.1.9) since they originate from low-energy nuclear deexcitation.

2.5. Deexcitation by nucleon emission

2.5.1. *Definition*

γ-deexcitation and/or internal conversion occur at low energy ($E^* < S_{n,p}$). In the case of highly excited nuclei ($E^* > S_{n,p}$), the mode of nuclear deexcitation by emission of γ-photon or conversion electron is in competition with the emission of proton p if ($E^* > S_p$), or neutron n if ($E^* > S_n$).

Let us study the particularly significant case of neutron emission that is responsible for the *emission of delayed neutrons* by nuclear fission products. To do this, we write the general equation for the fission of uranium-235 under the impact of a slow or thermal neutron:

$$n + {}^{235}_{92}\text{U} \rightarrow {}^{A}_{Z}\text{X} + {}^{A'}_{Z'}\text{Y} + 2\ or\ 3\ n\ + 2 \text{ or } 3 \text{ n} \qquad [2.100]$$

Let us identify several neutron qualifiers involved in a nuclear fission reaction:

– The neutron of reagents is called *slow neutron* or *thermal neutron*. This neutron is said to be slow because, if it is too slow, it rebounds on the uranium-235 nucleus, if it is too fast, it crosses it without being captured: fission does not occur in these cases. It is called a thermal neutron because it has a kinctic energy of the order of the thermal agitation energy of the gas molecules at the temperature of the nuclear reactor core where fission occurs. The kinetic energy of the slow or thermal neutron capable of inducing a uranium-235 fission is of the order of 0.025 MeV.

– The 2 or 3 neutrons appearing in the products of fission [2.100] are called *fission neutrons*. In the case where the fission fragment X is radioactive β^-, for example, because it has an excess of neutrons, it then breaks by emitting a so-called delayed neutron. The diagram of delayed neutron deexcitation is shown in Figure 2.19. The neutron is said to be delayed because it appears after the β^- decay of the fission product, X, and thus well after the appearance of the fission neutrons.

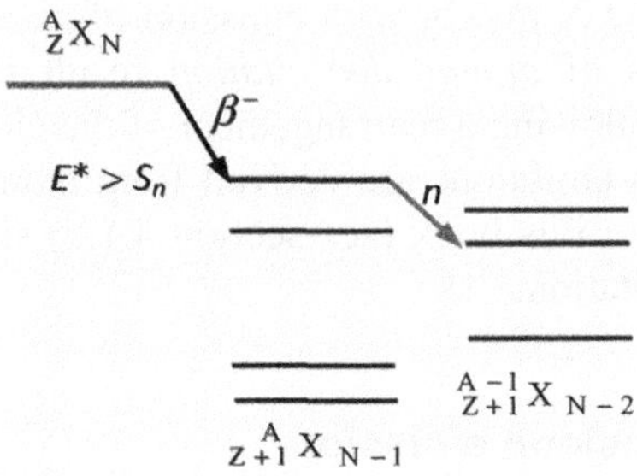

Figure 2.19. *General diagram of nuclear deexcitation by delayed-neutron emission following a β^- decay process*

Delayed neutron emission occurs according to the general equation:

$$^{A}_{Z}X_{N} \xrightarrow{\beta^-} {}^{A}_{Z+1}X^{*}_{N-1} \longrightarrow {}^{A-1}_{Z+1}X_{N-2} + {}^{1}_{0}n \qquad [2.101]$$

2.5.2. *Energy balance*

Consider the process of nuclear deexcitation by neutron emission according to [2.101]. We obtain:

$$^{A}_{Z+1}X^{*}_{N-1} \longrightarrow {}^{A-1}_{Z+1}X_{N-2} + {}^{1}_{0}n \qquad [2.102]$$

Let E^* be the excitation energy of the ${}_{Z+1}^{A}X^{*}_{N-1}$ nucleus. Determine the energies of the initial and final states of reaction [2.102]. We obtain:

$$- E_i \; E_i = E^* + m\,({}^{A}_{Z+1}X^{*}_{N-1})c^2$$

$$- E_f = m\,({}^{A-1}_{Z+1}X_{N-2})c^2 + m_n c^2 + E_c\,({}^{1}_{0}n) + E_{cr}\,({}^{A-1}_{Z+1}X_{N-2})$$

In the expression of E_f, E_{cr} refers to the recoil kinetic energy of the ${}^{A-1}_{Z+1}X_{N-2}$ nucleus. By making use of the principle of energy conservation, $E_I = E_f$, we obtain:

$$E^* + m\,({}_{Z+1}^{A}X^{*}_{N-1})c^2 = m\,({}^{A-1}_{Z+1}X_{N-2})c^2 + m_n c^2 + E_c\,({}^{1}_{0}n) + E_{cr}\,({}^{A-1}_{Z+1}X_{N-2})$$

That is:

$$E_c\left({}_{0}^{1}n\right) + E_{cr}\left({}_{Z+1}^{A-1}X_{N-2}\right) = E^* + m\left({}_{Z+1}^{A}X_{N-1}^{*}\right)c^2 - m\left({}_{Z+1}^{A-1}X_{N-2}\right)c^2 - m_n c^2 \qquad [2.103]$$

By definition, the *energy of deexcitation by* Q_n *neutron emission* is equal to:

$$Q_n = E_c\left({}_{0}^{1}n\right) + E_{cr}\left({}_{Z+1}^{A-1}X_{N-2}\right) \qquad [2.104]$$

Thus, using [2.103]

$$Q_n = E^* + m\left({}_{Z+1}^{A}X_{N-1}^{*}\right)c^2 - m\left({}_{Z+1}^{A-1}X_{N-2}\right)c^2 - m_n c^2 \qquad [2.105]$$

By considering the expression of the separation energy of a neutron [1.67a], the energy balance, Q_n, can take the form:

$$Q_n = E^* - S_n \qquad [2.106]$$

Knowing that $Q_n > 0$ according to equation [2.104], relationship [2.106] indeed shows that nuclear deexcitation by n neutron emission can only occur if $E^* > S_n$.

2.5.3. *Bound levels and virtual levels*

When studying the nuclear deexcitation process we differentiate between the *bound levels* and *virtual levels* of the nuclei. A bound level is a level whose excitation energy is insufficient to induce a deexcitation of the nucleus by emission of a particle such as a neutron, proton or α-particle. As a consequence, deexcitation of a bound nuclear level can only be achieved by the emission of a γ-photon or of an internal conversion electron in competition with the γ-emission. On the other hand, a virtual level is any level whose excitation energy is greater than the lowest dissociation energy of a nucleus in a lighter nucleus with emission of a particle (n, p, α, etc.).

A common example is the nitrogen-14 nucleus, the lowest dissociation energy of which is equal to 7.542 MeV. Figure 2.20 shows several excited levels of the nitrogen-14 nucleus, as well as mass differences, expressed in MeV, of various possible dissociation products. All levels above 7.542 MeV are virtual levels.

As an example, let us calculate the mass difference between the nitrogen-14 nucleus and its dissociation products, $^{13}C + {}^{1}H$, according to the equation:

$$ {}^{14}_{7}N^* \rightarrow {}^{13}_{6}C + {}^{1}_{1}H \qquad [2.107] $$

The masses of the nuclei involved in reaction [2.107] are equal to [EVA 61]:

$m\,(^{14}N) = 14.007515$ u; $m\,(^{13}C) = 13.007473$ u; $m\,(^{1}H) = 1.008142$ u.

Taking 1 u = 931.5 MeV/c^2, we obtain the mass difference:

$$ \Delta m = [m\,(^{13}C) + m\,(^{1}H)] - m\,(^{14}N) = 0.0081 \text{ u} = 7.54515 \text{ MeV}/c^2 \qquad [2.108] $$

Result [2.108] clearly shows that the energy level 7.545 MeV is a virtual level that can be dissociated by emission of a proton according to equation [2.107].

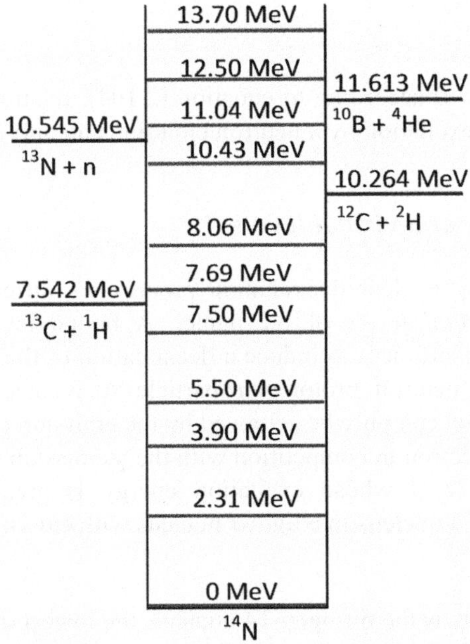

Figure 2.20. *Bound levels and virtual levels of the nitrogen-14 nucleus. All levels above 7.542 MeV are virtual levels that can be dissociated with emission of a particle: neutron (n), proton (p), particle α, deuteron (d). For clarity, the energy scale is not respected*

APPLICATION 2.11.– Account for the various deexcitation pathways of the virtual levels of the nitrogen-14 nucleus by emission of a particle, as shown in Figure 2.20.

Given data:

– $m\ (^{13}\mathrm{N}) = 13.009858$ u; $m\ (^{12}\mathrm{C}) = 12.003804$ u;

– $m\ (^{2}\mathrm{H}) = 2.014735$ u; $m\ (^{10}\mathrm{B}) = 10.016114$ u;

– $m_\alpha = 4.003873$ u; $m_\mathrm{n} = 1.008982$ u.

ANSWER.– The various deexcitation pathways of the virtual levels of the nitrogen-14 nucleus with emission of a neutron, a deuteron or an α particle (^{4}He) correspond to the following processes, respectively:

$$^{14}_{7}\mathrm{N}^* \rightarrow {}^{13}_{7}\mathrm{N} + {}^{1}_{0}\mathrm{n}$$

$$^{14}_{7}\mathrm{N}^* \rightarrow {}^{12}_{6}\mathrm{C} + {}^{2}_{1}\mathrm{H} \qquad [2.109]$$

$$^{14}_{7}\mathrm{N}^* \rightarrow {}^{10}_{5}\mathrm{B} + {}^{4}_{2}\mathrm{He}$$

The mass differences between the nitrogen-14 nucleus and its dissociation products then have the following values, respectively, according to [2.109]:

$$\Delta m = [m\ (^{13}\mathrm{N}) + m_\mathrm{n}] - m\ (^{14}\mathrm{N}) = 10.54924\ \mathrm{MeV}/c^2$$

$$\Delta m = [m\ (^{12}\mathrm{C}) + m\ (^{2}\mathrm{H})] - m\ (^{14}\mathrm{N}) = 10.26886\ \mathrm{MeV}/c^2 \qquad [2.110]$$

$$\Delta m = [m\ (^{10}\mathrm{B}) + m_\alpha] - m\ (^{14}\mathrm{N}) = 11.61767\ \mathrm{MeV}/c^2$$

Results [2.110] clearly justify the various deexcitation pathways of the virtual levels of the nitrogen-14 nucleus with emission of a particle (Figure 2.20).

In the general case, deexciting a virtual level of a nucleus, X, can leave the residual nucleus, Y, in an excited state, E^*_r. In the case of a nucleon, the maximum kinetic energy, E_cmax, of the nucleon emitted of separation energy, $S_{n,p}$ (S_n in the case of a neutron or S_p in the case of a proton) is given by the relationship:

$$E_\mathrm{cmax} = E^* - E^*_\mathrm{r} - S_{n,p} \qquad [2.111]$$

When the nucleon emission leaves the residual nucleus in its ground level, then $E^*_\mathrm{r} = 0$ and $E_\mathrm{cmax} = E^* - S_{n,p}$.

2.5.4. *Study of an example of delayed-neutron emission*

For a better understanding of the process of nuclear deexcitation by neutron emission, consider the specific case of bromine-87 decay. This is a decay process with *delayed-neutron emission* by the krypton-87 nucleus, a nuclear fission product of uranium-235. Using [2.103], in the case of bromine-87 with radioactive half-life, $T = 55.6$ s, we obtain:

$$ {}^{87}_{35}\mathrm{Br} \xrightarrow{\beta^-} {}^{87}_{36}\mathrm{Kr}^* \longrightarrow {}^{86}_{36}\mathrm{Kr} + {}^{1}_{0}\mathrm{n} \qquad [2.112] $$

In Figure 2.21, we have reproduced the diagram of nuclear deexcitation by delayed-neutron emission by the krypton-87 nucleus produced from the β^- decay of the bromine-87 nucleus, according to the notation $_Z X^A$ [EVA 61], for a nucleus with mass number A and charge number Z.

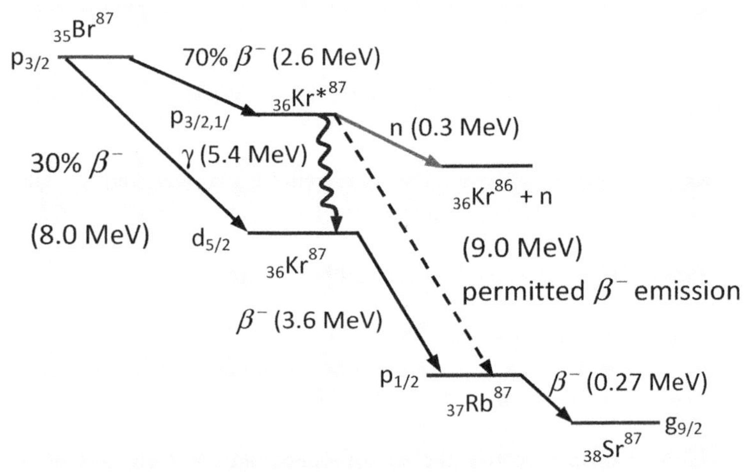

Figure 2.21. *Diagram of nuclear deexcitation by delayed-neutron emission by the krypton-87 nucleus produced from the β^- decay of the bromine-87 nucleus*

As shown in Figure 2.21, approximately 70% of the β^- decay of the ^{87}Br nucleus leads to the excited energy level 5.4 MeV of ^{87}Kr. This excited level is a virtual level whose energy is greater than the neutron separation energy in the ^{87}Kr nucleus, equal to approximately 5.1 MeV. It follows that the 5.4 MeV level is deexcited by emission of a neutron of 0.3 MeV energy, with production of the residual nucleus, ^{86}Kr, according to the last equation [2.112], which we will complete as follows:

$$ {}^{87}_{36}\text{Kr}^* \rightarrow {}^{86}_{36}\text{Kr} + {}^{1}_{0}\text{n} + 0.3 \text{ MeV} \quad [2.113] $$

APPLICATION 2.12.– Determine the deexcitation energy by neutron emission by the ^{86}Kr nucleus. Find the value of the separation energy of the neutron $S_n \approx 5.1$ MeV.

Given data: mass of ^{86}Kr: 85.93828 u.

ANSWER.– Using [2.104], we obtain:

$$ Q_n = E_c\,(\text{n}) + E_{cr}\,(^{86}\text{Kr}) \quad [2.114] $$

In [2.114] E_c (n) is known: E_c (n) = 0.3MeV. Let us then express E_{cr} (^{86}Kr) according to E_c (n) by making use of momentum conservation. For the sake of simplicity, let us write: E_c (n) = E_{cn} and E_{cr} (^{86}Kr) = E_{crKr}. By considering equation [2.113] for deexcitation of Kr* initially at rest, we obtain:

$$ \vec{0} = \vec{p}_{Kr} + \vec{p}_n \Rightarrow p_{Kr} = p_n \quad [2.115a] $$

Given that $E_c = p^2/2m$ and according to [2.115a], we obtain:

$$ m_{Kr} E_{cKr} = m_n E_{cn} \quad [2.115b] $$

By taking E_{crKr} from [2.115b] and applying the result in [2.114], we obtain:

$$ Q_n = \left(1 + \frac{m_n}{m_{Kr}}\right) E_{cn} \quad [2.116] $$

Numerically we find:

$$ Q_n = \left(1 + \frac{1.008982}{85.93828}\right) \times 0.3 = 0.3035 \text{ MeV} \quad [2.117] $$

Yet according to [2.106],

$$ S_n = E^* - Q_n \Rightarrow S_n = 5.4 - 0.3035 = 5.096 \text{ MeV} \quad [2.118] $$

We indeed find the value of the separation energy of the neutron $S_n \approx 5.1$ MeV.

2.6. Bethe–Weizsäcker semi-empirical mass formula

2.6.1. *Presentation of the liquid-drop model*

As specified in section 2.3.1, the nuclear models for studying the internal structure and properties of nuclei include the liquid-drop model, in which nucleons are highly correlated. This model allows us to explain nuclear phenomena that cannot be described within the framework of the shell model, such as phenomena linked *inter alia* to the energy of α and β decays, whose processes will be studied in Chapter 3. For this reason, we will complete this chapter with a detailed study of the properties of the nuclei within the framework of the *liquid-drop model* or *nuclear model with constant nucleon density* [EVA 61, STÖ 07, SEC 10, DUM 15, MAY 17, SAK 18a].

The liquid-drop model is based on the following hypotheses:

1) The nucleus assumed to be spherical of radius $R = r_0 A^{1/3}$ is likened to a drop of incompressible nuclear matter.

2) All nuclei have the same nucleon density; saturated nucleon shells are overlooked.

3) The forces between nucleons are independent of spin and charge.

4) Nuclear forces are very short-range, so that each nucleon only interacts with its closest neighbors, so the nucleons distributed in volume have more neighbors than those distributed on the surface.

On the basis of these hypotheses, we will present and discuss the physical contents of the *semi-empirical mass formula* proposed by Carl von Weizsäcker in 1935 and resumed by Hans Bethe in 1936. This then justifies the *Bethe–Weizsäcker semi-empirical mass formula* or the *Bethe–Weizsäcker formula*, in short.

2.6.2. *Bethe–Weizsäcker formula, binding energy*

The semi-empirical mass formula, known as the Bethe–Weizsäcker formula, expresses the $M(A, Z)$ of a nucleus according to the relationship:

$$\begin{aligned} M(A,Z)c^2 &= ZM_H c^2 + (A-Z)M_n c^2 \\ &\quad - a_v A + a_s A^{2/3} + a_c \frac{Z^2}{A^{1/3}} + a_a \frac{(A-2Z)^2}{A} \pm |\delta| \end{aligned} \qquad [2.119]$$

To clarify the physical meaning of the different terms in the Weizsäcker formula, let us first express the binding energy, E_l (A, Z), of a nucleus. To do this, let us consider expression [1.60] by replacing m_p with M_{H}:

$$E_l = \{[ZM_H c^2 + (A - Z)\, m_n c^2] - M\,(A, Z)\} c^2 \qquad [2.120]$$

By comparing [1.120] and [2.119], and knowing that $N = A - Z$, we obtain:

$$E_l(A, Z) = a_v A - a_s A^{2/3} - a_c \frac{Z^2}{A^{1/3}} - a_a \frac{(N - Z)^2}{A} \pm |\delta| \qquad [2.121]$$

The expression [2.121] of the binding energy enables us to obtain the precise physical meaning of the different terms in the Bethe–Weizsäcker formula [2.120]. It contains five terms having the dimension of an energy, whose physical meanings we will specify.

2.6.3. *Volume energy, surface energy*

The volume energy, noted E_v, is a term introduced by considering an infinite nucleus, in which the nucleons are bound by the strong nuclear force. It corresponds to the first term in expansion [2.121], i.e.:

$$E_v = a_v A \qquad [2.122]$$

In relationship [2.122], the constant a_v is called the *volume coefficient.* This coefficient is to be evaluated.

As explained in section 1.6.6, short-range nuclear forces only act on the close neighbors of a nucleon. A *saturation effect of nuclear forces* occurs, resulting in the binding energy being limited by the maximum number of nucleons that can surround the nucleon considered. For this reason, the binding energy is proportional to the number of nucleons contained in the drop and therefore to the mass number, A, of the nucleus. This then justifies the presence of the term $a_v A$ in the expression of the binding energy [2.121]. Knowing that for a bound system the binding energy is positive, then the volume energy [2.122] is counted positively. Physically, the volume coefficient, a_v, represents the *energy density per nucleon* in the nuclear matter assumed to be infinite.

The second term in [2.121] is negative *surface energy, E_s*:

$$E_s = - a_s A^{2/3} \qquad [2.123]$$

In relationship [2.123], a_s is called the positively-counted *surface coefficient*. This coefficient is also evaluated.

Surface energy is introduced to account for the fact that the nucleus is well-finished and that the nucleons on the surface are less bound than those within the nucleus. To theoretically justify expression [2.123], it can be considered that within the framework of the nuclear model with constant nucleon density, the nucleus assumed to be spherical of radius $R = r_0 A^{1/3}$ has a surface $S = 4\pi R^2 = 4\pi r_0^2 A^{2/3}$. Assuming the spherical nucleon has an average radius equal to r_0, it would occupy a volume $v_0 = (4/3)\pi r_0^3$. Each nucleon would then have a surface $s_0 = 4\pi r_0^2$. Since the nucleon density is constant, the number of nucleons distributed at the surface would then be of the order of $S/s_0 = 3A^{2/3}$. Considering a nucleon of surface πr_0^2, the number of nucleons distributed at the surface would be of the order of $4A^{2/3}$ [EVA 61]. These results justify the fact that the surface energy is proportional to $A^{2/3}$. The negative sign is justified by the fact that the volume energy corresponds to the energy of infinite nuclear matter. With the nucleus being finished, $E_l < E_v \Rightarrow E_s < 0$, according to [2.121].

2.6.4. *Coulomb energy*

The third term in the expression [2.121] is the Coulomb energy noted E_c, due to proton repulsion in the nucleus:

$$E_c = -a_c \frac{Z^2}{A^{1/3}} \qquad [2.124]$$

In [2.124], a_c refers to the *Coulomb coefficient* also to be evaluated.

Considering that the charge $+ Ze$ of the nucleus is uniformly distributed in volume, the *Coulomb energy of a sphere of R charged in volume* is given by the relationship (see exercise 2.5):

$$W_c = \frac{3}{5} k \frac{Z^2 e^2}{R} \qquad [2.125]$$

Within the framework of the liquid-drop model, the nucleus radius, $R = r_0 A^{1/3}$. The relationship [2.125] is then written in the form:

$$W_c = \frac{3ke^2}{5r_0} \times \frac{Z^2}{A^{1/3}} \qquad [2.126]$$

With the Coulomb term decreasing the energy volume, we obtain $E_c = -W_c$. This gives, according to [2.126]:

$$E_c = -\frac{3ke^2}{5r_0} \times \frac{Z^2}{A^{1/3}} = -a_c \times \frac{Z^2}{A^{1/3}} \quad [2.127]$$

The Coulomb coefficient, a_c, is then written:

$$a_c = \frac{3ke^2}{5r_0} \quad [2.128]$$

The expression [2.128] shows that the Coulomb coefficient can be evaluated theoretically if r_0 is known. The first theoretical expression of the Coulomb coefficient is determined using the expression of the Sakho unit nuclear radius [1.39]. We obtain [SAK 18a]:

$$a_c = \frac{3ke^2}{5\alpha^2 a_0}\left(1 + \frac{N}{Z}\right) \quad [2.129]$$

APPLICATION 2.13.– Show that in MeV, relationship [2.129] is written:

$$a_c = 0.307 \times \left(1 + \frac{N}{Z}\right) \text{ (MeV)} \quad [2.130]$$

Calculate the Coulomb coefficient for the nuclei ^{20}Ne ($Z = 10$) and ^{56}Fe ($Z = 26$).

Given data (CODATA recommended values):

$\alpha = 1/137.035999679$; $e = 1.602179487 \times 10^{-19}$ C, $a_0 = 0.52917720859 \times 10^{-10}$ m; 1eV $= 1.602179487 \times 10^{-19}$ J. Let $k = 1/4\pi\varepsilon_0 = 9 \times 10^9$ IS.

Let us call K the factor in [3.10]:

$$K = \frac{3ke^2}{5\alpha^2 a_0}$$

Numerically we find:

$$K = \frac{3 \times 9 \times 10^9 \times 1.602179487 \times 10^{-19}}{5 \times 5.325135412 \times 10^{-5} \times 5.291720859 \times 10^{-11}} = 3.07027 \times 10^5$$

Let $K \approx 0.307$ MeV. We then obtain [2.130]:

– for ^{20}Ne: $N/Z = 1 \Rightarrow a_c = 0.614$ MeV;

– for ^{56}Fe: $N/Z = 1 \Rightarrow a_c = 0.661$ MeV.

As we will see in Table 2.8, for various authors, the semi-empirical values of the Coulomb coefficient vary between 0.580 MeV and 0.710 MeV.

2.6.5. *Asymmetry energy, pairing energy*

The fourth term in expression [2.121] of the binding energy represents the *asymmetry energy, E_a*:

$$E_a = -a_a \frac{(N-Z)^2}{A} \qquad [2.131]$$

In this relationship, a_a is called the *asymmetry coefficient* to be evaluated.

The negative asymmetry energy, E_a, is a term introduced to account for excess neutrons ($N - Z$) compared to $N = Z$ symmetry for the number of nuclei, as observed experimentally.

As shown in Segrè diagram 1.18, the most stable nuclei ($Z < 30$) are located in the valley of stability near the first bisector, $N = Z$, also known as the *line of stability*. This shows that attractive nuclear interaction is stronger where the nucleus has an equal number of protons and neutrons. For heavy nuclei, the excess of neutrons partially compensates for the Coulomb repulsion between protons. This then justifies the fact that the asymmetry energy, which is a purely quantum contribution, is proportional to the excess of neutrons ($N - Z$), which is reflected by a decrease in the binding energy. Note that the asymmetry term is more significant for light nuclei than for heavy nuclei outside the valley of stability.

Finally, the fifth term in expression [2.121] of the binding energy represents the *pairing energy, E_p (pairing term)*:

$$E_p = \pm |\delta| \qquad [2.132]$$

The term for pairing energy is introduced to take into account the fact that, while even-even nuclei exist, there are several even-odd and odd-even nuclei, as well as odd–odd nuclei numbering five (see Table 2.1). This corrective term, also known as a *pairing term*, expresses the fact that the binding energy increases when identical nucleons of a nucleus are paired. As a result, even-even nuclei are more stable than odd-even or even-odd nuclei, which themselves are more stable than odd-odd nuclei because of the inability to pair an *unpaired nucleon*. If we use a_p to designate the *pairing coefficient*, we obtain:

$\delta = 0$ for odd A (even Z-odd N or odd Z-even N) [2.133a]

$\delta = + a_p A^{-1/2}$ for even Z-even N nuclei [2.133b]

$\delta = - a_p A^{-1/2}$ for odd Z-odd N nuclei [2.133c]

In [2.133], the pairing coefficient, a_p = 12 MeV [DUM 15] or (11.2 ± 09) MeV [BUS 13; MAT 46].

Table 2.5 shows the stable nucleus numbers according to the parity of the nucleon number, A, the proton number, Z, and the neutron number, N.

Z	*N*	*A*	**Number of stable nuclei**
even	even	even	166
even	odd	odd	55
odd	even	odd	51
odd	odd	even	5 (*)

(*): the nuclei concerned are: ^{1}H, ^{6}Li, ^{10}Be, ^{14}N and ^{180}Ta (tantalum)

Table 2.5. *Number of stable nuclei with respect to the parity of Z and N*

2.6.6. *Principle of semi-empirical evaluation of coefficients in Bethe–Weizsäcker form*

To determine the coefficients in the Bethe–Weizsäcker formula [2.119], we consider nuclei of odd-A nucleon number. This allows the pairing energy to be canceled according to [2.133a]. The binding energy [2.121] is then written in the form:

$$E_l(A,Z) = a_v A - a_s A^{2/3} - a_c \frac{Z^2}{A^{1/3}} - a_a \frac{(N-Z)^2}{A} \quad [2.134]$$

Relationship [2.134] is an equation with four unknowns, a_v, a_s, a_c , and a_a. These coefficients can then either be determined from the experimental data on four nuclide binding energies, or from the experimental data on the released energies of four nuclear reactions. The average values of the coefficients a_v, a_s, a_c and a_a

evaluated thus serve as a basis for calculation for a large number of nuclides. Nevertheless, it should be noted that formula [2.119] does not give very good results for magic nuclei, which are more stable than nuclei whose masses are determined with high precision using formula [2.119].

– Determining a_c and a_a.

The Coulomb coefficient, a_c, can be determined directly from equation [2.130] by considering all stable nuclei $N = Z$. Which then gives:

$$a_c = 0.613 \text{ MeV} \tag{2.135}$$

Historically, the empirical determination of the five coefficients in the Bethe–Weizsäcker formula is based on experimental data. In the general case, the expression [2.128] of the Coulomb coefficient is considered. Then to evaluate a_c, the experimental value of the unit nuclear radius, r_0, is used, usually chosen as being specific to the Bethe–Weizsäcker formula with [EVA 61]:

$$r_0 = (1.45 \pm 0.05) \times 10^{-13} \text{ cm}; \; 10 < A < 240 \tag{2.136}$$

Using the value [2.136] of r_0, the Coulomb coefficient has the following value, according to [2.128]:

$$a_c = \frac{3 \times 9 \times 10^9 \times 1.602179487 \times 10^{-19}}{5 \times 1.45 \times 10^{-15}} = 5.9667374 \times 10^5 \text{ eV}$$

$\alpha = 1/137.035999679$; $e = 1.602179487 \times 10^{-19}$ C, $a_0 = 0.52917720859 \times 10^{-10}$ m; $1\text{eV} = 1.602179487 \times 10^{-19}$ J. Let $k = 1/4\pi\varepsilon_0 = 9 \times 10^9$ IS.

Let:

$$a_c = 0.597 \text{ MeV} \tag{2.137}$$

To evaluate a_a, the values of the ratio a_a/a_c are determined for several nuclides and an average value is then deduced. This then allows the asymmetry coefficient to be evaluated using the empirical value [2.137] of the Coulomb coefficient.

As shown in equation [2.134], for fixed A, the binding energy is a parabola presenting a minimum. By definition, the term *nuclear charge of the most stable isobar* is used for the $Z_{\min}$ value of Z, for which the mass, $M(A, Z)$ is minimal:

$$\left. \frac{\delta M(A,Z)}{\delta Z} \right|_{Z = Z_{\min}} = 0 \tag{2.138}$$

By making use of condition [2.138], the volume, a_v, and surface a_s, coefficients are eliminated. This then gives an equation where only coefficients a_c and a_a remain. This enables us to express the ratio a_a/a_c.

Using [2.134], condition [2.138] leads to the ratio:

$$\frac{a_a}{a_c} = \frac{1}{2}\frac{Z_{\min} A^{2/3}}{A-2Z_{\min}} - \frac{A}{A-2Z_{\min}} \times \frac{(M_n - M_H)c^2}{4a_c} \qquad [2.139]$$

In equation [2.139], $Z_{\min}$ is generally non-integer. The charge is therefore fictitious. Knowing that the atomic number is an integer. The integer Z closest to $Z_{\min}$ is then taken for the calculations.

Let us determine the mass-energy difference $(M_n - M_H)\, c^2$ by considering the mass energies of the hydrogen atom and the neutron to be equal to: $M_H c^2 = 1.007825\ uc^2$; $M_n c^2 = 1.008665\ uc^2$. Knowing that: 1 u = 931.5 MeV/c^2, we then obtain:

$$(M_n - M_H)\, c^2 = 0.78246 \text{ MeV} \qquad [2.140]$$

Considering [2.137] and [2.140], we find the results gathered in Table 2.6, obtained using equation [2.139] by writing $Z_{\min} = Z$.

X element	*Z*	*A*	a_a/a_c
Astate (As)	33	75	29.874
Bromine (Br)	35	79	32.924
Bromine (Br)	35	81	27.370
Niobium (Nb)	41	93	35.484
Rhodium (Rh)	45	103	35.434
Iodine (I)	53	127	29.902
Cesium (Cs)	55	133	29.259
Terbium (Tb)	65	159	31.095
Holmium (Ho)	67	165	30.765
Thulium (Tm)	69	169	32.232
Tantalum (Ta)	73	181	31.675
Iridium (Ir)	77	191	32.819
Iridium (Ir)	77	193	31.347
Gold (Au)	79	197	32.636
Bismuth (Bi)	83	209	32.397
Average of the ratio a_a/a_c			31.681

Table 2.6. *Calculating the average value of the ratio a_a/a_c*

Looking at the last row of Table 2.6, we obtain:

$$\overline{(a_a / a_c)} = 31,681 = 31.681 \qquad [2.141]$$

The value of the asymmetry coefficient is given by the relationship:

$$a_a = a_c \times \overline{(a_a / a_c)} \qquad [2.142]$$

Using [2.137] and [2.141], we ultimately obtain:

$$a_c = 18.914 \text{ MeV} \qquad [2.143]$$

APPLICATION 2.14.– Then demonstrate relationship [2.139] and determine the absolute errors Δa_I on the evaluations of coefficients a_c and a_a. Now rewrite results [2.135] and [2.141], taking the absolute errors into account.

ANSWER.– Using [2.119], we obtain:

$$\frac{\partial M(A,Z)}{\partial Z} = M_H - M_n + \frac{1}{c^2}\left\{2a_c \frac{Z}{A^{1/3}} - 4\,a_a \frac{(A-2Z)}{A}\right\} \qquad [2.144a]$$

By minimizing equation [2.144a] according to condition [2.138], we obtain:

$$M_H - M_n + \frac{1}{c^2}\left\{2a_c \frac{Z_{\min}}{A^{1/3}} - 4\,a_a \frac{(A-2Z_{\min})}{A}\right\} = 0 \qquad [2.144b]$$

Given that $N = A - Z$, equation [2.144b] is written:

$$M_H - M_n + \frac{1}{c^2}\left\{2a_c \frac{Z_{\min}}{A^{1/3}} - 4\,a_a \frac{(A-2Z_{\min})}{A}\right\} = 0 \qquad [2.144c]$$

Let us multiply [2.144c] by $A/(A - 2Z_{\min})$. We obtain:

$$2a_c \frac{Z_{\min} A^{2/3}}{A-2Z_{\min}} - 4\,a_a - \frac{A}{A-2Z_{\min}}(M_n - M_H)c^2 = 0$$

That is:

$$a_a = \frac{a_c}{2}\frac{Z_{\min} A^{2/3}}{A-2Z_{\min}} - \frac{A}{A-2Z_{\min}}\frac{(M_n - M_H)c^2}{4} \qquad [2.145]$$

By dividing [2.145] by a_c on both sides, we obtain [2.139].

To determine the absolute errors, Δa_I, on the evaluations of coefficients a_c and a_a, let us consider [2.128] and [2.142] to obtain, respectively:

$$\begin{cases} \dfrac{\Delta a_c}{a_c} = \dfrac{\Delta r_0}{r_0} \\ \dfrac{\Delta a_a}{a_a} = \dfrac{\Delta a_c}{a_c} \end{cases} \Rightarrow \begin{cases} \Delta a_c = \dfrac{a_c}{r_0}\Delta r_0 \\ \Delta a_a = \dfrac{a_a}{r_0}\Delta r_0 \end{cases} \quad [2.146]$$

Given that $r_0 = 1.45 \times 10^{-13}$ cm, $\Delta r_0 = 0.05 \times 10^{-13}$ cm [2.136], $a =$ 18.914 MeV [2.137] and $a_a = 18.914$ MeV [2.143], numerically we obtain:

$$\Delta a_c = 0.021 \text{ MeV}; \Delta a_a = 0.652 \text{ MeV} \quad [2.147]$$

Taking into account the absolute errors, we ultimately obtain:

$$a_c = (0.597 \pm 0.021) \text{ MeV}; a_a = 18.914 \pm 0.652 \text{ MeV} \quad [2.148]$$

– **Determining a_v and a_p.**

To empirically determine the coefficients of volume a_v and pairing a_p, let us express the binding energy per nucleon using [2.134]. By replacing the neutron number, *N*, with $A - Z$, we obtain:

$$\frac{E_l(A,Z)}{A} = a_v - \frac{a_s}{A^{1/3}} - a_c\frac{Z^2}{A^{4/3}} - a_a\left(1 - \frac{2Z}{A}\right)^2 \quad [2.149]$$

Given that coefficients a_c and a_a are known [2.148], the equation is solved if two experimental nucleon binding values are known for any two nuclides of an odd number of nucleons. Consider the experimental values of (E_l/A) for ^{65}Cu ($Z = 29$) and ^{127}I ($Z = 53$) equal to 8.75 MeV/nucleon and 8.43 MeV/nucleon, respectively [EVA 61]. Using [2.148], equation [2.149] gives:

$$8.75 = a_v - 0.248711317a_s - 2.140468896$$

$$8.43 = a_v - 0.198944571a_s - 3.144113565$$

By solving the above system of equations, we find:

$$a_v = 14.307 \text{ MeV}; a_s = 13.737 \text{ MeV} \quad [2.150]$$

In summary, the following semi-empirical values are obtained:

$$a_c = 0.597\text{ MeV};\ a_a = 18.914\text{ MeV}$$
$$a_v = 14.307\text{ MeV};\ a_s = 13.737\text{ MeV} \qquad [2.151]$$

These results can be compared with the more recent values from Roy and Basu [ROY 06]:

$$a_c = (0.695 \pm 0.002)\text{ MeV};\ a_a = 22.435 \pm 0.065)\text{ MeV};$$

$$a_v = (15.409 \pm 0.026)\text{ MeV};\ a_s = 16.873 \pm 0.080)\text{ MeV}; \qquad [2.152]$$

$a_\text{p} = (11.155 \pm 0.864)$ MeV, this value often being approximated to 12 MeV.

The liquid-drop model is a comprehensive approach to roughly determine the mass and binding energy of nuclei. Considering the semi-empirical mass formula [2.119], the $M(A, Z)$ of a neutral atom and a nucleus are written as a function of the binding energy, E_l [2.134], respectively in the form:

$$M(A, Z) = ZM_\text{H}c^2 + (A - Z)M_\text{n}c^2 - E_l \qquad [2.153a]$$

$$M(A, Z) = ZM_\text{H}c^2 + (A - Z)M_\text{n}c^2 - E_l \qquad [2.153b]$$

Using [2.151], the binding energy [2.134] is determined, then the atomic mass, $M(A, Z)$ [2.153a] is derived. The results obtained for several odd-A nuclei compared to the experimental data are presented in Table 2.7.

Element	Z	A	$(E_l/A)_\text{cal}$ (MeV)	$(E_l/A)_\text{exp}$ (MeV)	$M(A, Z)_\text{cal}$ (u)	$M(A, Z)_\text{exp}$ (u)
Oxygen (O)	8	17	8.02	7.75	17.1320	17.0045
Sulfur (S)	16	33	8.56	8.50	33.2633	32.9819
Manganese (Mn)	25	55	8.75	8.75	55.4462	54.9558
Copper (Cu)	29	65	8.75	8.75	65.5295	64.6484
Iodine (I)	53	127	8.43	8.43	128.0469	126.9453
Platinum (Pt)	78	195	7.97	7.92	196.6156	195.0264
Berkelium (Bk)	97	245	7.63	7.52	247.0333	245.142

Table 2.7. *Binding energy and atomic mass of some nuclides. The calculated values are obtained using formulas [2.149] and [2.153a] based on the values [2.152] of coefficients a_v, a_s, a_a, a_c, and a_p. The experimental values are taken from [EVA 61]*

The comparison of the results gathered in Table 2.7 shows that the semi-empirical mass formula reproduces the experimental data in a fairly satisfactory manner. The deviations observed are due to the imprecision of the empirical values of coefficients a_v, a_c, a_s and a_a. Several *families of coefficients* are found in the literature, as indicated in Table 2.8.

Authors	a_v	a_s	a_a	a_c	a_p
Bethe and Bacher[a]	13.86	13.20	19.50	0.580	
Bohr and Wheeler[b]		14.00		0.590	
Matauch and Fluegge[c]	14.66	15.40	20.50	0.602	
Feenberg[d]	14.10	13.10	18.10	0.585	
Friedlander and Kennedy[e]	14.10	13.10	18.10	0.585	
Fermi[f]	14.00	13.00	19.30	0.583	
Wapstra[g]	14.10	13.00	19.00	0.595	
Seeger[h]	16.11	20.21	20.65	0.806	
Roy and Basu[i]	15.41	16.87	22.43	0.695	12.00

[a] [BET 36], [b] [BOH 39], [c][MAT 46], [d][FEE 47], [e][FRI 49], [f] [FER 50]
[g][WAP 58], [h][SEE 61], [i][ROY 06]

Table 2.8. *Families of coefficients a_v, a_s, a_a, a_c, and a_p*

APPLICATION 2.15.– The [SAK 18a] tables present the atomic masses 15.99491 u and 25.98689 u, respectively, for the nuclides ^{16}O ($Z = 8$) and ^{26}Al (Z = 13). Deduct the masses of the nuclei corresponding to the two nuclides considered from these results. Compare the atomic masses obtained with the predictions of the liquid-drop model. We will use the values of coefficients a_v, a_c, a_s, and a_a obtained by Wapstra (Table 2.8).

We will take (Table 1.3): $m_0 = 0.0005486$ u; $m_p = 1.007276$ u, $m_n = 1.00866$ u.

ANSWER.–

– *Nucleus masses*

Let $M(A, Z)$ denote the atomic mass of a nuclide, ^AX, and $M_X(A, Z)$ the mass of the corresponding nucleus. The mass of an electron is noted m_0. The mass of the nucleus is thus given by the relationship:

$$M_X(A, Z) = M(A, Z) - Zm_0 \quad [2.154]$$

We thus obtain:

– for ^{16}O: M_O (16, 8) = (15.99491 – 8 × 0.0005486) u = 15.9905212 u

– for ^{26}Al: M_{Al} (26, 13) = (25.98689 – 8 × 0.0005486) u = 25.9797582 u

Thus, in summary:

$$M_O (16, 8) = 15.9905212 \text{ u}; M_{Al} (26, 13) = 25.9797582 \text{ u}. \qquad [2.155]$$

– Predictions of the liquid-drop model, comparison

The ^{16}O and ^{26}Al nuclides are even-even and odd-odd nuclides, respectively. Using the Bethe–Weizsäcker formula [2.119] and the expression of pairing energy [2.133b] and [2.133c], for even A, we obtain:

$$\begin{aligned} M(A,Z)c^2 &= ZM_H c^2 + (A-Z)M_n c^2 \\ &- a_v A + a_s A^{2/3} + a_c \frac{Z^2}{A^{1/3}} + a_a \frac{(N-Z)^2}{A} \pm a_p \frac{1}{A^{1/2}} \end{aligned} \qquad [2.156]$$

Which then gives:

– for ^{16}O (even Z-even N):

$$M(16, 8)c^2 = 8M_{H}c^2 + 8M_n c^2 - 16\, a_v + 6.34960\, a_s + 25.39842\, a_c + 0.25\, a_p \qquad [2.157a]$$

– for ^{26}Al (odd Z-odd N):

$$M(26.13)c^2 = 13M_{H}c^2 + 13M_n c^2 - 26\, a_v + 8.77638\, a_s + 228.18596\, a_c - 0.19612\, a_p \qquad [2.157b]$$

Given that $M_H c^2$ = 1.007825 uc^2, 1 u = 931.5 MeV/c^2, a_v = 14.1 MeV, a_s = 13.0 MeV, a_c = 0.595 MeV and a_p = 33.50 MeV, equations [2.156] give:

$$M(16, 8) = 16.00351 \text{ u}; M(26, 13) = 25.54304 \text{ u} \qquad [2.157c]$$

The consistency between the predictions [2.157c] derived from the drop model and the atomic masses, M (16, 8) = 15.99491 u and M (26, 13) = 25.98689 u given in the tables is satisfactory.

NOTE.– *Using [2.119], the mass, $M_X(A, Z)$ of a nucleus is written:*

$$\begin{aligned} M_X(A,Z)c^2 &= Zm_pc^2 + (A-Z)M_nc^2 \\ &- a_vA + a_sA^{2/3} + a_c\frac{Z^2}{A^{1/3}} + a_a\frac{(N-Z)^2}{A} \pm a_p\frac{1}{A^{1/2}} \end{aligned} \quad [2.158]$$

Given that the mass difference $(M_H - m_p) = m_0$ (electron mass) is obtained by comparing [2.156] and [2.158], we obtain:

$$M_X(A, Z) = M(A, Z) - Z(M_H - m_p) = M(A, Z) - Zm_0$$

Equation [2.154] is found. It is therefore not necessary to calculate the mass of a nucleus using the semi-empirical mass formula [2.119], if the atomic mass of the nuclide considered is known. It is sufficient to simply use [2.154].

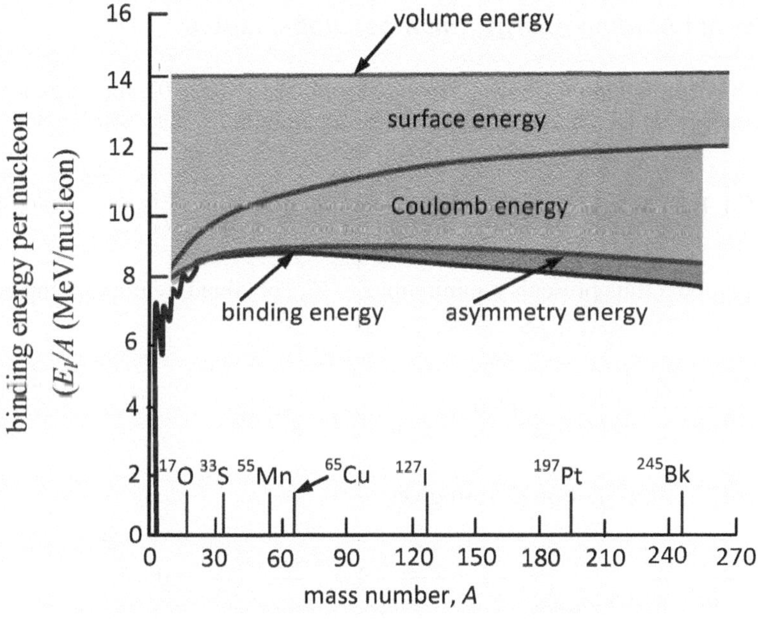

Figure 2.22. *Summary of the treatment of binding energy within the framework of the liquid-drop model*

In conclusion, let us summarize the treatment of binding energy within the framework of the liquid-drop model by including the contributions of the four corrective terms in the Bethe–Weizsäcker formula [2.119]. To do this, for the sake of convenience let us reverse the shape of the Aston curve shown in Figure 1.20. Considering the specific values of the binding energy nucleons of the nuclides listed in Table 2.7, the diagram shown in Figure 2.22 is obtained. This figure shows that for $A < 30$, the contribution of asymmetry energy to the decrease in volume energy is insignificant. For $A < 60$, the greatest decrease in binding energy per nucleon can be attributed to the surface energy. The Coulomb repulsion between protons becomes predominant for $A > 60$. Thus, for heavy nuclides, the greatest decrease in binding energy per nucleon can be attributed to the surface energy.

Lastly, for $A > 30$, note the significance of the contribution of asymmetry energy, which increases, albeit modestly compared to the contributions of the surface and Coulomb energies.

2.6.7. *Isobar binding energy, the most stable isobar*

As shown in equation [2.121], for fixed and odd A ($\delta = 0$), the variation in the binding energy, $E_l\,(A, Z)$ as a function of Z is a parabola ($N = A - Z$):

$$E_l(A,Z) = a_v A - a_s A^{2/3} - a_c \frac{Z^2}{A^{1/3}} - a_a \frac{(A-2Z)^2}{A} \qquad [2.159]$$

The parabola thus presents a minimum $Z = Z_{\min}$ obtained by minimizing equation [2.159], i.e.:

$$\left.\frac{\partial E_l(A,Z)}{\partial Z}\right|_{Z = Z_{\min}} = 0 \qquad [2.160]$$

Using [2.159], condition [2.160] gives:

$$\left.\frac{\partial E_l(A,Z)}{\partial Z}\right|_{Z = Z_{\min}} = -2\,a_c \frac{Z_{\min}}{A^{1/3}} + 4\,a_a \frac{(A-2Z_{\min})}{A} = 0$$

That is:

$$-a_c \frac{Z_{\min}}{A^{1/3}} + 2\,a_a\left(1 - \frac{2Z_{\min}}{A}\right) = 0$$

The minimum binding energy is then obtained for:

$$Z_{\min} = \frac{A}{2 + \dfrac{a_c}{2a_a} A^{2/3}} \qquad [2.161]$$

As an illustration, look for *the most stable isobar* among the nuclides ^{131}Te (Z = 52), ^{131}I (Z = 53), ^{131}Xe (Z = 54), and ^{131}Cs (Z = 55). Using the values of the coefficients a_c= 0.595 MeV and a_a = 19.0 MeV obtained by Wapstra, equation [2.161] gives the *nuclear charge of the most stable isobar*:

$$Z_{\min} = 54.4953 \Rightarrow Z = 54 \qquad [2.162]$$

The most stable isobar is actually xenon according to result [2.162].

Indeed, tellurium-131 and iodine-131 are all β^- radioactive. Cesium-131 (only isotope-133 is stable) decays by electron capture with the emission of X-ray energy in the range (29.5 – 33.5) keV. The equations of radioactive transformations are written (the complete equations will be presented in Chapter 3):

$$^{131}_{52}\text{Te} \xrightarrow{\beta^-} {}^{131}_{53}\text{I} \xrightarrow{\beta^-} {}^{131}_{54}\text{Xe} \qquad [2.163a]$$

$$^{131}_{55}\text{Cs} + e^- \rightarrow {}^{131}_{54}\text{Xe} + \nu + X \qquad [2.163b]$$

Given that a decay product (daughter nucleus) is more stable than the decaying nucleus (parent nucleus), equations [2.163] show that xenon-131 is indeed the most stable isobar.

NOTE.– *It is helpful to calculate Z_{min} using the values of a_c and a_a, obtained by the various authors whose results are presented in Table 2.8. The comparison shows that the accuracy of calculations using formula [2.161] depends on the accuracy of semi-empirical measurements on the values of the Coulomb coefficient, a_c, and the asymmetry coefficient, a. An imprecision is notably found in the evaluations of coefficients a_c and a_a by Seeger and by Matauch and Fluegge. The following sub-section shows that for A = 131, the most stable isobar is xenon, based on the mass parabola equation for constant A.*

In order to correctly interpret the results gathered in Table 2.9, the integer value just below $Z_{\min}$ is to be considered. Z = 54 is found, except for the results obtained from the evaluations of Seeger (Z = 52) and Matauch and Fluegge (Z = 55), for which the most stable isobar corresponds to ^{131}Te and ^{131}Cs, respectively, in contradiction to observations [2.164].

Authors	Mass number $A = 131$			
	a_c/a_a	Z_{min}	Z	isobar
Bethe and Bacher	0.02974	54.95997	55	Xenon
Matauch and Fluegge	0.02937	55.07022	55	Cesium
Feenberg	0.03232	54.20331	54	Xenon
Friedlander and Kennedy	0.03232	54.20331	54	Xenon
Fermi	0.03031	54.79099	55	Cesium
Wapstra	0.03132	54.49410	54	Xenon
Seeger	0.03903	52.32957	52	Tellurium
Roy Chowdhury and Basu	0.03102	54.58195	55	Cesium

Table 2.9. *Values of Z_{min} calculated from the values of a_c and a_a gathered in Table 2.8*

Taking into account the values of the ratio a_c/a_a recorded in Table 2.9, an average is found:

$$\overline{(a_c / a_a)} = 0.03184 \qquad [2.164]$$

Considering the inverse of the average [2.141], we find:

$$\overline{(a_c / a_a)} = 0.03156$$

This result is in accordance with [2.164].

If we now apply the average [2.165] in [2.161], we find:

$$Z_{\min} = \frac{A}{2 + 0.01592\, A^{2/3}} \qquad [2.165]$$

With stable nuclei being the nuclei with a minimum of binding energy, the formula [2.165] also corresponds to the equation giving the nuclear charge of nuclei within the valley of stability (see section 2.8.6). This equation shows that for light nuclei ($Z < 30$), $Z = Z_{\min} \approx A/2$, which corresponds to nuclei located in the valley of stability near the line of stability, $N = Z$ (see the Segré diagram, Figure 1.18).

APPLICATION 2.16.– Determine the most stable isobar for $A = 177$ and $A = 179$.

Given data: lutetium ($Z = 71$), hafnium ($Z = 72$), and tantalum ($Z = 73$) elements.

ANSWER.– Using [2.162] and the average value [2.164], we obtain:

– for $A = 177$, $Z_{min} = 70.7471 \Rightarrow Z = 71$: the most stable isotope, ${}^{177}_{71}\text{Lu}$;

– for $A = 179$, $Z_{min} = 71.4387 \Rightarrow Z = 71$: the most stable isotope,. ${}^{179}_{71}\text{Lu}$.

2.7. Mass parabola equation for odd *A*

2.7.1. *Expression*

As the Bethe–Weizsäcker formula [2.119] shows, for fixed A, the variation in atomic mass, M (A, Z), with atomic number Z is a parabola. For the sake of convenience, let us rewrite this formula by expanding the factor $(A - 2Z)^2$ for the asymmetry term. We obtain:

$$\begin{aligned} M(A,Z)c^2 &= ZM_Hc^2 + (A-Z)M_nc^2 - a_vA + a_sA^{2/3} \\ &+ a_c\frac{Z^2}{A^{1/3}} + a_aA + a_a\frac{4Z^2}{A} - a_a4Z \pm |\delta| \end{aligned}$$

By arranging this equation, we obtain the following *mass parabola equation*:

$$\begin{aligned} M(A,Z)c^2 &= \left(\frac{a_c}{A^{1/3}} + \frac{4a_a}{A}\right)Z^2 + (M_Hc^2 - M_nc^2 - 4a_a)Z \\ &+ \left(M_nc^2 + a_a - a_v + \frac{a_s}{A^{1/3}}\right)A \pm |\delta| \end{aligned} \quad [2.166]$$

Let us introduce the ratio a_c/a_a into the first term of the right-hand member of equation [2.166]. We obtain:

$$\begin{aligned} M(A,Z)c^2 &= \left(\frac{a_c}{A^{1/3}} + \frac{4a_a}{A}\right)Z^2 + (M_Hc^2 - M_nc^2 - 4a_a)Z \\ &+ \left(M_nc^2 + a_a - a_v + \frac{a_s}{A^{1/3}}\right)A \pm |\delta| \end{aligned} \quad [2.167]$$

Let us simplify the writing of the parabolic equation [2.167] as follows:

$$M(A, Z)c^2 = a\,Z^2 + b\,Z + d \pm |\,\delta| \qquad [2.168]$$

In equation [2.168]:

$$\begin{cases} a = \left(\dfrac{a_c}{A^{1/3}} + \dfrac{4a_a}{A}\right) \\ b = (M_H c^2 - M_n c^2 - 4a_a) \\ d = \left(M_n c^2 + a_a - a_v + \dfrac{a_s}{A^{1/3}}\right)A \end{cases} \qquad [2.169]$$

For odd A, the pairing energy is zero ($\delta = 0$) in [2.167] and [2.169]. The nuclear charge of the most stable isobar corresponding to the minimum value of Z is given by condition [2.138]. By applying it to parabolic equation [2.168], we obtain:

$$\left.\frac{\delta M(A,Z)}{\partial Z}\right|_{Z = Z_{\min}} = 2a\,Z_{\min} + b = 0$$

Which then gives:

$$Z_{\min} = -\frac{b}{2a} \qquad [2.170]$$

APPLICATION 2.17.– Show that [2.170] is equivalent to equation [2.139].

ANSWER.– In [2.170] replace a and b with their expressions taken from [2.169]. We obtain:

$$2\left(\frac{a_c}{A^{1/3}} + \frac{4a_a}{A}\right)Z_{\min} = -(M_H c^2 - M_n c^2 - 4a_a)$$

That is:

$$-4a_a\left(1 - \frac{2}{A}Z_{\min}\right) = (M_n - M_H)c^2 - 2\frac{a_c}{A^{1/3}}Z_{\min}$$

Thus:

$$4\,a_a\,(A - 2Z_{\min}) = -\,2a_c\,Z_{\min}\,A^{2/3} - A(\,M_n - M_H)\,c^2 \qquad [2.171]$$

By dividing [2.171] by $4a_c$, × $(A - 2Z_{min})$ on both sides, we find relationship [2.139].

2.7.2. *Determining the nuclear charge of the most stable isobar from the decay energy*

As explained in section 1.6.2, referring to the Segré diagram, the blue area occupied by nuclei with excess neutrons compared to stable nuclei with the same mass number, A, decay by β^- radioactivity, while the green area occupied by nuclei with excess protons compared to stable nuclei with the same mass number, A, decay by β^+ radioactivity. Although these two decay modes are studied in Chapter 3, it is interesting to see how, based on β decay energies, the nuclear charge of the most stable isobar can be determined from equation [2.170]. To do this, let us first express the *parabolic relationship between the mass of the isobars of odd A*, defined by the mass difference, $M(A, Z) - M(A, Z_{min})$. Using the parabolic mass equation [2.168], let us express the fictitious mass, $M(A, Z_{min})$:

$$M(A,Z_{min})c^2 = a\,Z_{min}^2 + b\,Z_{min} + d$$

By replacing b with its expression taken from [2.170] in the last term of the right-hand member of the above equation, we find:

$$M(A,Z_{min})c^2 = -\,a\,Z_{min}^2 + d \qquad [2.172]$$

Let us now express the parabolic relationship between the mass of the isobars of odd A. Using [2.168] and [2.172], we obtain:

$$[M(A,Z) - M(A,Z_{min})]c^2 = a(Z^2 + Z_{min}^2 - 2ZZ_{min})$$

Thus by leveraging the identities of note:

$$[M(A,Z) - M(A,Z_{min})]c^2 = a(Z - Z_{min})^2 \qquad [2.173]$$

Equation [2.173] reflects the parabolic relationship between the mass of the isobars of odd A.

Let us now express the β decay energies as a function of the nuclear charge, Z_{min}, using [2.173].

– Transition between isobars of odd A by β^- emission: $Z \rightarrow Z + 1$

During a β^- transition, the nuclear charge of the final nucleus has one more proton than the initial nucleus: it is therefore a $Z \rightarrow Z + 1$ transition. Using [2.168], the β^- *decay energy* is written:

$$Q_{\beta^-} = [M(A,Z) - M(A,Z+1)]c^2 = a(Z - Z_{\min})^2 - a(Z+1-Z_{\min})^2$$

Which then gives:

$$Q_{\beta^-} = a\,(Z - Z_{\min} + Z + 1 - Z_{\min})(Z - Z_{\min} - Z - 1 + Z_{\min})$$

That is, ultimately:

$$Q_{\beta^-} = -2a\left(Z - Z_{\min} + \frac{1}{2}\right) = 2a\left(Z_{\min} - Z - \frac{1}{2}\right) \qquad [2.174]$$

– Transition between isobars of odd A by β^+ emission: $Z \rightarrow Z - 1$

During a β^+ transition, the nuclear charge of the final nucleus has one less proton than the initial nucleus: it is therefore a $Z \rightarrow Z - 1$ transition. The β^+ *decay energy* is written using [2.168]:

$$Q_{\beta^+} = [M(A,Z) - M(A,Z-1)]c^2 = 2a\left(Z - Z_{\min} - \frac{1}{2}\right) \qquad [2.175]$$

Using [2.174] and [2.175], we can summarize the two expressions giving the β decay energies, that is:

$$Q_{\beta^\pm} = 2a\left[\pm(Z - Z_{\min}) - \frac{1}{2}\right] \qquad [2.176]$$

Let us now plot the curve translating the mass parabola, at the lowest point of which is the stable isobar of odd mass number A. By including the quantities [2.173] and [2.176], the curve shown in Figure 2.25 is obtained.

As shown in Figure 2.23, for odd and fixed A, the $Z \rightarrow Z \pm 1$ transitions between isobars give a sequence of β decays cascading toward the most stable final isobar of nuclear charge Z equal to the integer closest to $Z_{\min}$.

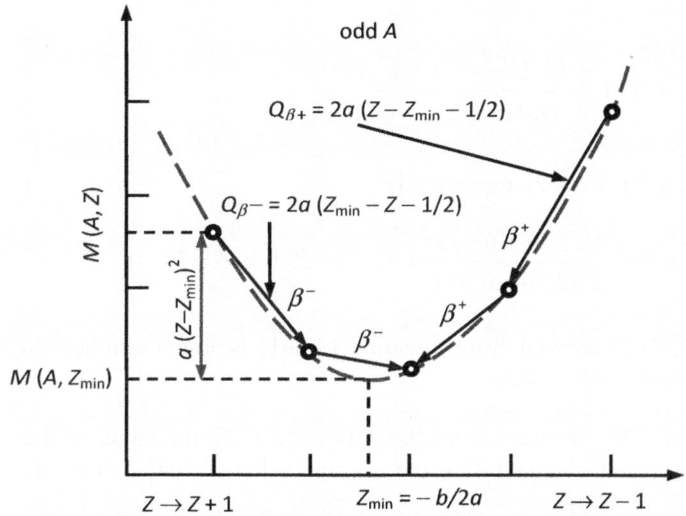

Figure 2.23. *Mass parabola for the isobars of odd A*

APPLICATION 2.18.– Let us consider the two β^- transitions [2.163a] cascading toward the xenon, which we will rewrite featuring the decay energy for each transition:

$$ {}^{131}_{52}\text{Te} \to {}^{131}_{53}\text{I} + \beta^- + (2.16 \pm 0.10)\ \text{MeV}\ {}^{131}_{52}\text{Te} \to {}^{131}_{53}\text{I} \qquad [2.177a] $$

$$ {}^{131}_{53}\text{I} \to {}^{131}_{54}\text{Xe} + \beta^- + (0.97 \pm 0.01)\ \text{MeV} \qquad [2.177b] $$

Deduce from the data in [2.177] the most stable isobar.

ANSWER.– Let us write, according to [2.177]:

$$ Q_\beta(131.52) = (2.16 \pm 0.10)\ \text{MeV} \qquad [2.178] $$

$$ Q_\beta(131.53) = (0.97 \pm 0.01)\ \text{MeV} $$

Using [2.174], we obtain:

$$ Q_\beta(131, 52) + Q_\beta(131, 53) = 4a\,(Z_{min} - 53) \qquad [2.179a] $$

$$ Q_\beta(131, 52) - Q_\beta(131, 53) = 2a \qquad [2.179b] $$

By applying the ratio of the two equations [2.179], we find:

$$Z_{\min} = 53 + \frac{1}{2}\frac{Q_\beta(131, 52) + Q_\beta(131, 52)}{Q_\beta(131, 52) - Q_\beta(131, 52)} \quad [2.180]$$

Using [2.178], we find, numerically:

$$Z_{\min} = 53 + \frac{1}{2}\frac{(2.16 + 0.97)}{(2.16 - 0.97)} = 54.3151260 \Rightarrow Z = 54 \quad [2.181]$$

Result [2.162] derived from equation [2.161] is found another way: xenon is indeed the most stable isobar.

APPLICATION 2.21.– Deduce from the data in [2.177], the value of the ratio a_c/a_a. Compare the result obtained with the average value [2.164]. Let us consider result [2.1].

ANSWER.– Using [2.169], we obtain:

$$\begin{cases} a = \left(\dfrac{a_c}{A^{1/3}} + \dfrac{4a_a}{A}\right) \\ b = (M_H c^2 - M_n c^2 - 4a_a) \end{cases} \quad [2.182]$$

It is therefore sufficient to calculate a and b and then to deduce therefrom the values of a_c and a_a using equations [2.182].

Using [2.179], we determine a, that is:

$$a = [Q_\beta(131, 52) - Q_\beta(131, 53)]/2 = 0.595 \text{ MeV} \quad [2.183a]$$

To determine b, let us use expression [2.170] of $Z_{\min}$ and its value [2.181]. We find:

$$b = -2a\, Z_{\min} = -64.635 \text{ MeV} \quad [2.183b]$$

By applying [2.183] in [2.182] and taking into account the mass difference [2.140], the system of equations for $A = 131$ is obtained:

$$\begin{cases} 0.595 = 0.196899a_c + 0.030534a_a \\ -64.535 = 0.78246 - 4a_a \end{cases} \quad [2.184]$$

Solving [2.184] gives:

$$a_a = 16.329 \text{ MeV}; \; a_c = 0.489 \text{ MeV} \quad [2.185]$$

Using [2.185], we ultimately obtain:

$$a_c/a_a = 0.029947 \quad [2.186]$$

Result [2.186] is consistent with the average value [2.164] equal to 0.03184.

2.7.3. *Mass parabola equation for even A*

The isobars for even A concern the even Z–even N and odd Z–odd N isobars. Since the transitions between isobars leave N constant, we will only present atomic number Z in the equations to be established.

For even A, equations [2.168] and [2.172] give:

– For even Z

$$M(A, Z)\, c^2 = a\, Z^2 + b\, Z + d - \delta \quad [2.187]$$

– For odd Z

$$M(A, Z)\, c^2 = a\, Z^2 + b\, Z + d + \delta \quad [2.188]$$

For even A, the nuclear charge of the most stable isobar is given by condition [2.138], i.e. by using [2.187]:

$$\left.\frac{\delta M(A, Z)}{\partial Z}\right|_{Z = Z_{\min}} = 2a\, Z_{\min} + b = 0 \Rightarrow Z_{\min} = -\frac{b}{2a}$$

Result [2.170] is thus found. Thus the value of $Z_{\min}$ is independent of the parity of mass number A.

Let us choose the pairing energy, such that $M(A, Z_{\min})$ has the smallest possible value for even A. We will thus consider equation [2.187], which is written taking [2.170] into account:

$$M(A, Z_{\min})c^2 = -a\, Z_{\min}^2 + d - \delta \quad [2.189]$$

Let us now establish the parabolic relationship between the mass of the isobars of even A.

– For even Z

Using [2.187] and [2.189], we obtain relationship [2.173], which we assign another number for the calculations to follow for even Z and odd Z:

$$[M(A,Z) - M(A,Z_{\min})]c^2 = a(Z - Z_{\min})^2 \quad [2.190a]$$

– For odd Z

Using [2.188] and [2.189], we obtain:

$$[M(A,Z) - M(A,Z_{\min})]c^2 = a(Z - Z_{\min})^2 + 2\delta \quad [2.190]$$

Let us now express the decay energies as a function of $Z_{\min}$, as in the case of the isobars of odd A.

– Transition between isobars of even A by β^- emission: $Z \rightarrow Z + 1$

Using [2.190], the β^- decay energy is written:

– For even Z

Since Z is even, mass $M(A, Z)$ verifies equation [2.187]. For even Z, $Z + 1$ is then odd. [2.188] is then used to express the mass $M(A, Z + 1)$. Which gives:

$$M(A, Z + 1)\, c^2 = a\,(Z + 1)^2 + b\,(Z + 1) + d + \delta \quad [2.191]$$

The β^- decay energy for even Z is then written according to [2.187] and [2.191]:

$$Q_{\beta^-} = [M(A, Z) - M(A, Z + 1)]\, c^2 = a\, Z^2 + b\, Z - a\,(Z + 1)^2 - b\,(Z + 1) - 2\delta$$

That is, taking into account [2.170]:

$$Q_{\beta^-} = -a - 2aZ - b - 2\delta = -a - 2aZ + 2aZ_{\min} - 2\delta$$

That is, ultimately:

$$Q_{\beta^-} = -2a\left(Z - Z_{\min} + \frac{1}{2}\right) - 2\delta \quad [2.192]$$

– For odd Z

Mass $M(A, Z)$ is given by [2.188]. Given that $Z + 1$ is odd if Z is even, then mass $M(A, Z + 1)$ is obtained using [1.187]. We thus obtain:

$$M(A, Z+1)\,c^2 = a\,(Z+1)^2 + b\,(Z+1) + d - \delta \qquad [2.193]$$

Using [2.188] and [2.193], the β^- decay energy for odd Z is written:

$$Q_{\beta-} = [M(A, Z) - M(A, Z+1)]\,c^2 = a\,Z^2 + b\,Z - a\,(Z+1)^2 - b\,(Z+1)\;\; 2\delta$$

That is, by considering [2.170]:

$$Q_{\beta^-} = -\,2a\left(Z - Z_{\min} + \frac{1}{2}\right) + 2\delta \qquad [2.194]$$

– Transition between isobars of even A by β^+ emission: $Z \rightarrow Z - 1$

Taking an approach analogous to the previous one, we obtain:

– For even Z (odd $Z - 1$)

Mass $M(A, Z–1)$ is obtained using [1.188]. We thus obtain:

$$M(A, Z-1)\,c^2 = a\,(Z-1)^2 + b\,(Z-1) + d + \delta \qquad [2.195]$$

The β^+ decay energy for even Z is then written according to [2.187], [2.195] and [2.170]:

$$Q_{\beta+} = [M(A, Z) - M(A, Z-1)]\,c^2 = 2aZ - a - 2aZ_{\min} - 2\delta$$

That is, ultimately:

$$Q_{\beta^+} = 2a\left(Z - Z_{\min} - \frac{1}{2}\right) - 2\delta \qquad [2.196]$$

– For odd Z (even $Z - 1$)

Mass $M(A, Z - 1)$ is obtained using [1.187]. We obtain:

$$M(A, Z-1)\,c^2 = a\,(Z-1)^2 + b\,(Z-1) + d - \delta \qquad [2.197]$$

The β^+ decay energy is written according to [2.188], [2.197] and [2.170]:

$$Q_{\beta+} = [M(A, Z) - M(A, Z-1)]\,c^2 = 2aZ - a - 2aZ_{\min} + 2\delta$$

Thus:

$$Q_{\beta^+} = 2a\left(Z - Z_{\min} - \frac{1}{2}\right) + 2\delta \qquad [2.198]$$

Let us give a summary of the β decay energies that we have just determined:

– Transition between isobars of even A by β^- emission: $Z \to Z+1$

$$\text{even } Z\text{: } Q_{\beta}^{-} = 2a\,(Z_{\min} - Z - 1/2) - 2\delta$$
$$\text{odd } Z\text{: } Q_{\beta}^{-} = 2a\,(Z_{\min} - Z - 1/2) + 2\delta \qquad [2.199]$$

Using the preceding results, we get the mass parabola for the isobars of even A (Figure.2.24).

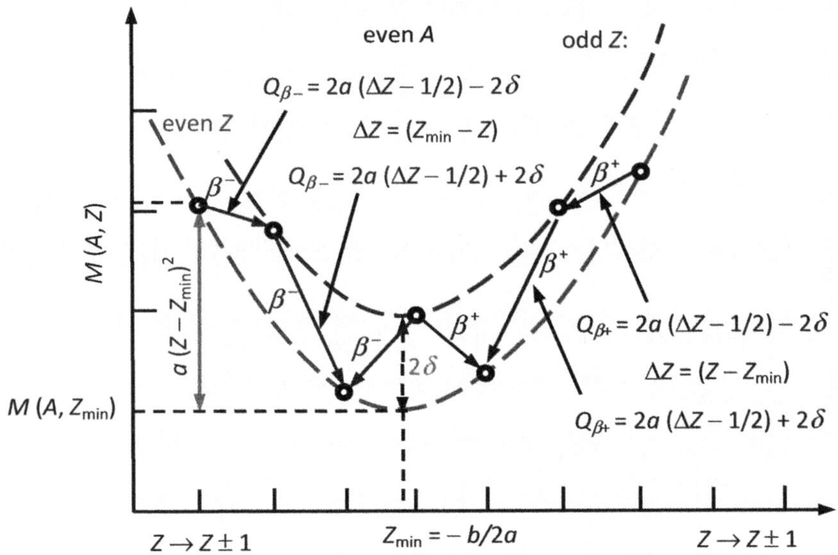

Figure 2.24. *Mass parabola for the isobars of even A. Note that there are two parabolas, one for even Z, even N and the other for odd Z, odd N; the two parabolas are offset by 2 δ*

– Transition between isobars of even A by β^+ emission: $Z \rightarrow Z - 1$

$$\text{even } Z\text{: } Q_\beta^+ = 2a\,(Z - Z_{\min} - 1/2) - 2\delta$$
$$\text{odd } Z\text{: } Q_\beta^+ = 2a\,(Z - Z_{\min} - 1/2) + 2\delta \qquad [2.200]$$

APPLICATION 2.20.– Show that by choosing the pairing energy, such that $M\,(A, Z_{\min})$ has the greatest possible value for even A, the parabola bottom for even Z is located at 2δ below the parabola bottom for odd Z.

ANSWER.–

For the largest possible value of $M\,(A, Z_{\min})$ for even A, equation [2.189] becomes:

$$M(A,Z)c^2 = -\,a Z_{\min}^2 + d + \delta \qquad [2.201]$$

Using [2.187] and [2.188], we obtain, respectively:

– For even Z

$$[M(A,Z) - M(A,Z_{\min})]c^2 = a(Z - Z_{\min})^2 - 2\delta \qquad [2.202]$$

For odd Z

$$[M(A,Z) - M(A,Z_{\min})]c^2 = a(Z - Z_{\min})^2 \qquad [2.203]$$

Results [202] and [203] clearly show that by choosing the pairing energy, such that $M\,(A, Z_{\min})$ has the greatest possible value for even A, the parabola bottom for even Z is located at 2δ below the parabola bottom for odd Z.

Carl Friedrich Von Weizsäcker was a German physicist and philosopher. In physics he is especially famous for having proposed, in 1935, the semi-empirical mass formula initially named the Weizsäcker formula in his honor.

Hans Albrecht Bethe was an American physicist of German origin. He was the winner of the Nobel Prize in Physics 1967 for his contribution to stellar nucleosynthesis. In 1936, he simplified the Weizsäcker formula. Thus, the semi-empirical mass formula is often called the Bethe–Weizsäcker formula.

Box 2.3. *Bethe (1906–2005); Weizsäcker (1912–2007)*

APPLICATION 2.21.– Determine the most stable neighboring isobars for $A = 120$.

ANSWER.– For $A = 120$, equation [2.165] gives:

$$Z_{\min} = \frac{120}{2 + 0.01592 \times 120^{2/3}} = 50.26568 \qquad [2.204]$$

Given that there are two parabolas, we will choose the integer value immediately below Z_{min}, i.e. $Z = 50$ (even Z isobar) and the integer value immediately above Z_{min}, i.e. $Z = 51$ (odd Z isobar). These are the isobars $^{120}_{50}Sn$ (tin-120) and $^{120}_{51}Te$ (tellurium-120).

2.8. Nuclear potential barrier

2.8.1. *Definition, model of the rectangular potential well*

Let us consider the diffusion process for a particle of charge *ze*, by a nucleus of charge *Ze*. At infinity, the potential energy of the system {particle-nucleus} is zero. At a distance *r* from the nucleus, the potential energy of the system {*Ze* nucleus – *ze* particle} is equal to the potential electrostatic energy, $U(r)$, given by the relationship:

$$U(r) = k\frac{Zze^2}{r} \qquad [2.205]$$

When *r* decreases to the order of Fermi, the nuclear forces begin to make themselves felt. If *r* decreases again, it arrives at an instant where the attractive nuclear forces are equal in intensity to the repulsive Coulomb forces between protons. At this moment, the charge particle, *ze*, is no longer subjected to any force. If *r* decreases further, the nuclear forces predominate and the particle of charge *ze*, is captured by the nucleus to form another stable nucleus. For example, a (^{4}He)-α particle can fuse with a ^{204}Hg mercury nucleus to give a stable ^{208}Pb lead nucleus.

The simplest representation model of the interaction potential of the system {*Ze* nucleus – *ze* particle} is that of the rectangular potential well of depth U_0 for $0 \leq r \leq R$ and presenting a discontinuity for $r = R$. The curve $U(r)$ from $r = R$ to $r = \infty$ is called the *nuclear potential barrier*. The value $U(R) = B$ is called the *nuclear potential barrier height*, also called the *Coulomb barrier height* given by the relationship:

$$B = k\frac{Zze^2}{R} \qquad [2.206]$$

The profile of the nuclear potential barrier is illustrated in Figure 2.25.

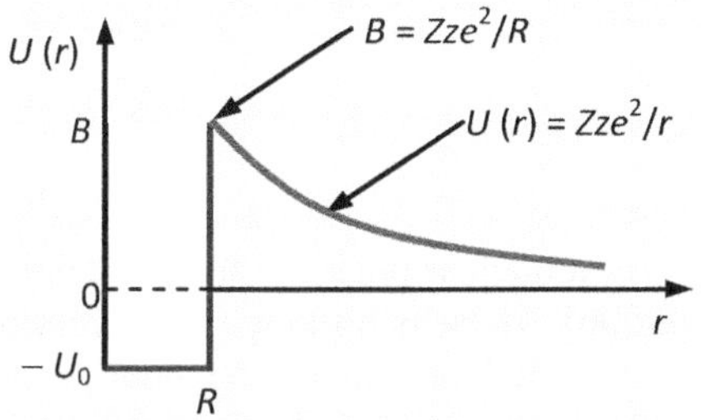

Figure 2.25. *Profile of the nuclear potential barrier between a nucleus of charge Ze and radius R and a particle of charge ze, r is the distance between the centers of the nucleus and the particle*

2.8.2. *Modifying the model of the rectangular potential well*

In reality, just like the nucleus, the charge particle, *ze*, has a nuclear radius. Let us note using R_Z, the radius of the nucleus ($R_Z = R$), and using R_z, the radius of the particle. As long as the distance $r >$ $(R_z + R_Z)$, the potential energy is of the Coulomb type and is given by [2.205]. When $r = (R_z + R_Z)$, the surfaces of the spherical envelopes of the nucleus and the particle are in contact (Figure 2.26).

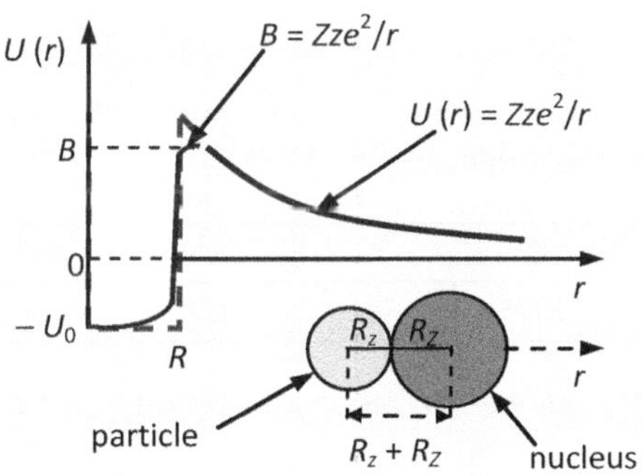

Figure 2.26. *Modified profile of the nuclear potential barrier between a nucleus of charge Ze and radius R_Z and a particle of charge ze and radius R_z. The value of the height of the potential barrier is a function of r with $R_Z < r < (R_z + R_Z)$*

At a distance $r < (R_z + R_Z)$, the attractive nuclear forces become predominant with respect to the Coulomb repulsion forces. The height of the nuclear potential barrier is no longer defined for $r = R$, but rather at a distance, r such that:

$$B = k\frac{Zze^2}{r}; \quad R_Z < r < (R_z + R_Z) \qquad [2.207]$$

Expression [2.207] gives a nuclear potential barrier height that is much closer to reality than expression [2.206]. There is no longer a discontinuity of the potential at $r = R$ and we obtain a modified profile of the nuclear potential barrier with a rounded vertex and bottom (Figure 2.26). When the radius, R_z, of the particle is not known, the height of the nuclear potential barrier corresponds to the height of the Coulomb barrier given by expression [2.206] with a nuclear radius, $R = r_0 A^{1/3}$.

2.9. Exercises

EXERCISE 2.1.– Experimental measurement of the radioactive half-life of vanadium-52

Using a meter, we measure the activity of a source of vanadium-52, which is a β^- emitter. Let $\langle N \rangle$ be the decay number measured for a constant duration, $\tau = 5$s. The results obtained are gathered in Table 2.10.

a) Determine the activity, A_0, of the source at the initial instant, $t = 0$.

b) Determine, in table form, the values of the ratio $A_0/A(t)$.

c) Use the experimental results to determine the radioactive half-life of vanadium-52. Express it in minutes.

t (min)	0	2	4	6	8	10	12
$\langle N \rangle$	1586	1075	741	471	355	235	155

Table 2.10. *Decay number $\langle N \rangle$ measured during a constant duration of 5s*

EXERCISE 2.2.– Shell structure, J^π of the ground state and of excited states

The following nuclide list is given: ^{15}N, ^{11}B and ^{27}Al⊳. Figure 2.27 shows the nuclear levels of platinum-188 ($Z = 78$) and thorium-228 ($Z = 90$).

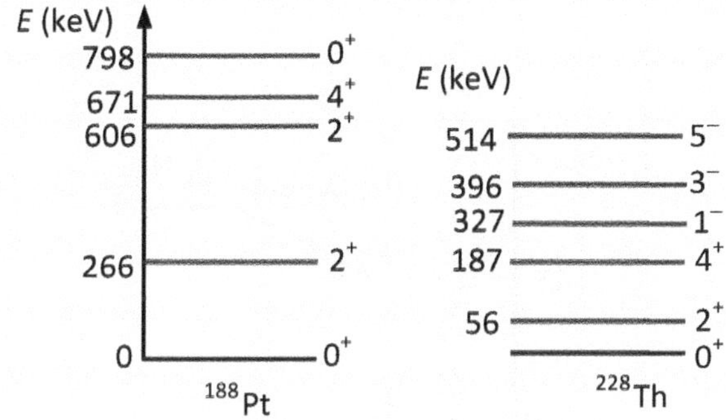

Figure 2.27. *Nuclear levels of platinum-188 and thorium-228. For clarity, the energy scale is not respected*

a) We will consider the nuclides ^{15}N, ^{11}B and ^{25}Mg.

i) For each nuclide, map out the shell structure derived from a harmonic potential and the distribution of nucleons on shells resulting from the Woods–Saxon potential with spin-orbit coupling.

ii) Deduce from the previous shell structures, the J^π of the ground state, and the J^π of the two excited states of lower energies.

b) Let us consider Figure 2.27.

i) Theoretically justify the J^π of the ground state of each of the nuclei.

ii) Specify the multipole transition type for the $5^- \rightarrow 4^+$ and $4^+ \rightarrow 2^+$ deexcitations and the forbidden transition. Derive therefrom the wavelengths of photons that could be emitted.

Given data: $c = 3 \times 10^{8}$ m · s^{-1}; $h = 6.63 \times 10^{-34}$ J · s; 1 eV = 1.6×10^{-19} J.

EXERCISE 2.3.– Multipole order, probability of γ-deexcitation of neon-22

We propose to determine the γ-deexcitation probability of neon-22. Neon-22 is the decay product of radiosodium-22. Figure 2.28 shows the excited level of neon-22 fed by β^+ decay and the electronic capture (EC) of sodium-22.

a) Theoretically justify the values of J^π shown in Figure 2.28.

b) Determine the multipole order and the type of multipole transition, E_l or M_l, corresponding to the deexcitation observed.

c) Estimate the corresponding transition probability. We will use the data given in Table 2.4.

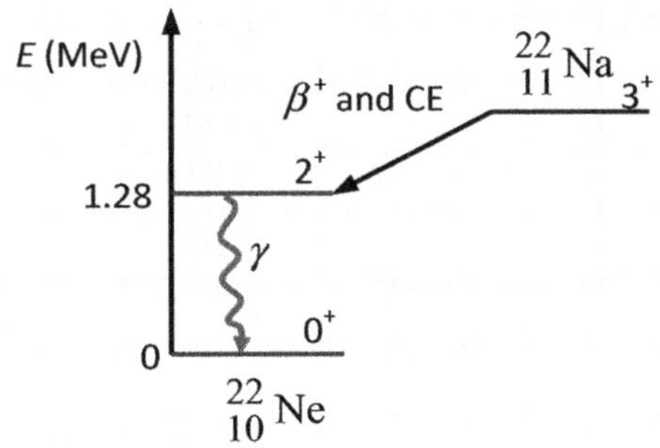

Figure 2.28. *γ-transitions to the ground level of the neon-22 nucleus*

EXERCISE 2.4.– Separation energy of the last proton and last neutron of dysprosium-161 (^{161}Dy).

In this exercise, we propose to determine the separation energy of the last $S_{n,p}$ nucleon (A, Z). To do so, we will consider the 66-proton dysprosium-161 nucleus. We are seeking to determine the energy required to extract a neutron or a proton from the ^{161}Dy nucleus.

a) A ^{160}Dy nucleus absorbs a neutron. Deduce the energy released during this reaction. What does this energy represent?

b) A ^{160}Tb (terbium-160) nucleus with atomic number $Z = 65$ absorbs a proton. Deduce the energy released during this reaction. What does this energy represent?

c) Calculate the binding energies of the ^{160}Dy, ^{161}Dy and ^{160}Tb nuclei. Then classify the nuclides in order of increasing stability.

d) Deduce therefrom the separation energies of the last neutron, S_n (161, 66) and the last proton, S_p (161, 66) of the dysprosium-161 nucleus. Now find the results of 4.1 and 4.2.

Given data: [m (^{A}X): atomic mass; m_0: electron mass):

– $m_n = 1.00866$ u; $m_p = 1.00728$ u; m_0 (electron) $= 5.486 \times 10^{-4}$ u;

– M(^{160}Dy) = 159.9251975 u; M(^{161}Dy) =160.9269334 u;

– M(^{160}Tb) = 159.9271676; 1 u = 931.5 MeV/c2.

EXERCISE 2.5.– Determining the Coulomb energy of a charged sphere from the potential created by a charge placed in the center of the nucleus.

We propose to establish expression [2.125] representing the Coulomb energy of a volume-charged sphere. To do this, let us consider a nucleus, assumed to be spherical, of radius R, whose charge is uniformly distributed in volume. We will assume that the charge volume density, ρ, is constant throughout the nuclear volume.

a) Express ρ as a function of the atomic number, Z, the elementary charge, e, and R.

b) A point M is located at a distance r from the center, O, of the nucleus ($r < R$). Draw a diagram then express the potential, $V(r)$, created at M by the point charge, $Q\ (r)$, placed at O and contained in the nuclear envelope of radius r. We will express it as a function of Z, e, r, and R.

c) Express the potential energy, dW, of the elementary charge, dq, immersed in the potential, $V\ (r)$. Deduce from this result, the expression of the Coulomb energy of a sphere of nuclear radius $R = r_0\, A^{1/3}$, uniformly charged in volume.

EXERCISE 2.6.– Determining the Coulomb energy of a charged sphere from the electromagnetic energy density.

We propose to establish expression [2.125] by using another method and considering the *electromagnetic energy density* stored in a sphere of radius R.

As in exercise 2.5, it is assumed that the charge of the nucleus, Q, is uniformly distributed in a sphere of center O and radius $R = r_0\, A^{1/3}$. The charge volume density is considered constant throughout the sphere.

a) Let us consider a point, M, located at a distance r from the center, O, of the nucleus. Represent the external electric field, $\vec{E}_{ext}$, created by the electric charge, Q, of the nucleus at a point M' ($r > R$) as a function of Q and R, as well as the internal electric field, $\vec{E}_{\text{int}}$, created by the internal charge, $Q\ (r)$, at a point M ($r < R$) as a function of Q, r and R.

b) Let us give the expression of the electromagnetic energy density [SAK 18b].

$$\omega = \frac{\varepsilon_0 E^2}{2} + \frac{\varepsilon_0 B^2}{2\mu_0} \qquad [2.208]$$

Deduce from [2.208], with supporting justification, the shape at which the electromagnetic energy density is summarized in the calculation of the Coulomb energy of a volume-charged sphere. Deduce therefrom the expressions of ω on the inside (ω_{int}) and outside (ω_{ext}) of the sphere of radius R.

c) Now express the elementary Coulomb energy, dW_c, stored on the inside and outside of the sphere of radius R as a function of ω_{int} and ω_{ext}. Now find expression [2.125] of the Coulomb energy of a volume-charged sphere.

EXERCISE 2.7.– Determining the height of the nuclear potential barrier: theory and experiment.

Let us study a problem of α particle diffusion by a nucleus of radius $R = r_0 A^{1/3}$.

a) We do not know the nuclear radius of the α particle. Show that within the framework of the liquid-drop model, the height of the nuclear potential barrier, B_{0theo} (the "0" index included for zero correction) can be theoretically written as:

$$B_{0theo} = B_{0théo} = \frac{2ke^2}{\alpha^2 a_0} A^{2/3} \quad [2.209]$$

b) The radius, Rz, of the α particle is taken into account. Let r be the distance between the centers of the particle and nucleus of radius R_Z. To establish a corrected expression ($B_{theocorr}$) of the height of the nuclear potential barrier, such that $R_Z < r < (R_z + R_Z)$, we will write, arbitrarily, for a nucleus other than the α particle:

$$r = R_Z + \frac{R_z}{A} \quad [2.210]$$

Using hypothesis [2.210], express $B_{theocorr}$ analogously to [2.209], where A will be the mass number of the nucleus, with that of the α particle replaced by its value.

c) Table 2.11 presents the experimental results obtained on the measurement of the height of the nuclear potential barrier for eight light elements: He, Li, Be, B, C, N, Mg, and Al. Using [2.209] and [2.210], compare the experimental results with the theoretical predictions by completing Table 2.11. Conclude.

Given data:

– $\alpha^2 = 5.325 \times 10^{-5}$; $a_0 = 5.29 \times 10^{-11}$ m; $k = 9 \times 10^9$ SI;

– $e = 1.602 \times 10^{-19}$ C; 1 eV $= 1.602 \times 10^{-19}$ J.

Element		B (MeV)		
$_{Z}X$	A	(B_{exp})[a]	(B_{theo0})[b]	$B_{theocorr}$[c]
$_{2}$He	4	2.4		
$_{3}$Li	7	3.3		
$_{4}$Be	9	4.0		
$_{5}$B	11	4.5		
$_{6}$C	12	5.1		
$_{7}$N	14	5.6		
$_{12}$Mg	24	8.5		
$_{13}$Al	27	9.0		

[a] Experimental values, [POL 35].

[b] Theoretical values, formula [2.209].

[c] Theoretical values, formula [2.210].

Table 2.11. *Comparison of the experimental (B_{exp}) and theoretical (B_{theo}) values of the height of the nuclear potential barrier, B, for several light elements*

EXERCISE 2.8.– Experimentally determining the Coulomb coefficient by measuring the variation in binding energy of neighboring isobars

In this exercise, we propose to experimentally determine the Coulomb coefficient, a_c, by measuring the variation in binding energy between neighboring isobars. To do this, we consider the experimental values for the measurement of the variation in binding energies of the following neighboring isobars:

(^{37}Ar and ^{37}K); (^{31}P and ^{31}S); (^{23}Mg and ^{23}Al); and (^{15}N and ^{15}O).

a) Recall the definition of mirror nuclei. Deduce therefrom the value of $A - 2Z$ for these nuclei. What can then be said about each of the above pairs of isobars?

b) Using the Bethe–Weizsäcker formula, express the mass difference, ΔM, and the binding energy difference, ΔE_l, given by the following relationships:

$$\Delta M = M(A, Z) - M(A, A - Z) \quad [2.211a]$$

$$\Delta E_l = E_l(A, Z) - E_l(A, A - Z) \quad [2.211b]$$

c) Now show that the variation in binding energy between neighboring isobars is written:

$$\Delta E_{lexp} = a_c A^{2/3} \quad [2.212]$$

d) Using the expression of the Sakho unit nuclear radius, the Coulomb coefficient is written:

$$a_c = \frac{3}{5}\frac{ke^2}{\alpha^2 a_0}\left(1+\frac{N}{Z}\right) = \frac{3}{5}\frac{ke^2}{\alpha^2 a_0}\frac{A}{Z}$$

We will write, for a pair of isobars (^AX_Z and $^AX_{Z+1}$):

$$a_c(Z) = \frac{3}{5}\frac{ke^2}{\alpha^2 a_0}\frac{A}{Z} \qquad [2.213a]$$

$$a_c(Z+1) = \frac{3}{5}\frac{ke^2}{\alpha^2 a_0}\frac{A}{Z+1} \qquad [2.213b]$$

The variation in binding energy between neighboring isobars is given theoretically with a good approximation by:

$$\Delta E_{ltheo} = \langle a_c \rangle A^{2/3} \qquad [2.214a]$$

In this relationship, $\langle a_c \rangle$ is the average value of a_c defined by:

$$\langle a_c \rangle = \frac{a_c(Z) + a_c(Z+1)}{2} \qquad [2.214b]$$

Complete Table 2.12 below using [2.213] and [2.214]. The numerical data from exercise 2.7. should also be used.

Isobar			**Coulomb coefficient (MeV)**			**Difference in binding energy**
A	$_Z$**X**	$_{Z+1}$**X**	a_c **(Z)**	a_c **(Z+1)**	$\langle a_c \rangle$	ΔE_{ltheo} **(MeV)**
37	$_{18}$Ar	$_{19}$K				
31	$_{15}$P	$_{16}$S				
23	$_{12}$Mg	$_{13}$Al				
15	$_7$N	$_8$O				

Table 2.12. *Theoretical values of the Coulomb coefficient and of the difference in binding energy of neighboring isobars*

e) The experimental results obtained by Marmier and Sheldon [MAR 69] are given in Table 2.10. Furthermore, using [2.228], we can estimate the average value of the unit nuclear radius for the neighboring isobars by the relationship:

$$\langle r_0 \rangle = \frac{3}{5} \frac{ke^2}{\langle a_c \rangle} \quad [2.215]$$

Reproduce, then complete Tables 2.13 and 2.14, taking into account the results presented in Table 2.12.

(A, Z)	**$(A, A-Z)$**	**Experiment**			**Theory**
$_Z$X	**$_{Z+1}$X**	$E_l\,(A, Z)$	$E_l\,(A, A-Z)$	ΔE_l	ΔE_l
$_{18}$Ar	$_{19}$K	315.510	08.587		
$_{15}$P	$_{16}$S	262.916	256.688		
$_{12}$Mg	$_{13}$Al	186.565	181.726		
$_7$N	$_8$O	115.494	111.952		

Table 2.13. *Comparison of the experimental [MAR 69] and theoretical values of the difference in binding energy of neighboring isobars. The results are expressed in MeV*

(A, Z)	**$(A, A-Z)$**	**Experiment**		**Theory**	
$_Z$X	**$_{Z+1}$X**	**a_c (MeV)**	**r_0 (fm)**	**$\langle a_c \rangle$ (MeV)**	**$\langle r_0 \rangle$ (fm)**
$_{18}$Ar	$_{19}$K				
$_{15}$P	$_{16}$S				
$_{12}$Mg	$_{13}$Al				
$_7$N	$_8$O				

Table 2.14. *Comparison of the experimental [MAR 69] and theoretical values of the Coulomb coefficient and the unit nuclear radius*

EXERCISE 2.9.– Expressions of Weisskopf estimates

In section 2.5.4, we presented the Weisskopf estimates given by relationships [2.80] and [2.81]. In this exercise we propose to establish these estimates based on Weisskopf's general formula, giving the probability of γ-deexcitation per unit time for electric and magnetic transitions.

For a process of γ-deexcitation between two nuclear levels, the transition probability per unit time, noted $T\ (\ell,\ m)$ for electric (E_l) and magnetic (M_l) transitions, is given by Weisskopf's general formula [WEI 51]:

$$T(\ell,m)=\frac{8\pi(\ell+1)}{\ell[(2\ell+1)!!]^2}\frac{\kappa^{2\ell+1}}{\hbar}\left|A(\ell,m)+A'(\ell,m)\right|^2 \qquad [2.216]$$

In formula [2.216]:

– $(2\ell+1)!!$ corresponds to the product of odd positive integers:

$(2\ell+1)\times(2\ell-1)\times(2\ell-3)\times(2\ell-5)\times\ldots\times 1;$

– ℓ and m are the orbital quantum number and the orbital magnetic quantum number, respectively;

– κ is the *wave number* of the transition considered: $\kappa=2\pi\nu/c=2\pi/\lambda$;

– A and A' represent the matrix elements of the multipole matrix induced by the electric current and the magnetic field.

By explaining the matrix elements, A and A', Weisskopf estimated the probabilities of electric and magnetic multipole transitions using the expressions [WEI 51]:

$$T_E(\ell)\approx\frac{4{,}4(\ell+1)}{\ell[(2\ell+1)!!]^2}\left(\frac{3}{\ell+3}\right)^2\left(\frac{\hbar\omega}{197MeV}\right)^{2\ell+1}R^{2\ell}\times 10^{21}\,\mathrm{s}^{-1} \qquad [2.217a]$$

$$T_M(\ell)\approx\frac{1{,}9(\ell+1)}{\ell[(2\ell+1)!!]^2}\left(\frac{3}{\ell+3}\right)^2\left(\frac{\hbar\omega}{197MeV}\right)^{2\ell+1}R^{2\ell-2}\times 10^{21}\,\mathrm{s}^{-1} \qquad [2.217b]$$

In formulas [2.217], $\hbar\omega=E_\gamma$ is the energy of the γ-photon emitted during the transition, expressed in MeV, and R is the radius of the nucleus considered: $R=1.2A^{1/3}$ fm.

We will write:

$$\lambda_\gamma(E_\ell)=T_E(\ell)\quad;\quad\lambda_\gamma(M_\ell)=T_M(\ell) \qquad [2.217c]$$

a) Using [2.217], express, as a function of the mass number, A, and the energy, E_γ, of the γ-photon, the deexcitation probabilities per unit time corresponding to the first four multipole transitions of each type.

b) Deduct from [2.217] the general expressions of the coefficients $C_\ell(E)$ and $C_\ell(M)$ involved in the Weisskopf estimates [2.80] and [2.81].

c) Determine the values for $C_\ell(E)$ relating to the first five electric multipole transitions.

d) Show that

$$C_\ell(M) = \frac{1900}{6336} C_\ell(E) \qquad [2.217d]$$

e) Then, determine the values for $C_\ell(M)$ relating to the first five electric multipole transitions.

Verify that the values of coefficients $C_\ell(E)$ and $C_\ell(M)$ presented in Table 2.4 are found.

2.10. Solutions to exercises

SOLUTION 2.1.– Experimentally determining the radioactive half-life of vanadium-52

a) Determining the activity, A_0, of the source

At the initial instant, $t = 0$, the activity, A_0, of the source is given by the relationship:

$$A_0 = \langle N \rangle_0 / \tau \Rightarrow A_0 = 1586/5 = 317.2 \text{ Bq} \qquad [2.218]$$

b) Determining the values of the ratio $A_0/A(t)$

Using relationship [2.218], we find the results gathered in Table 2.15.

t (min)	0	2	4	6	8	10	12
A (t)	317.2	215.0	148.2	94.2	71.0	47.0	31.0
A_0/A (t)	1	1.475	2.140	3.367	4.468	6.749	10.232
ln [A_0/A (t)]	0	0.389	0.761	1.214	1.497	1.909	2.326

Table 2.15. *Values of the ratio $A_0/A(t)$ of the initial and instantaneous activities of the source of vanadium-52*

We have added a last row to Table 2.15 in advance, to list the values of ln $[A_0/A(t)]$, which are useful for the exploitation of experimental data. This eliminates the need to draw up another table of values.

c) Usage: radioactive half-life of vanadium-52

The decay law of the source activity is written:

$$A(t) = A_0 e^{-\lambda t} \Rightarrow \ln [A_0/A(t)] = \lambda t \qquad [2.219]$$

Let us then trace the curve ln $[A_0/A(t)] = f(t)$. The resulting graph is shown in Figure 2.26. A slope line is obtained:

$$a = \lambda = \ln 2/T \Rightarrow T = \ln 2/a \qquad [2.220]$$

Numerically: $a = 0.192 \text{ min}^{-1} \Rightarrow T = 3.6$ min. This result is consistent with the min. value 3.743 (5) shown in the isotope tables.

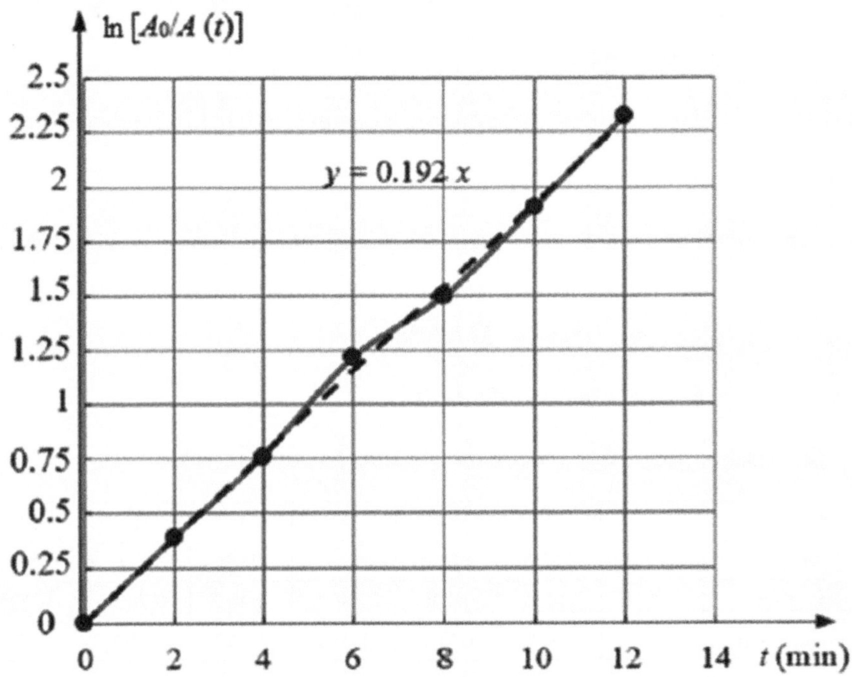

Figure 2.29. *Curve indicating the variation in the Napierian logarithm of the ratio $A_0/A(t)$ of the initial and instantaneous activities of the source of vanadium-52*

SOLUTION 2.2.– Shell structure, J^π of the ground state and of excited states

a) Shell structure of nuclides ^{15}N, ^{11}B and ^{25}Mg

i) Diagrams of nucleon distribution

The nucleon distribution over the nuclide shells ^{11}B (Z = 5), ^{15}N (Z = 7), and ^{25}Mg (Z = 12) is shown in Figures 2.30.

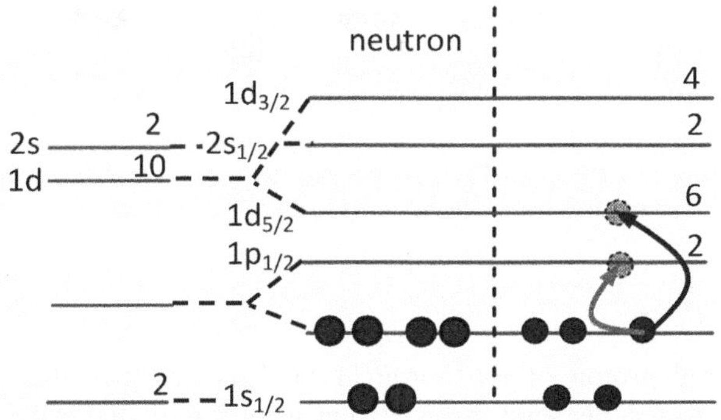

Figure 2.30a. *Shell structure derived from a harmonic potential (a) and distribution derived from the Woods–Saxon potential with spin-orbit coupling (b) for boron-11. The arrows indicate two possible of excitation of the unpaired proton*

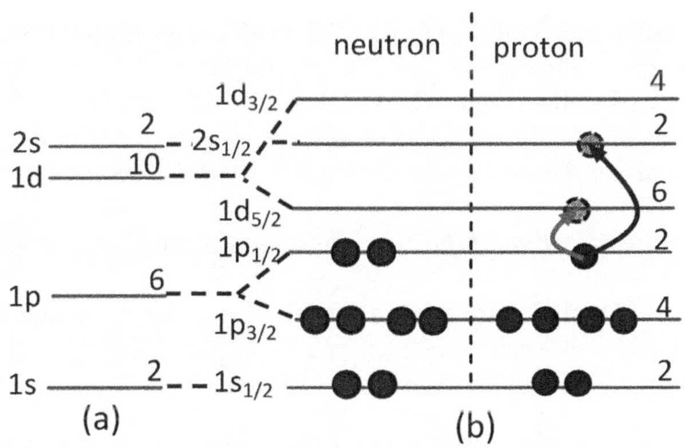

Figure 2.30b. *Shell structure derived from a harmonic potential (a) and distribution derived from the Woods–Saxon potential with spin-orbit coupling (b) for nitrogen-15. The arrows indicate two possible of excitation of the unpaired proton*

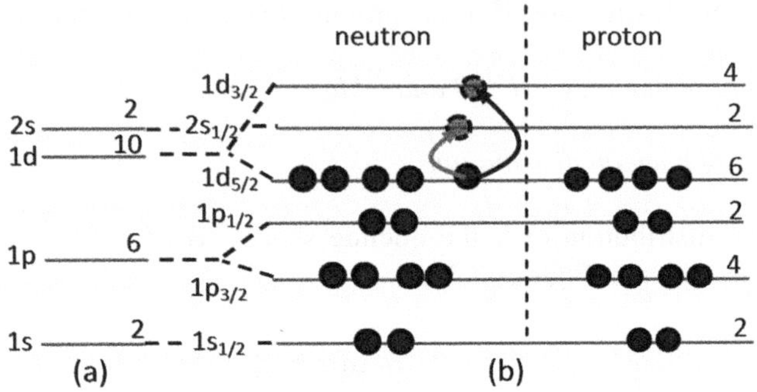

Figure 2.30c. *Shell structure derived from a harmonic potential (a) and distribution derived from the Woods–Saxon potential with spin-orbit coupling (b) for magnesium-25. The arrows indicate two possible of excitation of the unpaired proton*

ii) Values of J^π

– The unpaired proton of the boron-11 nucleus occupies the $1p_{3/2}$ level; i.e. $j = 3/2$. The parity for this state $\pi = -1$. Hence for the ground state, $J^\pi = 3/2^-$. The excitation of this proton can induce two transitions: $1p_{3/2} \rightarrow 1p_{1/2}$ and $1p_{3/2} \rightarrow 1d_{5/2}$ (Figure 2.34a). This gives two excited states: $1p_{1/2}$, $\pi = -1$ and $J^\pi = 1/2^-$, and $1d_{5/2}$, $\pi = +1$ and $J^\pi = 5/2^+$.

– Similarly, for the nitrogen-15 nucleus, the proton occupies the $1p_{1/2}$ level. This gives, for the ground state, $J^\pi = 1/2^-$ and for the two lower-energy excited states (Figure 2.34b): $1d_{5/2}$, $J^\pi = 5/2^+$ and $2s_{1/2}$, $J^\pi = 1/2^+$.

– For the magnesium-25 nucleus, it is the neutron occupying the $1d_{5/2}$ level that should be considered. This gives, for the ground state, $J^\pi = 5/2^+$ and for the two lower-energy excited states (Figure 2.34): $2s_{1/2}$, $J^\pi = 1/2^+$ and $1d_{3/2}$, $J^\pi = 3/2^+$.

b) Transitions between the energy levels of the ^{188}Pt and ^{228}Th nuclei

i) Theoretical justification of the J^π of the ground state

The ^{188}Pt and ^{228}Th nuclei are even-*A* nuclei. $J^\pi = 0^+$ for the ground state.

ii) Transition type, forbidden transition, wavelengths

– Multipole transition type, forbidden transition

– $5^- \rightarrow 4^+$ transition: there is a change in parity. This is an *M*-type transition. The quantum number, ℓ, takes the values from 9 to 1. The dominant transition corresponds to $\ell = 1$ and the multipole order $k_1 = 2$: a magnetic dipole transition, M_1, is thus obtained;

– $4^+ \rightarrow 2^+$ transition: parity is conserved. This is an E-type transition. The quantum number, ℓ, takes the values from 6 to 2. The dominant transition corresponds to $\ell = 2$ and the multipole order $k_2 = 4$: an electric quadrupole transition, E_4, is thus obtained;

– The electric monopole transition (E_0) $0^+ \rightarrow 0^+$ is forbidden since $\ell = 0$.

– *Wavelengths*

The wavelengths are given by the relationship:

$$\Delta E = \frac{hc}{\lambda} \Rightarrow \lambda = \frac{hc}{\Delta E} \approx \frac{124.3}{\Delta E} \text{ (pm)} \qquad [2.221]$$

In the last equality [2.121], ΔE is expressed in keV.

NOTE.–

– $5^- \rightarrow 4^+$ transition: $\Delta E = 327$ keV $\Rightarrow \lambda = 0.38$ pm;

– $4^+ \rightarrow 2^+$ transition: $\Delta E = 131$ keV $\Rightarrow \lambda = 0.95$ pm.

SOLUTION 2.3.– Multipole order, probability of γ-deexcitation of neon-22

a) Justification of the values of J^π

– For the ^{22}Ne nucleus, A is even, and for the ground state we obtain $J^\pi = 0^+$.

– For the ^{22}Na nucleus, A is even, but Z and N are odd. 10 protons and 10 neutrons are paired. Referring to Figure 2.30c, for example, we can see that the 11th proton and the 11th unpaired neutron are located on the $1d_{5/2}$ level ($\pi = +1$). This gives $j_1 = 5/2$ and $j_2 = 5/2$ for the 11th neutron. The total parity $\pi = (+1) \times (+1) = +1$. The total angular momentum takes the values between 5 ($j_1 + j_2$) and 2 ($|j_1 - j_2|$). This then gives, for J^π, the values 5^+, 4^+, 3^+ and 2^+. Only the level $J^\pi = 3^+$ is shown in Figure 2.32. The other values correspond to excited states.

b) Determining the multipole transition type

The deexcitation corresponds to the γ $2^+ \rightarrow 0^+$ transition. The parity is conserved and the multipole order is equal to 4 ($\ell = 2$): it is therefore an electric quadrupole transition, E_2.

c) Estimating the transition probability

Using the Weisskopf estimates, the probability of electric quadrupole transition, E_2, is given by the relationship according to [2.80]:

$$\lambda_\gamma(E_2) = C_2(E) A^{4/3} E_\gamma^5 \qquad [2.222]$$

NOTE.– $A = 22$, $C_2(E) = 7.4 \times 10^7$; $E_\gamma = 1.28$ MeV.

$$\lambda_\gamma(E_2) = 7.4 \times 10^7 \times 22^{4/3} \times 1.28^5 \approx 1.5 \times 10^{10}\ \mathrm{s}^{-1}$$

SOLUTION 2.4.– Separation energy of the last proton and last neutron of dysprosium-161 (^{161}Dy)

In this exercise, we propose to determine the separation energy of the last $S_{n,p}$ nucleon (A, Z). To do so, we will consider the 66-proton dysprosium-161 nucleus. We are seeking to determine the energy required to extract a neutron or a proton from the ^{161}Dy nucleus.

a) Energy released by the absorption of a neutron, meaning

Let us write the transformation equation for the ^{160}Dy nucleus following the absorption of a neutron:

$$^{1}_{0}\mathrm{n} + {}^{160}_{66}\mathrm{Dy} \rightarrow {}^{161}_{66}\mathrm{Dy} \qquad [2.223]$$

According to [2.223], the released energy, Q_l, is given by the relationship:

$$Q_l = \Delta mc^2 = [m_\mathrm{n} + m\,(^{160}\mathrm{Dy}) - m\,(^{161}\mathrm{Dy})]c^2 \qquad [2.224]$$

The released energy [2.224] therefore represents the separation energy of the last neutron, S_n (161, 66) of the ^{161}Dy nucleus.

NOTE.– *It is recalled that the mass of a nucleus is obtained by subtracting the mass of Z electrons from the atomic mass*

$$Q_l = [1.00866 + 159.8889899 - 160.8907258] \times 931.5 = 6.45 \text{ MeV} \qquad [2.225]$$

b) Energy released by the absorption of a proton, meaning

Let us write the transformation equation for the ^{160}Tb nucleus following the absorption of a proton:

$$^{1}_{1}\mathrm{H} + {}^{160}_{65}\mathrm{Tb} \rightarrow {}^{161}_{66}\mathrm{Dy} \qquad [2.226]$$

The released energy, Q_l, is then equal to:

$$Q_l = [m_p + m\,(^{160}\mathrm{Tb}) - m\,(^{161}\mathrm{Dy})]c^2 \qquad [2.227]$$

The released energy [2.224] therefore represents the separation energy of the last proton, S_p (161, 66) of the ^{161}Dy nucleus.

NOTE.–

$$Q_l = [1.00728 + 159.8915086 - 160.8907258] \times 931.5 = 7.51\ \text{MeV} \qquad [2.228]$$

c) Binding energies, classification

– *Binding energies*

Using the expression for the binding energy [1.60], we obtain:

$-E_l$ (^{160}Dy) = (66 × 1.00728 + 94 × 1.00866 – 159.8889899) × 931.5 = 1,309.251 MeV

$-E_l$ (^{161}Dy) = (66 × 1.00728 + 95 × 1.00866 – 160.8907258) × 931.5 = 1,315.701 MeV

$-E_l$ (^{160}Tb) = (65 × 1.00728 + 95 × 1.00866 – 159.8915086) × 931.5 = 1,308.191 MeV

In summary, this gives:

$$E_l\,(^{160}\mathrm{Dy}) = 1{,}309.251\ \text{MeV}$$

$$E_l\,(^{161}\mathrm{Dy}) = 1{,}315.701\ \text{MeV} \qquad [2.229]$$

$$E_l\,(^{160}\mathrm{Tb}) = 1{,}308.191\ \text{MeV}$$

– *Classification of nuclides in order of increasing stability*

Using [2.229], we obtain:

$$E_l\,(^{160}\text{Dy})/A = 8.18 \text{ MeV/nucleon}$$

$$E_l\,(^{161}\text{Dy})/A = 8.17 \text{ MeV/nucleon} \qquad [2.230]$$

$$E_l\,(^{160}\text{Tb})/A = 8.12 \text{ MeV/nucleon}$$

Hence:

$$E_l\,(^{160}\text{Dy})/A > E_l\,(^{161}\text{Dy})/A > E_l\,(^{160}\text{Tb})/A.$$

In order of increasing stability, we obtain: $^{160}\text{Tb} \rightarrow {}^{161}\text{Dy} \rightarrow {}^{160}\text{Dy}$.

d) Separation energy of the last nucleon

Using results [2.229], we obtain:

$$S_n\,(161, 66) = E_l\,(^{161}\text{Dy}) - E_l\,(^{160}\text{Dy}) = 6.45 \text{ MeV} \qquad [2.231]$$

$$S_n\,(161, 66) = E_l\,(^{161}\text{Dy}) - E_l\,(^{160}\text{Dy}) = 7.51 \text{ MeV}$$

Results [2.225] and [2.228] are indeed found.

SOLUTION 2.5.– Determining the Coulomb energy of a charged sphere from the potential created by a charge placed in the center of the nucleus

We propose to establish expression [2.225] representing the Coulomb energy of a volume-charged sphere. To do this, let us consider a nucleus assumed to be spherical, of radius R, whose charge is uniformly distributed in volume. We will assume that the charge volume density, ρ, is constant throughout the nuclear volume.

a) Expression of the charge volume density

The charge of the nucleus $Q = Ze$ and its volume $\tau\,(R) = (4/3)\pi R^3$. The charge volume density is therefore equal to:

$$\rho = \frac{Q}{\tau(R)} = \frac{4}{3}\frac{Ze}{\pi R^3} \qquad [2.232]$$

b) Expression of potential

Let us consider Figure 2.31.

Figure 2.31. *Charge Q (r) contained in the nuclear envelope with radius r*

The electrostatic potential created at *M* by the nuclear charge, *Q* (*r*), is given by the relationship (*the origin of the potentials is chosen as zero to infinity):*

$$V(r) = k\frac{Q(r)}{r} \quad [2.233]$$

Since the charge density is constant, then:

$$\rho = \frac{Q(r)}{\tau(r)} = \frac{Q}{\tau(R)} \Rightarrow Q(r) = \frac{r^3}{R^3}Q \quad [2.234]$$

Using [2.234], potential [2.233] is ultimately written:

$$V(r) = k\frac{Q}{R^3}r^2 = k\frac{Ze}{R^3}r^2 \quad [2.235]$$

c) Expression of potential energy and Coulomb energy

The potential energy, *dW*, of the elementary charge, *dq*, immersed in the potential, *V (r)* [2.235] is given by the relationship:

$$dW = dqV(r) = \rho d\tau V(r) \quad [2.236]$$

In spherical coordinates, the elementary volume $d\tau = r^2 dr \sin\theta d\theta\varphi$. Using [2.235], equation [2.236] is written:

$$dW = k\frac{Q}{\tau(R)} \times \left(\frac{Q}{R^3}r^2\right) r^2 dr \sin\theta \, d\theta \, d\varphi$$

Thus:

$$dW = \frac{3}{4\pi}k\frac{Q^2}{R^6}r^4 dr \sin\theta \, d\theta \, d\phi \qquad [2.237a]$$

By integrating equation [2.237], we obtain:

$$W = \frac{3}{4\pi}k\frac{Q^2}{R^6}\int_0^R r^4 dr \int_0^\pi \sin\theta \, d\theta \int_0^{2\pi} d\varphi \qquad [2.237b]$$

Which gives:

$$W = \frac{3}{4\pi}k\frac{Q^2}{R^6} \times \frac{R^5}{5} \times 4\pi \qquad [2.237c]$$

That is, ultimately:

$$W = \frac{3}{5}k\frac{Q^2}{R} = \frac{3}{5}k\frac{Z^2 e^2}{r_0 A^{1/3}} \qquad [2.238]$$

This corresponds to expression [2.225] of the Coulomb energy, W_c, of sphere with nuclear radius, $R = r_0 A^{1/3}$ uniformly charged in volume.

SOLUTION 2.6.– Determining the Coulomb energy of a charged sphere from the electromagnetic energy density

a) Expressions of internal and external electric fields

Let us consider Figure 2.32.

The external electric field, $\vec{E}_{ext}$, created by the electric charge, Q, of the nucleus at a point M' ($r > R$) and the internal electric field, $\vec{E}_{int}$, created by the internal charge, Q (r), at a point M ($r < R$) are given by the expressions:

$$\vec{E}_{ext} = k\frac{Q}{r^2}\vec{u}\,; \; \vec{E}_{int} = k\frac{Q(r)}{r^2}\vec{u} = k\frac{Q}{R^3}r\vec{u} \qquad [2.239]$$

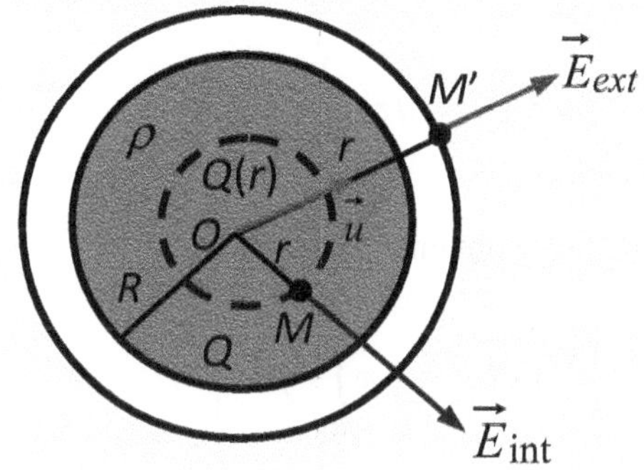

Figure 2.32. *Internal and external electric fields created by charges Q (r) and Q at M and M', respectively*

b) Shape of electromagnetic energy density

In the calculation of the Coulomb energy of a volume-charged sphere, the protons are assumed to be fixed. Therefore, there is no volume current. The electromagnetic energy density is summarized in the electrical term, i.e.:

$$\omega = \omega_{el} = \frac{\varepsilon_0 E^2}{2} \Rightarrow \begin{cases} \omega_{\text{int}} = \dfrac{\varepsilon_0 E_{\text{int}}^2}{2} \\ \omega_{ext} = \dfrac{\varepsilon_0 E_{ext}^2}{2} \end{cases} \qquad [2.240]$$

c) Expression of the Coulomb energy of a volume-charged sphere

The elementary Coulomb energy, dW_c stored inside and outside the sphere of radius R is written:

$$dW_c = \omega_{el} d\tau \Rightarrow dW_c = \omega_{\text{int}} d\tau_{\text{int}} + \omega_{ext} d\tau_{ext} \qquad [2.241]$$

Using [2.240] and [2.239], equation [2.241] is written:

$$dW_c = k^2 \frac{\varepsilon_0 Q^2}{2R^6} r^2 d\tau_{\text{int}} + k^2 \frac{\varepsilon_0 Q^2}{2r^4} d\tau_{ext}$$

That is:

$$dW_c = k^2 \frac{\varepsilon_0 Q^2}{2}\left(\frac{r^2}{R^6} d\tau_{\text{int}} + \frac{1}{r^4} d\tau_{ext}\right) \quad [2.242]$$

Let us integrate [2.242]. We obtain:

$$W_c = k^2 \frac{\varepsilon_0 Q^2}{2}\left(\frac{4\pi}{R^6}\int_0^R r^4 dr + 4\pi\int_R^\infty \frac{1}{r^2} dr\right) \quad [2.243]$$

This then gives:

$$W_c = 2\pi k^2 \varepsilon_0 Q^2\left(\frac{1}{5R} + \frac{1}{R}\right) = 2\pi k \times \frac{1}{4\pi\varepsilon_0}\frac{\varepsilon_0 Q^2}{R} \times \frac{6}{5} = \frac{3}{5} k \frac{Q^2}{R}$$

We indeed find [2.225].

SOLUTION 2.7.– Determining the height of the nuclear potential barrier: theory and experiment

a) Theoretical expression of the height of the nuclear potential barrier

For a problem of α particle diffusion by a nucleus of radius $R = r_0 A^{1/3}$, the height of the nuclear potential barrier is given by the general expression:

$$B = k\frac{Zze^2}{R} = k\frac{2Ze^2}{r_0 A^{1/3}} \quad [2.244]$$

Let us rewrite the expression of the Sakho unit nuclear radius [1.39] as follows:

$$r_0 = \frac{\alpha^2}{(1+N/Z)} a_0 = \alpha^2 a_0 \frac{Z}{A} \quad [2.245]$$

By applying expression [2.244] in [2.245], we obtain:

$$B_{0theo} = k\frac{2Ze^2}{\alpha^2 a_0 A^{1/3}}\frac{A}{Z} = \frac{2ke^2}{\alpha^2 a_0} A^{2/3} \quad [2.246]$$

b) Corrected expression of the height of the nuclear potential barrier

For $r = R_Z + R_z/A$, the height of the nuclear potential barrier is written:

$$B = k\frac{Zze^2}{r} = k\frac{2Ze^2}{(R_Z + R_z / A)} = k\frac{2Ze^2}{(r_{0A}A^{1/3} + r_{0\alpha}4^{1/3} / A)} \quad [2.247]$$

According to [2.245], we obtain:

$$r_{0A} = \alpha^2 a_0 \frac{Z}{A};\, r_{0\alpha} = \frac{\alpha^2 a_0}{2} \quad [2.248]$$

Using [2.248], equation [2.247] is then written:

$$B_{theocorr} = \frac{2ke^2}{\alpha^2 a_0}\frac{Z}{\left(\frac{Z}{A^{2/3}} + \frac{4^{1/3}}{2A}\right)} = \frac{2ke^2}{\alpha^2 a_0}\frac{Z}{\frac{Z}{A^{2/3}}\left(1 + \frac{4^{1/3}}{2A}\frac{A^{2/3}}{Z}\right)}$$

That is, ultimately:

$$B_{theocorr} = \frac{2ke^2}{\alpha^2 a_0}\frac{A^{2/3}}{\left(1 + \frac{1}{(2A)^{1/3}}\frac{1}{Z}\right)} \quad [2.249]$$

c) Comparison of experimental results and theoretical predictions

Let us calculate the factor $2ke^2/\alpha^2 at_0$ in equations [2.246] and [2.248]. We find:

$$\frac{2ke^2}{\alpha^2 a_0} = \frac{2\times 9\times 10^9 \times 1.602\times 10^{-19}}{5.325\times 10^{-5}\times 5.29\times 10^{-11}} = 1.02367 \text{ MeV} \quad [2.250]$$

Taking into account result [2.250], equations [2.246] and [2.249] are written:

$$B_{0\text{theo}} = 1.02367 \times A^{2/3} \text{ MeV} \quad [2.251]$$

$$B_{theocorr} = 1.02367 \times \frac{A^{2/3}}{\left(1 + \frac{1}{(2A)^{1/3} Z}\right)} \text{ MeV} \quad [2.252]$$

Using [2.251] and [2.252], we find the theoretical results gathered in Table 2.16 compared with the experimental data.

Element		***B* (MeV)**		
$_Z$X	*A*	$(B_{exp})^a$	$(B_{theo0})^b$	$B_{theocorr}{}^c$
$_2$He	4	2.4	2.579	-
$_3$Li	7	3.3	3.746	3.291
$_4$Be	9	4.0	4.429	4.043
$_5$B	11	4.5	5.063	4.726
$_6$C	12	5.1	5.365	5.072
$_7$N	14	5.6	5.946	5.679
$_8$O	16		6.500	6.254
$_9$F	19		7.289	7.056
$_{10}$Ne	20		7.542	7.328
$_{12}$Mg	24	8.5	8.517	8.326
$_{13}$Al	27	9.0	9.213	9.039
$_{14}$Si	28		9.439	9.266
$_{15}$P	31		10.102	9.935
$_{16}$S	32		10.318	10.159
$_{17}$Cl	35		10.953	10.799
$_{18}$Ar	40		11.973	11.821

[a] Experimental values, [POL 35].

[b] Theoretical values, formula [2.251].

[c] Corrected theoretical values, formula [2.252].

Table 2.16. *Comparison of the experimental (B_{exp}) and theoretical (B_{theo}) values of the height of the nuclear potential barrier, B, for several light elements*

Overall, there is very good agreement between the theoretical and experimental results when the theoretical expression of the height of the nuclear potential barrier is corrected by choosing $r = R_Z + R_z/A$.

NOTE.– *The experimental values 4.5 MeV and 8.5 MeV are probably less accurate than the other measured values.*

SOLUTION 2.8.– Experimentally determining the Coulomb coefficient by measuring the variation in binding energy of neighboring isobars

The following pairs of neighboring isobars are considered:

$$(^{37}\text{Ar and }^{37}\text{K});\ (^{31}\text{P and }^{31}\text{S});\ (^{23}\text{Mg and }^{23}\text{Al});\ \text{and }(^{15}\text{N and }^{15}\text{O}) \qquad [2.253]$$

a) Reminder, value of $A-2Z$

Mirror nuclei are pairs of nuclei with the same mass number, A, with their numbers of protons Z and neutrons N exchanged; this is reflected in the relationship: $Z-N=\pm 1$. It follows that:

$$A-2Z=Z+N-2Z=N-Z \Rightarrow A-2Z=\pm 1 \qquad [2.254]$$

Let us calculate the values of $A-2Z$ for the pairs of nuclei [2.253]. We obtain:

– ^{37}Ar ($Z=18$): $A-2Z=1$; ^{37}K ($Z=19$): $A-2Z=-1$

– ^{31}P ($Z=15$): $A-2Z=1$; ^{31}S ($Z=16$): $A-2Z=-1$

– ^{23}Mg ($Z=12$): $A-2Z=1$; ^{23}Al ($Z=13$): $A-2Z=-1$

– ^{15}N ($Z=7$): $A-2Z=1$; ^{15}O ($Z=8$): $A-2Z=-1$

Each of the pairs of nuclei [2.253] verifies the last equality [2.254]: they are therefore pairs of mirror nuclei.

b) Expressions of mass difference and binding energy difference

– Expression of mass difference

Using the Bethe–Weizsäcker formula, we obtain ($A-Z=A-(A-Z)=Z$):

$$\begin{cases} M(A,Z) = Z\,M_H + (A-Z)M_n \\ \qquad - a_v\,A + a_s\,A^{2/3} + a_c\,\dfrac{Z^2}{A^{1/3}} + a_a\,\dfrac{(Z-N)^2}{A} \pm |\delta| \\ M(A,A-Z) = (A-Z)M_H + ZM_n \\ \qquad - a_v\,A + a_s\,A^{2/3} + a_c\,\dfrac{(A-Z)^2}{A^{1/3}} + a_a\,\dfrac{(Z-N)^2}{A} \pm |\delta| \end{cases} \qquad [2.255]$$

Using [2.255], the difference in mass, $\Delta M = M(A,Z) - M(A,A-Z)$ is written:

$$\Delta M = [ZM_H + (A-Z)M_n] - [(A-Z)M_H + ZM_n]$$
$$+\frac{a_c}{A^{1/3}}[Z^2 + (A-Z)^2]$$

That is:

$$\Delta M = [A(M_n - M_H) - 2Z(M_n - M_H)] - \frac{a_c}{A^{1/3}}[(A-2Z)]A$$

That is, ultimately:

$$\Delta M = (A-2Z)[(M_n - M_H) - a_c A^{2/3}] \quad [2.256a]$$

– *Expression of the binding energy difference*

The difference in binding energy, $\Delta E_l = E_l(A, Z) - E_l(A, A-Z)$ is equal to:

$$\Delta E_l = [ZM_H + (A-Z)M_n] - [(A-Z)M_H + ZM_n] + [M(A, A-Z) - M(A,Z)]$$

i.e.:

$$\Delta E_l = (A-2Z)[M_n - M_H]\,\Delta M \quad [2.256b]$$

c) Demonstration

By applying [2.256a] to [2.256b], we obtain:

$$\Delta E_l = (A-2Z)(M_n - M_H) - (A-2Z)[(M_n - M_H) - a_c A^{2/3}]$$

That is:

$$\Delta E_l = (A-2Z)a_c A^{2/3} \quad [2.256c]$$

Let us write $\Delta E_{lexp} = |\Delta E_l|$. The variation in binding energy between neighboring isobars is written taking the result [2.254] into account:

$$\Delta E_{l\exp} = a_c A^{2/3} \quad [2.257]$$

d) Theoretical expression of the variation of the binding energy between neighboring isobars

We will write, for a pair of isobars (${}^{A}X_Z$ and ${}^{A}X_{Z+1}$):

$$a_c(Z) = \frac{3}{5}\frac{ke^2}{\alpha^2 a_0}\frac{A}{Z}\,;\; a_c(Z+1) = \frac{3}{5}\frac{ke^2}{\alpha^2 a_0}\frac{A}{Z+1}$$

Let us rewrite these expressions by multiplying the top and bottom of their right-hand members by 2, with a view to using result [2.250]. We obtain:

$$a_c(Z) = \frac{3}{10}\frac{2ke^2}{\alpha^2 a_0}\frac{A}{Z}\,;\, a_c(Z+1) = \frac{3}{10}\frac{2ke^2}{\alpha^2 a_0}\frac{A}{Z+1}$$

Taking [2.250] into account, we obtain:

$$a_c(Z) = 0.307101 \times \frac{A}{Z}\ \text{MeV} \qquad [2.258a]$$

$$a_c(Z+1) = 0.307101 \times \frac{A}{Z+1}\ \text{MeV} \qquad [2.258b]$$

The average value $\langle a_c \rangle$ is defined by the relationship:

$$\langle a_c \rangle = \frac{\langle a_c \rangle (Z) + \langle a_c \rangle (Z+1)}{2}\ \text{MeV} \qquad [2.259]$$

Furthermore, the variation in binding energy between neighboring isobars is theoretically given by:

$$\Delta E_{ltheo} = \langle a_c \rangle A^{2/3} \qquad [2.260]$$

Using [2.258], [2.259] and [2.260], the theoretical predictions gathered in Table 2.17 are obtained.

Pair of neighboring isobars			**Coulomb coefficient (MeV)**			**Binding energy difference**
A	$_Z$**X**	$_{Z+1}$**X**	a_c **(Z)**	a_c **(Z+1)**	$\langle a_c \rangle$	ΔE_{ltheo} **(MeV)**
37	$_{18}$Ar	$_{19}$K	0.631	0.598	0.615	6.829
31	$_{15}$P	$_{16}$S	0.635	0.595	0.615	6.069
23	$_{12}$Mg	$_{13}$Al	0.589	0.543	0.566	4.588
15	$_{7}$N	$_{8}$O	0.658	0.576	0.617	3.753

Table 2.17. *Theoretical values of the Coulomb coefficient and of the difference in binding energy of neighboring isobars*

Let us now compare the theoretical values of the difference in binding energy of neighboring isobars with the experimental data. The results presented in Table 2.18 are obtained.

(*A*, *Z*)	**(*A*, *A* – *Z*)**	**Experiment**			**Theory**
$_{Z}$X	$_{Z+1}$X	$E_l\,(A, Z)$	$E_l\,(A, A-Z)$	ΔE_l	ΔE_l
$_{18}$Ar	$_{19}$K	315.510	308.587	6.923	6.829
$_{0.15}$P	$_{16}$S	262.916	256.688	6.228	6.069
$_{12}$Mg	$_{13}$Al	186.565	181.726	4.839	4.588
$_{7}$N	$_{8}$O	115.494	111.952	3.542	3.753

Table 2.18. *Comparison of the experimental [MAR 69] and theoretical values of the difference in binding energy of neighboring isobars. The results are expressed in MeV*

The comparison of the values listed in the last two columns of Table 2.18 shows agreement between the theoretical predictions and the experimental results. The maximum deviation of 0.251 MeV between the theoretical and experimental results is due to the imprecision on the average value $\langle a_c \rangle$ of the Coulomb coefficient.

e) Average value of the unit nuclear radius

The average value of the unit nuclear radius for the neighboring isobars is given by the relationship:

$$\langle r_0 \rangle = \frac{3}{5}\frac{ke^2}{\langle a_c \rangle} = \frac{1.38585816\times 10^{-28}}{\langle a_c \rangle} \qquad [2.261]$$

Examples of calculations

$$\Delta E_{l\text{exp}} = a_{c\text{exp}}\, A^{2/3} \Rightarrow a_{c\text{exp}} = \Delta E_{l\text{exp}} / A^{2/3} \qquad [2.262]$$

For $\Delta E_{l\text{exp}} = 6.923$ MeV, we obtain, for the pair ($_{18}$Ar and $_{19}$K) with $A = 37$:

$$a_{c\text{exp}} = 6.923/37^{2/3} = 0.623\ \text{(MeV)}$$

For the unit nuclear radius,

$$r_{0\text{exp}} = 1.38585816 \times 10^{-28}/a_{c\text{exp}} \qquad [2.263]$$

Thus, for the pair of isobars ($_{18}$Ar and $_{19}$K):

$$r_{0\exp} = 1.38585816 \times 10^{-28}/(0.623 \times 10^6 \times 1.602 \times 10^{-19})$$

$$= 1.38857 \times 10^{-15}\,\text{m} = 1.39\ \text{fm}$$

By adopting an approach analogous to the previous one, we find the results gathered in Table 2.19.

(*A*, *Z*)	**(*A*, *A* – *Z*)**	**Coulomb coefficient**		**Unit nuclear radius**	
$_Z$**X**	$_{Z+1}$**X**	$a_{c\exp}$ **(MeV)**	$\langle a_c\rangle_{theo}$ **(MeV)**	$r_{0\exp}$ **(fm)**	$\langle r_0\rangle_{theo}$ **(fm)**
$_{18}$Ar	$_{19}$K	0.623	0.615	1.389	1.407
$_{15}$P	$_{16}$S	0.631	0.615	1.371	1.407
$_{12}$Mg	$_{13}$Al	0.598	0.566	1.447	1.530
$_7$N	$_8$O	0.582	0.617	1.486	1.404

Table 2.19. *Comparison of the experimental [MAR 69] and theoretical values of the Coulomb coefficient and the unit nuclear radius*

The comparison of the results gathered in Table 2.19 shows sound agreement between theory and experiment.

SOLUTION 2.9.– Expressions of Weisskopf estimates

a) General expressions of coefficients $C_\ell(E)$ and $C_\ell(M)$

In relationships [2.217], let us replace R with 1.2 $A^{1/3}$ fm and $\hbar\omega = E_\gamma$. We obtain:

$$T_E(\ell) \approx \frac{4.4(\ell+1)\times(1.2)^{2\ell}\times 10^{21}}{\ell[(2\ell+1)!!]^2}\left(\frac{3}{\ell+3}\right)^2\left(\frac{1}{197}\right)^{2\ell+1} A^{2\ell/3}E_\gamma^{2\ell+1} \qquad [2.264]$$

$$T_M(\ell) \approx \frac{1.9(\ell+1)\times(1.2)^{2\ell-2}\times 10^{21}}{\ell[(2\ell+1)!!]^2}\left(\frac{3}{\ell+3}\right)^2\left(\frac{1}{197}\right)^{2\ell+1} A^{(2\ell-2)/3}E_\gamma^{2\ell+1} \qquad [2.265]$$

Comparing [2.264] and [2.265], respectively, with the Weisskopf estimates [2.80] and [2.81], we see that:

$$C_\ell(E) \approx \frac{4.4(\ell+1)\times(1.2)^{2\ell}\times10^{21}}{\ell[(2\ell+1)!!]^2}\left(\frac{3}{\ell+3}\right)^2\left(\frac{1}{197}\right)^{2\ell+1} \quad [2.266]$$

$$C_\ell(M) \approx \frac{1.9(\ell+1)\times(1.2)^{2\ell-2}\times10^{21}}{\ell[(2\ell+1)!!]^2}\left(\frac{3}{\ell+3}\right)^2\left(\frac{1}{197}\right)^{2\ell+1} \quad [2.267]$$

Note that $C_\ell(E)$ is expressed in $(\text{fm})^{2\ell}$ and $C_\ell(M)$ in $(\text{fm})^{2\ell-2}$.

b) Values of coefficient $C_\ell(E)$

– *Case of electric multipole transitions*

Using [2.266], we obtain:

– for $\ell = \mathbf{1}$:

$$C_1(E) \approx \frac{4.4\times2\times(1.2)^2\times10^{21}}{[3!!]^2}\left(\frac{3}{4}\right)^2\left(\frac{1}{197}\right)^3 = \frac{9.323282124\times10^{14}}{[3!!]^2}$$

$= 1.03592 \times 10^{14}$

$([3!!] = [(2\times\mathbf{1}+1)\times(2\times\mathbf{1}-1)] = 3$

– for $\ell = \mathbf{2}$:

$$C_2(E) \approx \frac{4.4\times3\times(1.2)^4\times10^{21}}{2\times[5!!]^2}\left(\frac{3}{5}\right)^2\left(\frac{1}{197}\right)^5 = \frac{1.66050539}{[5!!]^2}\times10^{10}$$

$= 7.380024 \times 10^7$

$([5!!] = [(2\times\mathbf{2}+1)\times(2\times\mathbf{2}-1)\times(2\times\mathbf{2}-3)] = 15$

– for $\ell = \mathbf{3}$:

$$C_3(E) \approx \frac{4.4\times4\times(1.2)^6\times10^{21}}{3\times[7!!]^2}\left(\frac{3}{6}\right)^2\left(\frac{1}{197}\right)^7 = \frac{3.803253862}{[7!!]^2}\times10^5$$

$= 34.49663$

$([7!!] = [(2\times\mathbf{3}+1)\times(2\times\mathbf{3}-1)\times(2\times\mathbf{3}-3)\times(2\times\mathbf{3}-5)] = 105$

– for $\ell = \mathbf{4}$:

$$C_4(E) \approx \frac{4.4\times 5\times(1.2)^8\times 10^{21}}{4\times[9!!]^2}\left(\frac{3}{7}\right)^2\left(\frac{1}{197}\right)^9 = \frac{9.719928088}{[9!!]^2}$$

$= 1{,}088427322 \times 10^{-5}$

$([9!!] = [(2\times\mathbf{4}+1)\times(2\times\mathbf{4}-1)\times(2\times\mathbf{4}-3)\times(2\times\mathbf{4}-5)\times(2\times\mathbf{4}-7)] = 945$

– for $\ell = \mathbf{5}$:

$$C_5(E) \approx \frac{4.4\times 6\times(1.2)^{10}\times 10^{21}}{5\times[11!!]^2}\left(\frac{3}{8}\right)^2\left(\frac{1}{197}\right)^{11} = \frac{2.65082\times 10^{-4}}{[11!!]^2}$$

$= 2{,}45319037 \times 10^{-12}$

$([11!!] = [(2\times\mathbf{5}+1)\times(2\times\mathbf{5}-1)\times(2\times\mathbf{5}-3)\times(2\times\mathbf{5}-5)\times(2\times\mathbf{5}-7)$

$\times(2\times\mathbf{5}-9)] = 10{,}395$

In summary:

$$C_1(E) = 1.0\times 10^{14};\ C_2(E) = 7.4\times 10^{7};\ C_3(E) = 34.5$$

$$C_4(E) = 1.1\times 10^{-5};\ C_5(E) = 2.5\times 10^{-12} \qquad [2.268]$$

c) Expression and values of coefficients $C_\ell(M)$

Let us express the ratio $C_\ell(M)/C_\ell(E)$. Using [2.266] and [2.267], we obtain, after simplification:

$$C_\ell(M) = \frac{1.9\times(1.2)^{2\ell-2}}{4.4\times(1.2)^{2\ell}}C_\ell(E) \Rightarrow C_\ell(M) = \frac{1900}{6336}C_\ell(E) \qquad [2.269]$$

Using [2.268] and the untruncated values of $C_\ell(E)$ above, we find, for the first five magnetic multipole transitions:

$$- \; C_1(M) = \frac{1900}{6336} \times C_1(E) = \frac{1900}{6336} \times 1.03592 \times 10^{14} = 3.10645202 \times 10^{13} \qquad [2.270]$$

$$- \; C_2(M) = \frac{1900}{6336} \times C_2(E) = \frac{1900}{6336} \times 7.380024 \times 10^{7} = 2.213075379 \times 10^{7}$$

$$- \; C_3(M) = \frac{1900}{6336} \times C_1(E) = \frac{1900}{6336} \times 34.49663 = 10.34463336$$

$$- \; C_4(M) = \frac{1900}{6336} \times C_4(E) = \frac{1900}{6336} \times 1.088427322 \times 10^{-5} = 3.263907689 \times 10^{-6}$$

$$- \; C_5(M) = \frac{1900}{6336} \times C_5(E) = \frac{1900}{6336} \times 2.45319037 \times 10^{-12} = 7.356773521 \times 10^{-13}$$

In summary:

$$C_1\,(M) = 3.1 \times 10^{13};\; C_2\,(M) = 2.2 \times 10^{7};\; C_3\,(\mathrm{M}) = 10.3$$

[2.271]

$$C_4\,(M) = 3.3 \times 10^{-6};\; C_5\,(M) = 7.4 \times 10^{-13}$$

The values recorded in Table 2.4 are clearly found by comparing them with results [2.268] and [2.271].

3

Alpha Radioactivity

Overall objective	
To know the general properties of α and β radioactivities and electron capture	
Specific objectives	
To interpret Becquerel's observations that led to the discovery of natural radioactivity	To write the balanced equation for α decay
To know the experimental facts that led to the discovery of α radioactivity	To apply the mass number and charge number conservation laws
To define the activity of a radioactive source	To define the mass loss in a nuclear reaction
To know the conventional unit of activity	To define the energy released during α decay
To define radioactivity	To calculate the α decay energy
To define radioactive decay	To define the α decay energy diagram
To define radioactive radiation	To interpret the fine structure of α lines
To distinguish between a parent nucleus and a daughter nucleus	To know the Gerger–Nuttall law
To give examples of α emitting radioelements or radionuclides	To interpret the quantum model of α emission by tunnel effect
To establish the relationship between radioactive half-life and decay constant of a radioelement	To establish the relationship between radioactive half-life and decay constant

For a color version of all of the figures in this chapter, see www.iste.co.uk/sakho/nuclear1.zip.

To highlight α, β and γ radiation experimentally	To describe the process of creating an electron-positron pair
To theoretically justify the classical nature of α emission	To describe the materialization process of a *g* photon interacting with a nucleus
To theoretically justify the relativistic character of β emission	To describe the annihilation process of an electron-positron pair
To establish Rutherford and Soddy's empirical law of exponential decay	To know the nature of the four interactions within the framework of the *standard model*
To define the probability of decay per unit time of a radionuclide	To know the properties of the *u* and *d* quarks
To know the Gamow–Gurney–Condon quantum model of α emission	To know the properties of the W^- and W^+ bosons
To know the hypotheses of Gamow's theory of α emission	To explain the emission of the *neutrino* and the *antineutrino* from the weak interaction
To make use of Gamow's quantum theory to estimate the radioactive half-life	To schematically describe the decay of the neutron by low interaction
To define exotic radioactivity	To schematically describe the decay of the proton by low interaction
To distinguish between α emission and cluster emission	To describe the Cowan and Reines experiment that led to the discovery of the neutrino
To theoretically justify the highly improbable nature of cluster emission	To write the internal structure of the neutron and the proton according to the Gell-Mann and Zweig quark model
Prerequisites	
Energy and momentum conservation laws	Nucleon shell model
Properties of a relativistic particle	Nuclear deexcitation
Angular momenta and parity of a system	Radioactive decay law

3.1. Experimental facts

3.1.1. *Becquerel's observations, radioactivity*

In 1896, Henri Becquerel experimented with phosphorescent materials that emit light into the dark after exposure to light. To do this, he sealed a photographic plate in black paper and put this package in contact with various phosphorescent materials. The experimental observations were negative except those involving uranium salts, which then left an imprint on the photographic plate through the layer of paper used. However, it soon transpired that the impression of the photographic emulsion was not related to the *phenomenon of phosphorescence* since it occurred even if the uranium was not previously exposed to light. Interpreting his observations, Becquerel thus assumed that a mysterious radiation had just been revealed by chance. The discovery of this phenomenon marked the birth of radioactivity. Not having any idea as to the nature of radiation, for six hours he somewhat imprudently left several decimeters of a radium salt in his vest pocket. Twenty days later, he noticed that his skin was falling off and a wound was festering at the precise spot where the radiation, unknown at that time, had struck.

Henri Becquerel was a French physicist. He discovered natural radioactivity by chance in 1896. He won the Nobel Prize in Physics 1903 (shared with the Curie couple, see Box 3.2). The unit of activity is named in his honor.

Box 3.1. *Becquerel (1852–1908)*

3.1.2. *Discovery of α radioactivity and β^- radioactivity*

As noted in Chapter 1, uranium has 26 isotopes, including ^{238}U (99.2739 to 99.2752%), ^{235}U (0.7198 to 0.7202%) and ^{234}U (0.0050 to 0.0059%). The decay of uranium-238 is a series of chain reactions leading to stable lead-206. The *decay chain* of uranium-238 leading to lead is written (the products involved in the global decay chain leading to ^{206}Pb are presented in section 3.3):

$$^{238}_{92}U \xrightarrow{\alpha} {}^{234}_{90}Th \xrightarrow{\beta^-} {}^{234m}_{91}Pa \xrightarrow{\beta^-} {}^{234}_{92}U \xrightarrow{\alpha}$$

$$^{230}_{90}Th \xrightarrow{\alpha} {}^{226}_{88}Ra \xrightarrow{\alpha} {}^{222}_{86}Rn \xrightarrow{\alpha} {}^{218}_{84}Po \ldots \xrightarrow{\beta^-} \quad [3.1]$$

$$^{210}_{83}Bi \xrightarrow{\beta^-} {}^{210}_{84}Po \xrightarrow{\alpha} {}^{206}_{82}Pb \text{ (stable)} + \text{several } \gamma$$

In 1897, Pierre Curie and Marie Curie entered the scene with the aim of finding substances with the same properties as the uranium salts handled by Becquerel. In 1898, they discovered polonium and then radium, two radioactive substances involved in the uranium-238 decay chain [3.1]. After four years of research, the Curies succeeded in processing several tons of pitchblende and obtained just one tenth of a gram of pure radium salt. The Curies called radioactivity the property of nuclei to emit radiation by decay [FER 64]. On July 4 1934, Marie Curie died of leukemia (blood cancer) induced by the same radiation that had caused Becquerel's wound. Today, we know that that mysterious radiation is the gamma (γ) photon [3.1], the most dangerous ionizing radiation of all (see Chapter 4, section 4.4), exposure to which can lead to slow death.

As the sequence of nuclear reactions [3.1] indicates, the decay chain of uranium-238 produces α, β^- and γ particles.

In 1898, Ernest Rutherford (Box 1.4) discovered that radioactivity in a uranium ore is a mixture of two distinct phenomena, as indicated in the decay chain [3.1] of uranium-238. Rutherford called these two decay modes α radioactivity and β radioactivity (here β^-).

From 1902, Rutherford and Frederick Soddy studied the phenomenon of radioactivity. They managed to condense radium emanations and concluded that helium had to be one of the decay products of radioactive substances. Rutherford determined the nature of the helium element emitted by radium: it is the He^{2+} nucleus, corresponding to α alpha radiation.

Beta β radioactivity was observed in the form of a radiation deflected by electric or magnetic fields in the opposite direction from alpha radiation. This proves that the β particles emitted spontaneously by radioactive substances are negatively charged. With the discovery of the electron in 1897 by Joseph John Thomson (see Chapter 1, section 1.1.2), β particles were identified as electrons. The β radioactivity mode named by Rutherford thus corresponds to β^- radioactivity.

The radioactivity corresponding to the α and β^- decay modes is called *natural radioactivity* (to be differentiated from *artificial radioactivity*, see section 3.2). By definition, natural radioactivity is the property of some unstable nuclei to spontaneously convert into other, more stable nuclei. This non-induced nuclear transformation is called *radioactive decay*. It is accompanied by the emission of a *radioactive radiation* consisting of the various particles emitted by radioactive substances, also called *radioelements* or *radionuclides*. The decaying nucleus is called the *parent nucleus* and the nucleus into which the parent nucleus converts is called the *daughter nucleus*.

Pierre Curie was a French physicist and Marie Sklodowska Curie was a French physicist and chemist of Polish origin. As a couple they discovered polonium in 1898 and then radium. They won the Nobel Prize in Physics 1903 for their work on radium radioactivity. Marie Curie was also the winner of the Nobel Prize in Chemistry 1911 for her work on polonium and radium.

Frederick Soddy was a British radiochemist. In 1902 he established, together with Rutherford, the exponential law of radioactive decay. Soddy was awarded the Nobel Prize in Chemistry 1921 for his contributions on radioactive substances and for his research on the nature of isotopes.

Box 3.2. *Marie Curie (1867–1934); Pierre Curie (1859–1906); Soddy (1877–1956)*

3.1.3. *Discovery of the positron*

The β radioactivity mode consists of two decay submodes, one of which is the β^- decay mode. It took until 1932 for the second β decay submode to be identified with the discovery of the *positive electron* called the positron (the *negative electron* then being called the *negatron*). In 1934, there followed the discovery of artificial radioactivity, which highlighted β^+ radiation, similar to β^- radiation.

The existence of the positron was postulated in 1928 by the British physicist Paul Adrien Maurice Dirac (1902–1984) through his relativistic wave equation (see [ASL 08] p. 889). In 1932 Carl Anderson studied the particles present in *cosmic radiation* using a *Wilson chamber*, or *fog chamber*. When a charged particle passes through the chamber, steam condenses into fine droplets along its path. The particle trace is thus visualized. Under the action of a magnetic field the trajectory of the particle is curved in a direction that depends on the sign of its charge. In 1933, Anderson [AND 33a] succeeded in photographing the first trajectory of a particle whose direction of curvature indicates that it is a positive particle that is not a proton, whose course would have been much shorter. He identified the particle with the positive electron, or β^+ positron, which does not exist in our environment. This electron is produced by a rare decay mode called β^+ radioactivity (see section 3.2). In addition, the β^+ particle appears during the process of simultaneous production of electron-positron (or negatron-positron) pairs when a γ photon of energy greater than or equal to 1.022 MeV interacts with a nucleus (Figure 3.1). However, it should be noted that during decay processes, few γ photons possess the energy to induce the *creation of electron-positron pairs*. Thus, the production of pairs plays a marginal role in our environment.

The discovery of the positive electron constituted the first experimental evidence of the existence of what became known as *antimatter*.

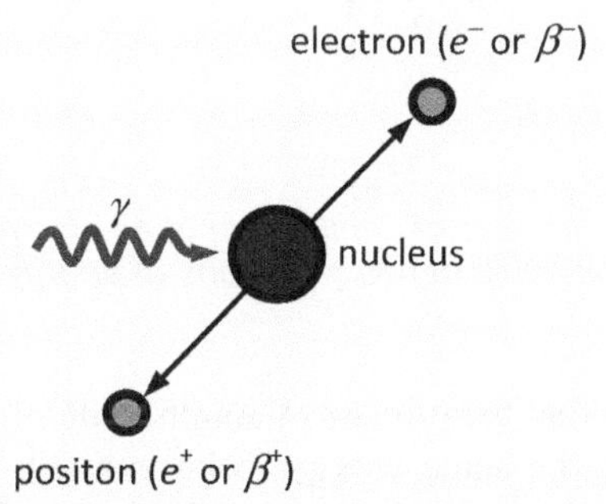

Figure 3.1. *Creation of electron-positron pair by a process of materialization of a γ photon of energy greater than 1.022 MeV interacting with a nucleus*

In 1933, Anderson and Seth Henry Neddermeyer [AND 33b] showed that the γ radiation of 2.62 MeV energy produced by the β decay of Th C″ (^{208}Tl, $Z = 81$), forms *positron-negatron pairs* when it is absorbed by heavy elements such as lead.

APPLICATION 3.1.– Show that the threshold energy of the γ photon for electron-positron pair creation is equal to 1.022 MeV. Also show that during the process of dematerialization of the γ photon, the positron and the electron formed move in directly opposite directions.

ANSWER.– The positron and electron have same resting energy, $E_0 = m_0c^2$. The threshold energy therefore has the value: $E_{\text{threshold}} = 2m_0c^2$. That is, numerically, $E_{\text{threshold}} = 2 \times 0.511$ MeV $= 1.022$ MeV.

The amount of motion of the {absorbed photon-nucleus} system is zero at the initial instant of photon absorption (the recoil of the nucleus is negligible). Thus, by applying the *principle of conservation of momentum*:

$$\vec{0} = \vec{p}_{e-} + \vec{p}_{e+} \Rightarrow \vec{p}_{e-} = -\vec{p}_{e+} \qquad [3.2]$$

As soon as it is created, a positron rapidly encounters an electron in matter. An *annihilation* reaction occurs with the production of two γ photons of the same energy and propagating in opposite directions (Figure 3.2). This annihilation process is used in nuclear medicine for cancer screening (see Volume 2, Chapter 3).

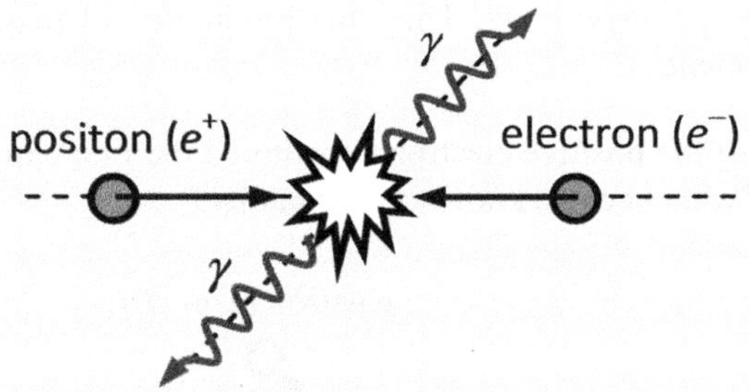

Figure 3.2. *Process of annihilation of an electron-positron pair with production of two γ photons of energy 1.022 MeV*

APPLICATION 3.2.– Show that the process of annihilation of an electron-positron pair causes two γ photons of the same energy to be emitted, propagating in opposite directions. The energy carried by each photon will be specified.

ANSWER.– During the annihilation process, the electron resting energy and that of the positron are each converted into electromagnetic energy. Thus, two photons, γ and γ' ($\gamma' = \gamma$), of the same energy, 0.511 MeV, are formed.

The momentum of the {electron-positron} system is zero at the initial instant of dematerialization. That is, by applying conservation of momentum:

$$\vec{0} = \vec{p}_{\gamma} + \vec{p}_{\gamma'} \Rightarrow \vec{p}_{\gamma} = -\vec{p}_{\gamma'} \qquad [3.3]$$

Carl David Anderson was an American physicist. He is famous for having discovered the positron in 1933, thus experimentally confirming the prediction of Paul Adrien Dirac in 1928. In 1936, together with Neddermeyer, his first student, he discovered the muon, a particle 207 times larger than the electron. Anderson received half of the Nobel Prize in Physics 1936 for his discovery of the positron.

Seth Henry Neddermeyer was an American physicist. In 1982, he received the Enrico Fermi Prize for his participation in the discovery of the positron and the muon.

Box 3.3. *Anderson (1905–1991); Neddermeyer (1907–1988)*

3.1.4. *Discovery of the neutrino, Cowan and Reines experiment*

The standard model describing the interaction between elementary particles is based on four fundamental interactions: strong interaction, electromagnetic interaction, weak interaction and gravitational interaction. The weak interaction carried by the W^+ and W^- vector bosons allows us to explain the emission of neutral and uncharged particles, called the neutrino and antineutrino. These bosons have the same mass, equaling (80,800 ± 2,700) MeV/c^2 and electrical charges equaling +e and –e, respectively. The neutrino and antineutrino accompany the β radioactive emission. By weak interaction, a nucleon of the nucleus can convert into another species. To compensate for the change in charge, an electron or a positron is then expelled from the nucleus. The emission of the electron is accompanied by an electronic antineutrino, $\overline{\nu_e}$, and that of the positron by an electronic neutrino, ν_e. The processes of nuclear transformations of a nucleon into another species are as follows:

$$\begin{cases} n \rightarrow p + e^- + \overline{\nu_e} \\ p \rightarrow n + e^+ + \nu_e \end{cases} \qquad [3.4]$$

Let us interpret the first transformation [3.4] within the framework of weak interaction, using the quark model.

The neutron and proton have quark structures *udd* and *udu*, respectively (Figure 1.16). We recall that the electrical charges of the *u* and *d* quarks are equal to +2/3*e* and −1/3*e* respectively. Under weak interaction, a *d* quark of the neutron converts into a *u* quark by emitting a W^- boson. The W^- boson subsequently decays into an electron and into an antineutrino These two processes of *d* quark transformation and W^- boson decay are described by the respective equations (we will note the conservation of the electrical charge):

$$\begin{cases} d \to u + W^- \\ W^- \to e^- + \bar{\nu}_e \end{cases} \qquad [3.5]$$

The overall process of a neutron converting into a proton via weak interaction is shown in Figure 3.3.

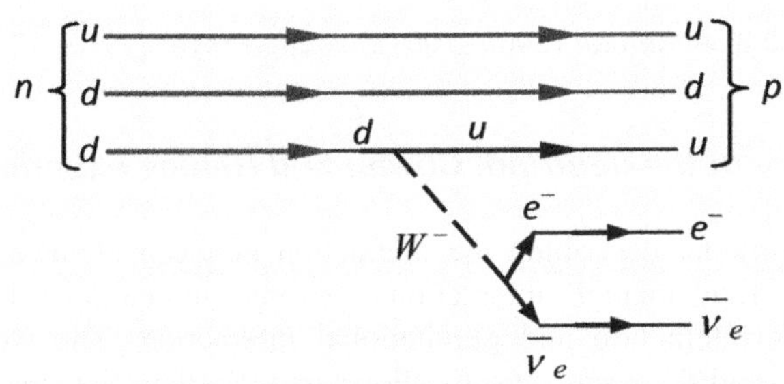

Figure 3.3. *Neutron decay by low interaction. It is the decay of the W^- boson into an electron and an antineutrino that allows the neutron to convert into a proton*

APPLICATION 3.3.– Using a diagram, reflect the overall process of the transformation of a proton into a neutron via weak interaction.

ANSWER.– Under weak interaction, a *u* quark of the proton converts into a *d* quark by emitting a W^+ boson. The W^+ boson subsequently decays into a positron and a neutrino (Figure 3.4).

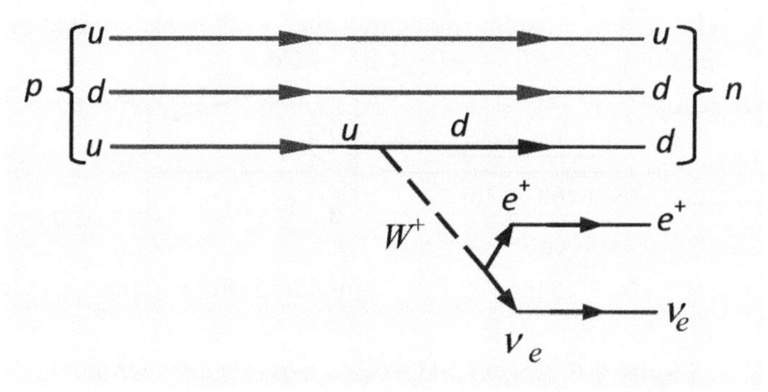

Figure 3.4. *Proton decay by low interaction. It is the decay of the W^+ boson into a positron and a neutrino that allows the proton to convert into a neutron*

The existence of the neutrino was first postulated in 1930 by the Austrian physicist Wolfgang Ernst Pauli (1900–1958) to satisfy the conservation of energy, essentially during β decay. Nevertheless, the word "neutrino" was not introduced by Pauli in the parlance of particle physicists.

Indeed, to generally satisfy the energy and momentum conservation laws during β radioactivity, Pauli suggested the existence of a new neutral particle of zero mass accompanying the β emission, which he called a neutron (which proved to be incorrect since the neutron is a nucleon discovered by Chadwick in 1932, see Chapter 1, section 1.3.2).

In a conversation with Enrico Fermi (see biography, box 4.3) at the Physics Institute of Via Panisperna in Rome, the Italian physicist Edoardo Amadi (1908–1989) gave the new particle suggested by Pauli the name "neutrino" in Italian, meaning "little neutron", in order to distinguish it from the neutron, which is much more massive. From 1932, one after another, Fermi and Pauli adopted the word neutrino, which has since been cited in the parlance of particle physicists.

Postulated in 1930, the existence of the neutrino was not proven experimentally until 26 years later, by means of what became known as the *neutrino experiment* carried out by Cowan and Reines.

In 1956, Clyde Cowan and Frederick Reines conducted the historical experiment aimed at detecting the neutrino. The *Cowan and Reines experimental set-up* is shown in Figure 3.5.

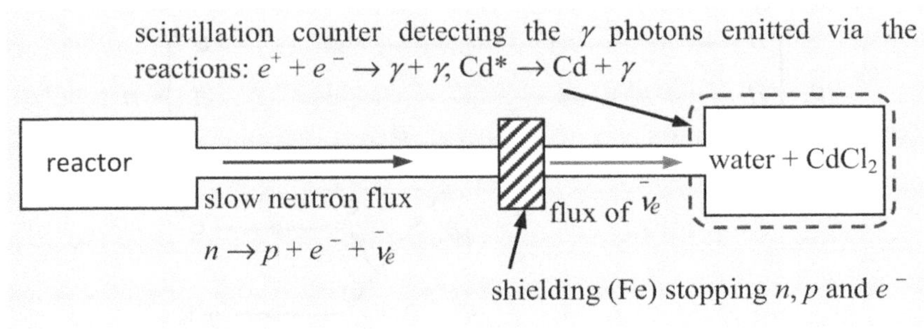

Figure 3.5. *Cowan and Reines experimental set-up*

The principle of the Cowan and Reines experiment is to use slow neutrons produced by a nuclear reactor in Los Alamos, USA. These neutrons, with an average lifetime of 887 s, decay into a proton, an electron and an antineutrino according to the first equation [3.4], which we recall here:

$$n \to p + e^- + \bar{\nu}_e$$

This reaction then provides an antineutrino flux of 5×10^{13} particles per second per square centimeter. To isolate the antineutrino flux, Cowan and Reines placed a shielding of thousands of tons of iron on the path of neutron decay products, which then stopped the protons, electrons and residual neutrons via the first equation [3.4] (Figure 3.5). The antineutrino flux passes through the shielding, practically without interacting, and then accesses two reservoirs containing a total of 200 L of water added to 40 kg of cadmium chloride ($CdCl_2$). The two reservoirs are surrounded by a *scintillation counter. Photomultipliers* coupled to the counter enable the conversion of light signals into electrical signals. When an emitted photon is received by the cathode of a photomultiplier, it induces the emission of an electron by photoelectric effect. This electron is then accelerated by an electric field and comes to strike the first electrode of the photomultiplier. This then induces a shower of accelerated electrons toward the second electrode and so on. A photon arriving on the cathode of a photomultiplier thus induces a brief current pulse. The pulses recorded are then enumerated by the counter.

While passing through the reservoirs, antineutrino capture reactions are produced by the water protons according to the equation:

$$\bar{\nu}_e + p \to e^+ + \text{n} \qquad [3.6]$$

The *antineutrino capture reaction* [3.6] thus produces a neutron and a positron.

Once formed, positrons interact, rapidly being annihilated with electrons of matter. This produces the emission of at least two photons in coincidence (annihilation with emission of more than two photons is extremely rare, see note below):

$$e^{+} + e^{-} \rightarrow \gamma + \gamma \qquad [3.7]$$

NOTE.– *Probability of emission of more than two photons by annihilation.*

In a very limited number of cases, the annihilation can lead to the emission of more than two photons. However, in practice, this type of marginal reaction is not taken into account since the probability of an annihilation of three photons occurring in water is estimated at less than 0.5% [NIC 10].

The introduction of cadmium (a good neutron absorber) into the water allows the detection of the neutron emitted by the antineutrino capture reaction [3.6]. Under the impact of a neutron, an atom of cadmium (Cd) is carried in an excited state, Cd*, with an average lifetime of 15 μs. There follows a γ-deexcitation process:

$$n + \text{Cd} \rightarrow \text{Cd}^{*} \rightarrow \text{Cd} + \gamma \qquad [3.8]$$

The emission of the two annihilation photons by coincidence [3.7] and the emission of the delayed photon [3.8] thus show the interaction of the antineutrino with matter (here water + cadmium). Using a scintillation counter (Figure 3.5) then enables detection of a time-delayed pulse with respect to the two pulses in coincidence resulting from the annihilation of an electron-positron pair.

The experiment of Cowan and Reines thus made it possible to highlight the electronic antineutrino, the *antiparticle* of the electronic neutrino just as the positron is the antiparticle of the electron. Since then, other experiments have proven the existence of the neutrino that appears during β^{+} disintegration, as indicated in the second equation [3.4]. In addition, the neutrino capture reaction produces positrons analogously to reaction [3.6]:

$$\nu_e + p \rightarrow e^{+} + n \qquad [3.9]$$

Since the effective interaction cross-section of a neutrino with matter is extremely small (σ = 3 × 10^{-48} m^2 = 3 × 10^{-20} barns), the matter is virtually transparent to neutrinos. On a flux of several billion neutrinos arriving on Earth, only one neutrino interacts with matter, the others crossing the Earth from either side without interacting. It should be noted that a flux of 3×10^{14} neutrinos per second and per square meter is continuously received from the Sun.

NOTE.– *In the categorization of elementary particles within the framework of the standard model, we can identify a family of particles called leptons, numbering six. Leptons are spin-1/2 fermions that are insensitive to strong interactions and devoid of internal structure. They consist of three particles of* $-e$ *charge: electron* e^- *(*m_0 *=* *0.511 MeV/*c^2*) muon* μ^- *(*m_0 *= 105.6 MeV/*c^2*) and tauon* τ^- *(*m_0 *= 1,777 MeV/*c^2*) associated with three neutrinos,* ν_e *(electronic neutrino),* ν_μ *(muonic neutrino) and* ν_τ *(tauonic neutrino), respectively. The six leptons are then grouped into three pairs: (*e^-*,* ν_e*), (*μ^-*,* ν_μ*) and (*τ^-*,* ν_τ*). For these six leptons there are six corresponding antiparticles. It is therefore important to differentiate between the neutrino accompanying* β *decay (here the electronic neutrino) and the muonic and tauonic neutrinos.*

Moreover, based on current knowledge, neutrinos have masses equal to:

– electronic neutrino, ν_e: $m_0 < 10^{-6}$ MeV/c^2;

– muonic neutrino, ν_μ: $m_0 < 0.19$ MeV/c^2;

– tauonic neutrino, ν_τ: $m_0 < 18.2$ MeV/c^2.

Therefore, only the electronic neutrino has a near-zero mass.

Clyde Lorrain Cowan Jr. and Frederick Reines were American physicists. They are known for their discovery of the neutrino, for which they shared the Nobel Prize in Physics 1995. Reines is also known for his work on the development of radiation detectors used in medicine to measure the total amount of radiation delivered to the human body during radiation treatments.

Box 3.4. *Cowan (1919–1974); Reines (1918–1998)*

3.1.5. *Highlighting α, β and γ radiation*

By studying the phenomenon of radioactivity, Marie Curie, Pierre Curie and Rutherford discovered that an electric or magnetic field can separate *uranic rays* into three separate beams made up of α, β and γ radiation.

Let us use a uniform electric field as an example to highlight α, β and γ radiations emitted by a mixture of natural and artificial radioactive substances.

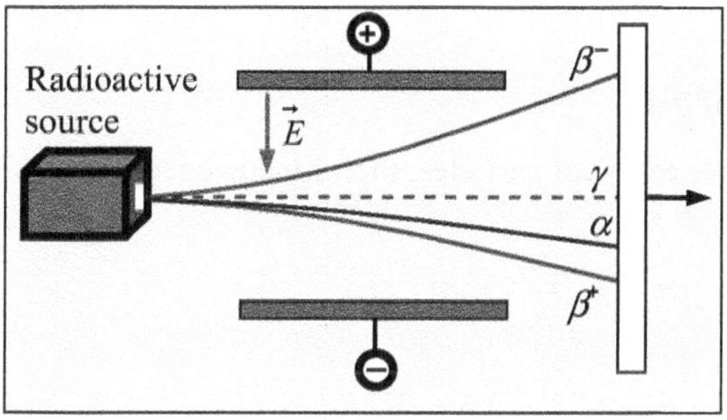

Figure 3.6. *Separation, by a uniform electric field, of α, β and γ radiations emitted by a radioactive source consisting of a mixture of natural and artificial radioactive substances*

Figure 3.6 shows the trajectories of α, β and γ particles. One can see that the β radiation is constituted of β^- and β^+ particles (emitted by artificial radioactive substances). The deviation direction is used to specify the particle's sign. The neutral photon is not deflected. Note that in the case of the magnetic field, the same deflections are observed except that the trajectories described by the α and β particles are circular arcs).

Moreover, the α particles emitted by radioactive sources are classical particles with ejection rates of the order of 20,000 km $\cdot$ s^{-1} while the β^+ and β^- particles are relativistic particles with ejection velocities of the order of 280,000 km $\cdot$ s^{-1}. The γ photon moves in a vacuum at the speed of light, 300,000 km $\cdot$ s^{-1}. It should be noted that the hazardous nature of radioactive radiations relates to the damage that they induce when passing through biological matter. Whereas α particles can be stopped by a simple sheet of paper, β particles can pass through 7 mm of aluminum, while the most hazardous, γ radiation, can pass through 20 cm of lead. These biological effects will be discussed in forthcoming volumes.

APPLICATION 3.3.– The experimental study of the decay of radium-226 reveals the emission of α particles of maximum kinetic energy 4.8 MeV. Similarly, the decay of sodium-24 is accompanied by the emission of β^- particles of maximum kinetic energy 1.39 MeV. Find the order of magnitude of the ejection velocities of α particles emitted by radioactive substances, then show that the β^- particles emitted by sodium-24 are relativistic particles.

Given data: $m_\alpha = 4.0015$ u; $m_0 = m_\beta = 0.00055$ u; 1u = 1.66×10^{-27} kg.

We will take 1 eV = 1.6×10^{-19} C and $m_0c^2 = 0.511$ MeV.

ANSWER.–

– **Case of α particles**

If α particles are classical particles, their kinetic energies verify the relationship:

$$E_c = \frac{1}{2} m_\alpha v^2 \Rightarrow v = \sqrt{\frac{2E_c}{m_\alpha}} \qquad [3.10]$$

Numerically, we find:

$$v = \sqrt{\frac{2 \times 4.8 \times 10^6 \times 1.6 \times 10^{-19}}{4.0015 \times 1.66 \times 10^{-27}}} = 1.52 \times 10^7 \text{ m} \cdot \text{s}^{-1} = 15{,}000 \text{ km} \cdot \text{s}^{-1}.$$

The order of magnitude (20,000 km · s^{-1}) of the ejection velocities of the α particles emitted by radioactive substances is indeed found.

– **Case of β particles**

The β^- particles emitted by sodium-24 are assumed to be classical particles. Numerically we obtain, using [3.10]

$$v = \sqrt{\frac{2 \times 1.39 \times 10^6 \times 1.6 \times 10^{-19}}{0.00055 \times 1.66 \times 10^{-27}}} = 6.98 \times 10^8 \text{ m} \cdot \text{s}^{-1} > c = 300{,}000 \text{ km} \cdot \text{s}^{-1}.$$

The above result is unacceptable. It follows that expression [3.10] is not verified for β^- particles. They are therefore relativistic particles whose kinetic energy is given by the relationship:

$$E_c = m_0 c^2 \left[\frac{1}{\sqrt{1 - v^2/c^2}} - 1 \right] \Rightarrow v = \sqrt{1 - \frac{1}{(E_c / m_0 c^2 + 1)^2}}\, c \qquad [3.11]$$

Numerically we find:

$$v = \sqrt{1 - \frac{1}{(1.39/0.511 + 1)^2}} \times 300\,000 \approx 289\,000 \times 300{,}000 \approx 289{,}000 \text{ km} \cdot \text{s}^{-1}$$

The order of magnitude (280,000 km · s^{-1}) of the ejection rates of the β particles is indeed found.

3.2. Radioactive decay

3.2.1. *Rutherford and Soddy's empirical law*

Let us consider a radioactive source, with λ the probability per unit time for a nucleus to decay in the time interval t and $t + dt$, and $N(t)$ the number of nuclei remaining at instant t. By definition:

$$\lambda dt = -\frac{dN}{N(t)} \quad [3.12]$$

The "–" sign in relationship [3.12] reflects the decrease in the number of nuclei over time.

If the radioactive source contains N_0 initial nuclei, the integration of [3.12] gives:

$$N(t) = N_0 e^{-\lambda t} \quad [3.13]$$

In relationship [3.13], λ is also called the *decay constant* and is expressed in seconds minus one (s^{-1}). This relationship reflects *Rutherford and Soddy's empirical law* established in 1902.

3.2.2. *Radioactive half-life*

By definition, the *radioactive half-life*, T (or (half-life period) is the time after which half of the initial nuclei, N_0, of a radioactive sample has decayed.

Thus, at $t = T$, $N(T) = N_0/2$. Using empirical law [3.13], we obtain:

$$N(T) = \frac{N_0}{2} = N_0 e^{-\lambda T} \Rightarrow \ln 2 = \lambda T \quad [3.14]$$

This then gives:

$$T = \frac{\ln 2}{\lambda} \quad [3.15]$$

In relationship [3.15], λ is expressed in s^{-1} and T in s.

Table 3.1 indicates the radioactive half-lives of several radionuclides. T varies from fractions of a second to billions of years.

Radionuclides	Half-life, T	Radionuclides	Half-life, T
${}^{204}_{82}Pb$ (α)	1.0×10^{19} years	${}^{137}_{55}Cs$ (β^-)	30.07 years
${}^{238}_{92}U$ (α)	4.5×10^{9} years	${}^{60}_{27}Co$ (β^-)	5.2 years
${}^{40}_{19}K$ (β^-)	1.3×10^{9} years	${}^{210}_{84}Po$ (α)	138 days
${}^{235}_{92}U$ (α)	7.0×10^{8} years	${}^{131}_{53}I$ (β^-)	8.0 days
${}^{239}_{94}Pu$ (α)	2.4×10^{4} years	${}^{212}_{83}Bi$ (α)	60 min
${}^{14}_{6}C$ (β^-)	5,730 years	${}^{19}_{10}Ne$ (β^+)	18 s
${}^{226}_{88}Ra$ (α)	1,622 years	${}^{212}_{84}Po$ (α)	3×10^{-7} s

Table 3.1. *Radioactive half-life of several radionuclides*

APPLICATION 3.4.– Show that at instant $t = nT$, the number N of non-decayed nuclei from a radioactive source containing initial N_0 nuclei is given by the relation:

$$N = \frac{N_0}{2^n} \quad [3.16]$$

ANSWER.– Using Rutherford and Soddy's law [3.13] and [3.15], we obtain:

$$N(nT) = N_0 e^{-\lambda nT} = N_0 e^{-n \ln 2} = N_0 2^{-n} \Rightarrow N = \frac{N_0}{2^n}$$

The graphical representation of the *exponential decay law* of the number of nuclei of a radioactive source is shown in Figure 3.7.

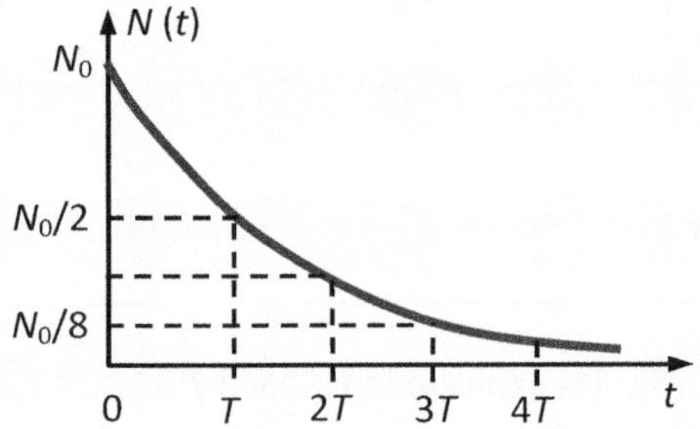

Figure 3.7. *Exponential decay of the number, N(t), of radioactive nuclei*

3.2.3. *Average lifetime of a radioactive nucleus*

The number of nuclei remaining at instant t is $N(t)$. Between instant t and instant $t + dt$, the number of decayed nuclei is equal to $-dN(t)$. Each of these nuclei lived for t seconds before decaying. The total lifetime of the $-dN(t)$ nuclei is therefore equal to $-tdN(t)$. By definition, the *average lifetime of a radioactive nucleus*, noted τ, is equal to the sum of the lifetimes of each nucleus, divided by the sum of the $-dN(t)$ nuclei. That is, given by the relationship:

$$\tau = \frac{\int_0^\infty t dN(t)}{\int_0^\infty dN(t)} \qquad [3.17]$$

Using [3.13], we obtain:

$$\tau = \frac{\int_0^\infty t\lambda N_0 e^{-\lambda t} dt}{\int_0^\infty \lambda N_0 e^{-\lambda t} dt} = \frac{\int_0^\infty t e^{-\lambda t} dt}{\int_0^\infty e^{-\lambda t} dt} \qquad [3.18]$$

The integrals in [3.18] are of the form:

$$\int_0^\infty x^n e^{-ax} dx = \frac{n!}{a^{n+1}} \qquad [3.19]$$

By writing [3.19], we obtain, according to [3.18]:

$$\tau = \frac{1!/\lambda^2}{0!/\lambda} = \frac{1}{\lambda} \qquad [3.20]$$

Taking into account the relationship [3.15], the average lifetime of a radioactive nucleus is ultimately written:

$$\tau = \frac{1}{\lambda} = \frac{T}{\ln 2} \qquad [3.21]$$

For example, for bismuth 212, with a half-life of 60 min (Table 3.1), the average lifetime is approximately 86 minutes 34 seconds. Note that, in the calculations, it is instead the radioactive half-life that is used. The average lifetime, τ, is used very little.

3.2.4. *Activity of a radioactive source*

By definition, *activity A* (*t*) of a radioactive source is the number of decay per second. Knowing that λ is the probability per unit time for a nucleus to decay in the time interval t and $t + dt$ and $N(t)$, the number of nuclei remaining at instant t, then:

$$A(t) = \lambda N(t) \qquad [3.22]$$

Taking account of [1.58] we obtain:

$$A(t) = A_0 e^{-\lambda t} \; ; \; A_0 = \lambda N_0 \qquad [3.23]$$

In [3.23], $A_0 = A(0)$ is the initial activity of the source.

Activity A is expressed in *becquerel* (Bq) with 1 Bq = 1 decay per second.

NOTE.– *The older unit of activity is the curie:* $1\ Ci = 3.7 \times 10^{10}\ Bq$.

APPLICATION 3.5.– We will consider a radioactive source of iodine-131 of half-life 8.1 days and of initial activity 2.2×10^5 Bq. Calculate the number of nuclei at $t_1 = 81$ days and then at $t_2 = 810$ days. Conclude.

ANSWER.– Note that $t_1 = 10\,T$ and $t_2 = 100\,T$

Using [2.4], [2.5], [2.7] and [2.8] we obtain:

$$N(t) = \frac{A_0 T}{\ln 2} e^{-\ln 2 (t/T)} \qquad [3.24]$$

NOTE.– $N(t_1) = 2.2 \times 10^8$; $N(t_2) = 1.7 \times 10^{-17} \approx 0$.

CONCLUSION.– The source is switched off at the instant of date $t = 810$ days.

3.3. α radioactivity

3.3.1. *Balanced equation*

During the α decay process, the parent nucleus, ${}^{A}_{Z}X$, spontaneously converts into a daughter nucleus, ${}^{A'}_{Z'}Y$, possibly with the emission of γ-photons. The *balanced equation for α disintegration* is thus written:

$$ {}_{Z}^{A}X \rightarrow {}_{Z'}^{A'}Y + {}_{2}^{4}He + \gamma + \gamma \qquad [3.25] $$

Equation [3.25] satisfies the principle of mass number and charge number conservation:

$$ \text{– } \textit{mass number conservation}: A = A' + 4 \Rightarrow A' = A - 4 \qquad [3.26a] $$

$$ \text{– charge number conservation: } Z = Z' + 2 \Rightarrow Z' = Z - 2 \qquad [3.26b] $$

By using [3.26], equation [3.25] is written as follows:

$$ {}_{Z}^{A}X \rightarrow {}_{Z-2}^{A-4}Y + {}_{2}^{4}He + \gamma \qquad [3.27] $$

Example, for the α emitter radium-228 (Table 3.1), the decay equation is written:

$$ {}_{88}^{226}Ra \rightarrow {}_{86}^{222}Rn + {}_{2}^{4}He + \gamma \qquad [3.28] $$

3.3.2. *Mass defect (loss of matter), decay energy*

By definition, the mass loss in a nuclear reaction is equal to the difference between the total mass of the nuclei before the reaction (m_b) and the total mass of the nuclei after the reaction (m_a). Using Δm to note the mass loss, we obtain:

$$ \Delta m = m_b - m_a \qquad [3.29] $$

APPLICATION 3.6.– Calculate the mass loss in the case of reaction [3.28].

Given data: m (^{226}Ra) = 225. 9771332 u; m (^{222}Rn) = 221.9703834 u; m (^{4}He) = 4.0015 u; 1 u = 931.5 MeV/c^2.

ANSWER.– The mass of the γ photon is zero. We thus obtain:

$$ \Delta m = m\,(^{226}\text{Ra}) - [m\,(^{222}\text{Rn}) + m\,(^{4}\text{He})] \qquad [3.30] $$

Numerically, we find:

$$ \Delta m = 0.0052498 \text{ u} = 4.890 \text{ MeV} \approx 4.9 \text{ MeV}/c^2 \qquad [3.31] $$

Let us now perform the energy balance of the decay reaction [3.27]. We obtain:

– Before decay (resting energy is $E_0 = m_0c^2$):

$$ E_i = E_0\,({}_{Z}^{A}X) = m_X c^2 \qquad [3.32a] $$

– After decay:

$$E_f = E_0\,({}^{A-4}_{Z-2}Y\;) + E_0\,({}^{4}_{2}He\;) + E_{c\alpha} + E_{cr}\,(\mathrm{Y}) + E_\gamma$$

That is:

$$E_f = m_{\mathrm{Y}}c^2 + m_\alpha c^2 + E_{c\alpha} + E_{cr}\,(\mathrm{Y}) + E_\gamma \qquad [3.32b]$$

Making use of the principle of energy conservation, we obtain, using [3.32]:

$$E_i = E_f \Rightarrow m_{\mathrm{X}}c^2 = m_{\mathrm{Y}}c^2 + m_\alpha c^2 + E_{c\alpha} + E_{cr}\,(\mathrm{Y}) + E_\gamma \qquad [3.33a]$$

By integrating equation [3.33a], we obtain:

$$E_{c\alpha} + E_{cr}\,(\mathrm{Y}) + E_\gamma = [m_{\mathrm{X}} - (m_{\mathrm{Y}} + m_\alpha)]c^2 \qquad [3.33b]$$

By definition, the *energy released during α decay* according to equation [3.27], also called *α decay energy*, noted Q_α, is equal to the sum of the kinetic energy of the α particle and the recoil energy of the daughter nucleus and the quantum energy of the γ photon, i.e. according to [3.33b]:

$$Q_\alpha = E_{c\alpha} + E_{cr}\,(\mathrm{Y}) + E_\gamma \qquad [3.34]$$

Taking into account [3.33b], another relationship is found, enabling the α decay energy to be calculated:

$$Q_\alpha = [m_{\mathrm{X}} - (m_{\mathrm{Y}} + m_\alpha)]c^2 \qquad [3.35]$$

The mass loss in equation [3.27] is written, according to [3.29]:

$$\Delta m\; = [m_{\mathrm{X}} - (m_{\mathrm{Y}} + m_\alpha)] \qquad [3.36]$$

The α decay energy [3.35] then verifies the relationship:

$$Q_\alpha\; = \Delta m\; c^2 \qquad [3.37]$$

We thus find the mass–energy equivalence relationship:

APPLICATION 3.7.– Calculate the energy released during the disintegration of radium-226 according to equation [3.28].

ANSWER.– Using [3.37] and [3.31], we find:

$$Q_\alpha \approx 4.9\ \mathrm{MeV} \qquad [3.38]$$

In practice, the recoil kinetic energy of the daughter nucleus is negligible in relation to the kinetic energy of the α particle. To verify this assertion, let us apply the principle of conservation of momentum to the decay process of the parent nucleus ^{A}X. We will use p_X to designate the momentum of the parent nucleus and p_Y and p_α for the respective momenta of the daughter nucleus and the α particle. With the momentum being conserved, we obtain:

$$\vec{p}_X = \vec{p}_Y + \vec{p}_\alpha = \vec{0} \Rightarrow \vec{p}_Y = -\vec{p}_\alpha \qquad [3.39]$$

Given that $E_c = p^2/2m$, we obtain, using the last relation [3.39]:

$$m_Y E_{cr}(\mathrm{Y}) = m_\alpha E_{c\alpha} \qquad [3.40]$$

By approximating $m = A$u, equation [3.40] is written, taking into account the conservation of the mass number [3.26a]:

$$A' E_{cr}(\mathrm{Y}) = 4\, E_{c\alpha} \Rightarrow \quad (A-4) E_{cr}(\mathrm{Y}) = 4\, E_{c\alpha}$$

Hence:

$$E_{cr}(Y) = \frac{4}{A-4} E_{c\alpha} \qquad [3.41]$$

Taking into account [3.41], the α decay energy [3.34] can be put in the form:

$$Q_\alpha = E_{c\alpha} + \frac{4}{A-4} E_{c\alpha} + E_\gamma = \left(1 + \frac{4}{A-4}\right) E_{c\alpha} + E_\gamma$$

That is, ultimately:

$$Q_\alpha = \frac{A}{A-4} E_{c\alpha} + E_\gamma \qquad [3.42]$$

APPLICATION 3.8.– Radium can decay into radon with or without the emission of one or more γ photons. Calculate the maximum kinetic energy of the α particles emitted during the decay of radium-226. Deduce the recoil kinetic energy of radon. Conclude.

ANSWER.– According to [3.42], the kinetic energy of the α particles verifies the equation:

$$\frac{A}{A-4}E_{c\alpha} = Q_\alpha - E_\gamma \qquad [3.43]$$

The kinetic energy of the α particles is maximum if there is no photon emission. Let:

$$E_{c\alpha} = \frac{A-4}{A}Q_\alpha \qquad [3.44]$$

Using [3.38], for radium-228 we obtain:

$$E_{c\alpha} = \frac{224}{228} \times 4.9 = 4.8 \text{ MeV} \qquad [3.45]$$

Using [3.41], the recoil kinetic energy of radon has a value of:

$$E_{cr}\,(\text{Rn}) = \frac{4}{224} \times 4.8 = 0.0857 \text{ MeV} \qquad [3.46]$$

CONCLUSION.– If high precision is not required, the recoil kinetic energy of the daughter nucleus can be overlooked.

3.3.3. *Decay energy diagram*

The *α decay energy diagram* gives an overall mapping of the decay process of an α emitting nucleus. Notably, it includes the level of the parent nucleus, the excited levels and the ground level of the daughter nucleus, the α transitions between the energy level of the parent nucleus and those of the daughter nucleus, and the γ decay transitions of the daughter nucleus.

Let us consider the simple case where a single γ photon is emitted. The α decay energy diagram is then mapped as shown in Figure 3.8.

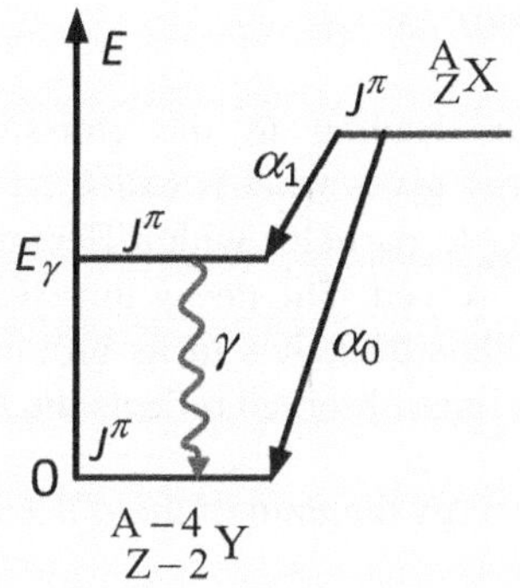

Figure 3.8. *α decay energy diagram*

Figure 3.8 shows two α transitions: a direct α_0 transition corresponding to a decay of the parent nucleus, leaving the daughter nucleus in its ground level, and a direct α_1 transition, feeding the excited level of the daughter nucleus. This level is deexcited to the ground level with emission of a γ photon of energy E_γ.

APPLICATION 3.9.– One of the α transitions resulting from the decay of bismuth-212 corresponds to α particles of kinetic energy equal to 5.481 MeV. Show that this transition is accompanied by the emission of a γ photon to the ground level of radon. The value of the transition energy is to be specified.

Given data: Energy released during decay of bismuth-212: 6.201 MeV.

ANSWER.– The energy released during decay of bismuth-212 is greater than the kinetic energy of the α particles:

$$Q_\alpha = 6.201 \text{ MeV} > E_{c\alpha} = 5.481 \text{ MeV}$$

The transition is therefore accompanied by the emission of a γ photon.

Using [2.59], we will take the energy of the γ photon emitted:

$$E_\gamma = Q_\alpha - \frac{A}{A-4} E_{c\alpha} \qquad [3.47]$$

NOTE.–

$$E_\gamma = 6.201 - \frac{212}{208} \times 5.481 = 0.615 \text{ MeV}$$

3.3.4. *Fine structure of α lines*

The decay of a nucleus leading to the emission of an α particle with well-determined kinetic energy gives what is called an α line. In the general case, several α lines corresponding to particles with different kinetic energies comprised between 4 and 9 MeV are observed. The decay thus feeds several excited levels of the daughter nucleus which thus deexcites to its ground level with an emission of several γ photons. The set of lines observed reflects the *fine structure of the α lines.*

Figure 3.9 indicates the energy diagram of the α decay of uranium-232.

We can identify the five excited levels of thorium-228 fed by the decay and six α transitions. The α_0 transition without photon emission occurs at 68%.

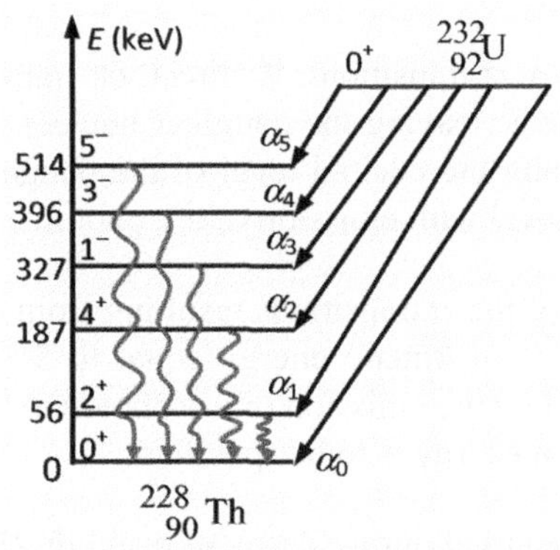

Figure 3.9. *Uranium-232 α decay energy diagram*

APPLICATION 3.10.– Using Figure 3.9, determine in tabular form the kinetic energies of the α particles corresponding to the six transitions shown. The γ photon energy emitted for each transition will be specified.

Given data:

– *mass of the uranium-232 nucleus*: 231.986685 u;

– *mass of the thorium-228 nucleus*: 227.9793671 u;

– *α particle mass*: 4.0015 u;

– $1\ \text{u} = 931.5\ \text{MeV}/c^2$.

ANSWER.– Using [3.47], the kinetic energy of an α particle is given by the relationship:

$$E_{c\alpha} = \frac{A-4}{A}(Q_\alpha - E_\gamma) \quad [3.48]$$

In the case of uranium-232, we obtain:

$$E_{c\alpha} = \frac{228}{234}(Q_\alpha - E_\gamma) \quad [3.49]$$

Considering [3.35], the α decay energy of uranium-232 is written as:

$$Q_\alpha = [m_{\text{U}} - (m_{\text{Th}} + m_\alpha)]c^2 \quad [3.50]$$

Numerically:

$$Q_\alpha = [231.986685 - (227.9793671 \; + 4.0015)]\ \text{u} = 5.41937385\ \text{MeV}$$

i.e.:

$$Q_\alpha = 5.419\ \text{MeV} \quad [3.51]$$

Using [3.48] and [3.50], we find the results gathered in Table 3.2.

α transition	**Energy of γ photons (in keV)**	**Kinetic energy $E_{c\alpha}$ of α particles (in MeV)**
α_0 transition	0	5.419
α_1 transition	56	5.225
α_2 transition	187	5.098
α_3 transition	327	4.961
α_4 transition	396	4.894
α_5 transition	514	4.779

Table 3.2. *Energies of the γ photons and α particles emitted during the decay of uranium-232 according to the decay diagram shown in Figure 3.9*

As an example, for the α_1 transition

$$E_{c\alpha} = \frac{228}{234}(5419 - 56) = 5225.49\ \text{keV} \approx 5.225\ \text{MeV}$$

3.3.5. *Geiger–Nuttall law*

Johannes Wilhelm Geiger and John Mitchell Nuttall studied the phenomenon of α radioactivity. In 1911, they showed that for even-even α emitters of the uranium family, the logarithm of the distance traveled (R) by the α particle was a linear function of the logarithm of the λ radioactive constant [CUR 32a, STÖ 07, LES 10, FOO 12, MAY 17]. The empirical relationship between R and λ, known as the Geiger–Nuttall law, is written:

$$\ln R = a + b \ln \lambda \qquad [3.52]$$

In relationship [3.52], a and b are empirical parameters.

In the air we have, approximately:

$$R\,(\text{cm}) = 0.325 E_\alpha(\text{MeV}) \qquad [3.53]$$

Knowing that the λ decay constant is linked to the radioactive half-life, T (T = ln2/λ), Geiger and Nuttall showed that the greater the kinetic energy, E_α (where the decay energy Q_α), the shorter the radioactive half-life, T, of the α emitter. Table 3.3 shows the correlation between T and E_α for several α emitters [FOO 12].

α emitter	**Radioactive half-life, *T***	**Kinetic energy $E_{c\alpha}$ of α particles (in MeV)**
^{147}Ce	5×10^{16} years	1.30
^{238}U	4.5×10^{9} years	4.20
^{239}Pu	24,110 years	5.15
^{242}Cm	162.8 days	6.21
^{212}At	0.314 s	7.83
^{212}Po	2.98×10^{-7} years	8.78

Table 3.3. *Correlation between the radioactive half-life, T, and kinetic energy, E_α, for some α emitters. In particular, it is noted that the greater the kinetic energy, the shorter the radioactive half-life of the α emitter*

For the even-even α-emitting nuclei of the uranium family, the variation law of the radioactive half-time, T, as a function of the kinetic energy, E_α, is written:

$$\ln T = a + \frac{b}{\sqrt{E_\alpha}} \qquad [3.54]$$

In relationship [3.54] another formulation of the Geiger–Nuttall law is shown. In this relationship, T is expressed in seconds, E_α in MeV, and a and b are empirical parameters dependent on the atomic number of the daughter nucleus. If Z is the atomic number of the daughter nucleus, we have:

$$a \approx -3.7 \times Z^{2/3} - 49.3; b = 3.7\,Z \qquad [3.55]$$

Using [3.51], law [3.54] is placed in the form:

$$\ln T = -3.7\,Z^{2/3} - 49.3 + \frac{3.7\,Z}{\sqrt{E_\alpha}} \qquad [3.56a]$$

For isotopes of a given α emitter, the curve $\ln T = f\,(E_\alpha^{1/2})$ is practically linear (Figure 3.10) in accordance with the Geiger–Nuttall law [3.56].

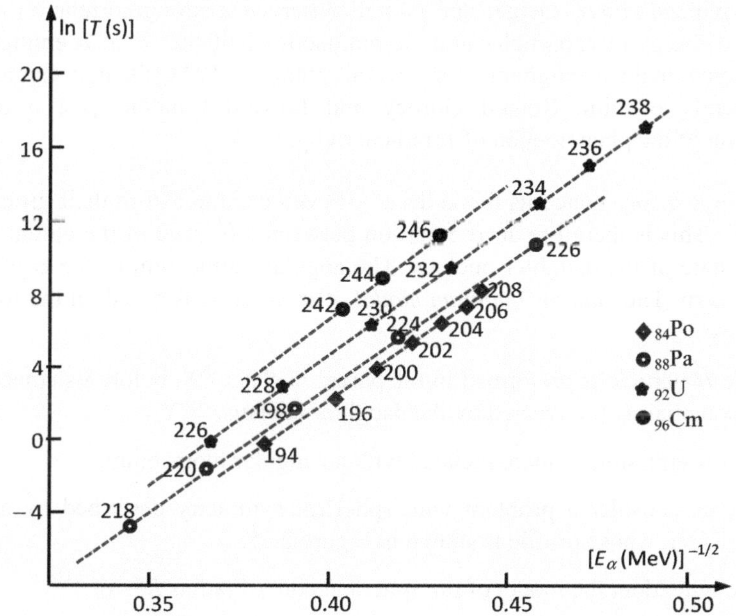

Figure 3.10. *Variation of curve* $LnT = f\,(E_\alpha^{1/2})$ *for several* α $_ZX$ *emitters (polonium* $_{84}Po$*, protactinium* $_{88}Pa$*, uranium* $_{92}U$ *and cerium* $_{96}Cm$*). The shapes obtained comply with the Geiger–Nuttall law*

As an illustrative example, let us numerically examine the validity of the Geiger–Nuttall law [3.56] in the case of uranium-238 decay. The daughter nucleus obtained is thorium with atomic number $Z = 90$. From Table 3.3 we will take, for uranium-238: $T = 4.5 \times 10^9$ years. Taking 1 year = 365.25 days, we find:

$$\ln T = 39.495 \approx 39 \quad [3.56b]$$

Moreover, using [3.56] we obtain:

$$\ln T = -\,3.7 \times 90^{2/3} - 49.3 + \frac{3.7 \times 90}{\sqrt{4.2}} = 38.880 \approx 39 \quad [3.57]$$

There is excellent agreement between results [3.56] and [3.57].

3.3.6. *Quantum model of α emission by tunnel effect*

As explained above, Geiger and Nuttall observed a strong correlation between the kinetic energy of α particles and the radioactive half-life, T, of α emitters. This observation remained enigmatic for several years. In 1928, George Gamow and, independently of him, Ronald Gurney and Edward Condon gave a quantum explanation of the phenomenon of α radioactivity.

Gamow's theory concerns the α decay of even-even nuclei in their ground state of $J^{\pi} = 0^+$. This is therefore an α emission between a 0^+ state of the parent nucleus and a 0^+ state of the daughter nucleus. The angular momentum of the α particle is therefore zero. The *Gamow quantum model of α emission* is based on the following hypotheses:

1) The α particle is preformed in the parent nucleus, ^{A}X, before its emission and lies in a potential, $U(r)$, created by the daughter nucleus, ^{A-4}Y.

2) The α emission is not associated with an angular momentum.

3) Let us consider a problem with spherical symmetry described by a central potential, $U(r)$, whose profile is shown in Figure 2.25.

4) We will adopt the model of the spherical core of radius $R = r_0 A^{1/3}$.

To model the daughter nucleus α particle interaction, the α particle is considered trapped in the potential of the daughter nucleus and is retained by attractive nuclear forces. By overlooking the skin effect of the daughter nucleus, the distance, a, between the centers of inertia of the daughter nucleus, ^{A-4}Y, and the α particle (He) in contact is defined by the relationship:

$$a = R\,(^{A-4}Y) + R\,(He) = r_0\,[(A-4)^{1/3} + 4^{1/3}] \quad [3.58]$$

In relationship [3.8], $r_0 = 1.1$ fm.

When the daughter nucleus ${}^{A-4}$Y and the α particle are in contact, the height of the nuclear potential barrier is equal to, according to [2.207]:

$$B = k\frac{2(Z-2)e^2}{a} \qquad [3.59]$$

Let us consider, for illustrative purposes, the specific case of the decay of radium-226 ($Z = 88$). The distance, a, between the centers of inertia of the daughter nucleus and the α particle in contact has the numerical value:

$$a = 1.1 \times 10^{-15} \times [(226-4)^{1/3} + 4^{1/3}] = 8.41 \text{ fm} \qquad [3.60]$$

The height of the nuclear potential barrier [3.59] then has the numerical value:

$$B = 9 \times 10^9 \frac{2(88-2)\times 1.602\times 10^{-19}}{8.41\times 10^{-15}} = 29.49 \text{ MeV} \qquad [3.61]$$

To escape from the daughter nucleus, *Rn*, the α particle must have a kinetic energy greater than 29.49 MeV. However, the kinetic energy of the α particle is equal to 4.8 MeV [3.38]. It follows that the α particle remains trapped in the potential of the daughter nucleus.

From a quantum point of view, the state of the α particle in the potential of the daughter nucleus is described by a wave function. Thus, in the *Gamow–Gurney–Condon interpretation*, the α particle is retained by the Coulomb barrier, striking the latter's walls approximately 10^{21} times per second. The wave describing the state of the α particle in the potential of the daughter nucleus is then a stationary wave (Figure 3.11). Then, after 10^{25} to 10^{44} impacts against this barrier, the α particle passes through the barrier by the tunnel effect. The wave describing the state of the α particle passing through the barrier is called a *fading wave*, which decreases exponentially (Figure 3.11).

In the general case, to describe the mechanism of tunnel effect emission, the potential of the daughter nucleus is modeled as shown in Figure 3.11.

For $r < a$, the potential is attractive and is equal to $-U_0$: the attractive forces are nuclear. For $r > a$, the potential is repulsive and at

$$U(r) = k\frac{2(Z-2)e^2}{r} \qquad [3.62]$$

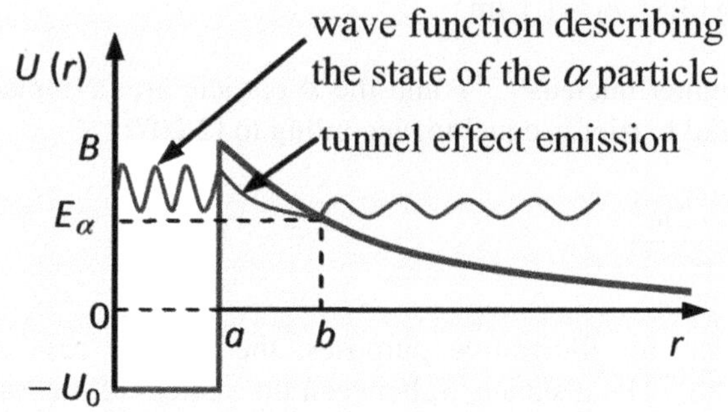

Figure 3.11. *Modeling of α emission by tunnel effect. The barrier width is equal to (b – a), b >> a (not to scale)*

The value of the radius, b, is defined by the condition $U(b) = E_\alpha$. That is, using [3.62]:

$$E_\alpha = k\frac{2(Z-2)e^2}{b} \Rightarrow b = k\frac{2(Z-2)e^2}{E_\alpha} \qquad [3.63]$$

Numerically, the specific case of radium-226 decay ($E_\alpha \approx 4.8$ MeV) gives:

$$b = 9\times 10^9\,\frac{2(88-2)\times 1.602\times 10^{-19}}{4.8\times 10^6} = 51.66 \text{ fm} \qquad [3.64]$$

Comparing [3.60] and [3.64] gives a ratio $b/a = 51.66/8.41 = 6.14$. The width of the barrier $(b - a) \approx 5a$ (in Figure 3.11, this width is not respected).

3.3.7. *Estimating the radioactive half-life, Gamow factor*

To estimate the radioactive half-life, as a first approximation it is considered that the λ decay constant is the product of the frequency, f, at which the α particle strikes the barrier by the probability, P, with which it crosses the barrier by tunnel effect. That is:

$$\lambda = f \times P \qquad [3.65]$$

To estimate the frequency, f, let us consider an α particle of kinetic energy $E_{c\alpha} = E_\alpha$ passing across the nucleus of radius R. We thus obtain (one can note that for a uniform circular motion, the frequency $f = v/2\pi r$):

$$f = \frac{v_\alpha}{2R};\, v_\alpha = \sqrt{\frac{2E_\alpha}{m_\alpha}}\ \text{fm} \qquad [3.66]$$

Furthermore, the tunnel effect decreases exponentially, as indicated by the fading wave shape in the barrier shown in Figure 3.10. Within the framework of the *Gamow–Gurney–Condon*, the probability, P, is given by the relationship:

$$P = e^{-G} \qquad [3.67]$$

In relationship [3.67], G is called "*Gamow factor*", equal to:

$$G = \frac{2}{\hbar}\int_a^b \sqrt{2m_\alpha\left(U(r) - E_\alpha\right)}\,dr \qquad [3.68]$$

It now remains to determine the Gamow factor, G, by integrating equation [3.68]. The calculations are long and drawn out. We will limit ourselves to giving the expression of G, allowing the half-life time, T, to be estimated using the Gamow formula. We then find, approximately, for a heavy nucleus of radius R [LES 10, FOO 12]:

$$\begin{cases} f \approx 3.47\times 10^{21}\times \dfrac{\sqrt{E_\alpha}}{R} \\ P \approx \exp\left(2.97\sqrt{Za} - \dfrac{3.96Z}{\sqrt{E_\alpha}}\right) \end{cases} \qquad [3.69]$$

In approximated expressions [3.69], Z is the atomic number of the daughter nucleus, E_α is expressed in MeV and a and R are expressed in Fermi. Generally, for an α emitting nucleus of mass number A, $a \approx 1.27 \times A^{1/3}$ and $R \approx 1.1 \times A^{1/3}$. Expressions [3.69] are then written in the form:

$$\begin{cases} f \approx 3.47\times 10^{21}\times \dfrac{\sqrt{E_\alpha}}{1.1\times A^{1/3}} \\ P \approx \exp\left(2.97\sqrt{Z\times 1.27\times A^{1/3}} - \dfrac{3.96Z}{\sqrt{E_\alpha}}\right) \end{cases} \qquad [3.70]$$

By comparing the expression of probability [3.70] and relationship [3.67], we obtain:

$$G \approx \left(\frac{3.96Z}{\sqrt{E_\alpha}} - 2.97\sqrt{Z \times 1.27 \times A^{1/3}} \right) \qquad [3.71]$$

For example, let us estimate the decay half-life of radium-226 using [3.65] and then compare the result with the experimental value, $T = 1{,}600$ years.

For radium-226, the precise value of the kinetic energy, $E_\alpha = 4.87$ MeV [LES 10, FOO 12]. The daughter nucleus is the radon with atomic number $Z = 86$. Using [3.73], we obtain:

$$G \approx \left(\frac{3.96 \times 86}{\sqrt{4.87}} - 2.97\sqrt{86 \times 1.27 \times 226^{1/3}} \right) = 77.72 \qquad [3.72]$$

The numerical value of the probability, P, is then, according to [3.67]:

$$P = e^{-77.72} = 1.76 \times 10^{-34} \qquad [3.73]$$

Using the first relationship [3.70], the frequency, f, has the numerical value:

$$f \approx 3.47 \times 10^{21} \times \frac{\sqrt{4.87}}{1.1 \times 226^{1/3}} = 1.14 \times 10^{21}\ \text{s}^{-1} \qquad [3.74]$$

Result [3.74] shows that in the Gamow quantum model, the α particle strikes the walls of the Coulomb barrier approximately 10^{21} times per second. The λ decay constant [3.6] is then equal to:

$$\lambda = 1.76 \times 10^{-34} \times 1.14 \times 10^{21} = 2.0 \times 10^{-13}\ \text{s}^{-1} \qquad [3.75]$$

Result [3.75] is already very different from the experimental value, 1.37×10^{-11} s^{-1} [LES 10]. This is due to the different approximations adopted to determine the frequency, f, at which the α particle strikes the barrier and the probability, P, with which it crosses the barrier by tunnel effect.

Using [3.75], the radioactive half-life of radium-226 is then:

$$T = \ln 2/\lambda = 3.47 \times 10^{13}\ \text{s} = 109{,}822 \text{ years} \qquad [3.76]$$

The value [3.76] of the decay half-life estimated based on the Gamow theory is about 69 times greater than the experimental value, $T = 1{,}600$ years. The estimation of the radioactive half-life by the Gamow model is therefore poor.

George Gamow was an American-Russian physicist and cosmologist. In nuclear physics, in 1928 he developed, independently of British physicist Ronald Wilfred (or Wilfrid) Gurney and American physicist Edward Uhler Condon, the quantum theory of α emission by tunnel effect, thus allowing the radioactive half-life of an α emitter to be estimated. This theory helped in the understanding of Geiger–Nuttall's observation of 1911. In addition, Gamow established, together with American-Hungarian physicist Edward Teller (Box 4.2), what are known as the Gamow–Teller selection rules, governing permitted and prohibited transitions β (see section 3.4.4).

Box 3.5. *Gamow (1906–1968); Gurney (1898–1953); Condon (1902–1974)*

NOTE.– *Cluster emission.*

Exotic radioactivity is a process of decay of atomic nuclei with *emission of clusters* such as ^{12}C and ^{14}C. We can cite the case of uranium-238, whose decay can occur with emission of ^{12}C and ^{14}C according to the equations:

$$^{238}_{92}U \rightarrow {}^{226}_{86}Rn + {}^{12}_{6}C \qquad [3.77]$$

$$^{238}_{92}U \rightarrow {}^{224}_{86}Rn + {}^{14}_{6}C \qquad [3.78]$$

Generally speaking, the decay of atomic nuclei with *cluster emission* is highly unlikely for at least the following two reasons.

1) Let m be the mass of the particle (α or clusters) emitted by tunnel effect. Considering equation [3.68], we obtain:

$$G = \rho\sqrt{m} \qquad [3.79]$$

with:

$$\rho = \frac{2}{\hbar}\int_a^b \sqrt{2\left(U(r) - E_\alpha\right)}\,dr$$

The probability, P, of tunnel effect emission is then written, according to [3.67]:

$$P = e^{-\rho\sqrt{m}} \qquad [3.80]$$

Equation [3.80] shows that the probability P of tunnel effect emission depends on the mass of the emitted particle and decreases sharply with it. For the α particle, this probability of the order of 10^{-34} [3.73] is already very low. It follows that for heavier clusters, the probability of tunnel effect emission is extremely low.

2) As shown in equation [3.59], the height of the Coulomb barrier increases with the atomic number, z, of the emitted particle ($z = 2$ for the α particle):

$$B = k\frac{z(Z-2)e^2}{a} \quad [3.81]$$

Radius b is then written, according to [3.63]:

$$b = k\frac{z(Z-2)e^2}{E_\alpha} \quad [3.82]$$

Thus, as the atomic number z of the emitted particle increases, the height of the potential barrier [3.81] and the width ($b - a$) of the barrier increase. Since the probability, P, decreases strongly with the mass of the emitted particle, the result is that the fading wave (Figure 3.11) disappears before the particle crosses the barrier by tunnel effect.

In conclusion, exotic radioactivity with cluster emission is highly improbable.

3.4. Exercises

EXERCISE 3.1.– Experimental determination of the half-life of an α emitter.

In this exercise we propose to determine the half-life of an α emitter experimentally. Experimental measurements are made using a Geiger–Müller counter (its principle is explained in Chapter 4, section 4.1.1).

We have a sample of mass m_0 containing N_0 α radioactive nuclei. Measurements made using a Geiger–Müller counter resulted in Table 3.4 being drawn up.

t (h)	0	3	4	5	7	9	11	15	20
$N(t)/N_0$	1	0.740	0.680	0.612	0.506	0.414	0.343	0.230	0.140

Table 3.4. *Variation as a function of time in the ratio of the number, N(t), of non-disintegrated nuclei to the initial number, N_0, of nuclei of an α emitter*

a) Propose a methodology for determining the probability per unit time of decay of the radionuclide studied.

b) Making use of the experimental data presented in Table 3.4, determine the probability per unit time of decay of the radionuclide studied. Deduce its half-life from this.

c) Deduce, from the plotted experimental curve, the activity of the α emitter studied at date $t = 14$ h if we initially have 2 moles of nuclei.

Given datum: $N_A = 6.02 \times 10^{23}$ mol^{-1}.

EXERCISE 3.2.– Decay of an α emitting parent nucleus.

Let us study the decay of a $^{A}_{Z}X$ nucleus α emitter to be identified.

Given data:

– *nuclear masses of nuclides X*;

Nuclide	$^{4}_{2}$He	$^{229}_{90}$Th	$^{233}_{92}$U	$^{235}_{92}$U
M_X **(u)**	4.002603	229.031762	233.039635	235.0439299

Table 3.5. *Atomic masses of several nuclides*

– *electron mass*: 0.00055 u;

– *atomic mass unit*: 1 u = 931.5 MeV/c^2;

– *J of the ground state*; ^{233}U: $5/2^+$; ^{229}Th: $5/2^+$.

a) Identify the α emitter simply by reading Table 3.5, considering that thorium-229 is the daughter nucleus. The conservation law used will be specified.

b) Calculate, in MeV, the energy released during the decay of the α emitter.

c) Calculate kinetic energies $E_{c\alpha}$ of the α particle and $E_{c\text{Th}}$ of thorium, assuming the α emitter to be initially immobile.

d) Analysis of the nuclear radiation reveals the existence of a γ radiation and that the spectrum of emitted α particles has a fine structure. Of particular note are two groups of α particles of kinetic energies, $E_{c\alpha 1} = 4.1$ MeV and $E_{c\alpha 2} = 4.7$ MeV.

e) Is a $5/2^+ \rightarrow 5/2^+$ nuclear transition observed? Substantiate the answer.

f) Then map out the energy diagram of the α decay studied. It will include the various nuclear transitions observed and, in MeV, the kinetic energies measured as well as the energies of the nuclear levels of the parent nucleus and the daughter nucleus.

g) Determine the energies of the gamma photons emitted.

EXERCISE 3.3.– Nuclear power produced by the decay of plutonium-239.

Plutonium, with atomic number 94, is α radioactive. Little of it exists in its natural state. It does, however, form in the core of nuclear reactors by a chain reaction [4.99] when a uranium-238 nucleus captures a neutron. We proposed to determine the nuclear power produced by the decay of one kilogram of plutonium-239.

Given data:

– *atomic masses*: $M_\alpha = 4.0026032$ u; $M_U = 235.0439299$ u;

– $M_{Pu} = 239.0521634$ u;

– radioactive half-life of plutonium-239: T = 24,410 years.

We will only consider the nuclear transition between the fundamental levels of plutonium and the daughter nucleus.

a) Calculate the kinetic energies of the particles produced by the decay of plutonium-239.

b) Let N_0 be the initial number contained in a sample of 1 kg of plutonium-239. Show that the activity, $A(t)$, of ^{239}Pu is almost equal to its initial activity, A_0.

c) Then express the nuclear power produced by the decay of ^{239}Pu as a function of N_0, M_{Pu}, M_u, M_α and T. Perform the numerical application.

3.5. Solutions to exercises

SOLUTION 3.1.– Experimental determination of the half-life of an α emitter.

a) *Methodology*

The exponential decay law of the number, $N(t)$, of radioactive nuclei is written:

$$N(t) = N_0 e^{-\lambda t} \Rightarrow -\ln [N(t)/N_0] = \lambda t \qquad [3.83]$$

According to [3.83], the curve $-\ln [N(t)/N_0] = f(t)$ is a slope line, λ. By plotting this curve, the probability per unit time of λ decay is determined experimentally.

b) *Probability per unit time of decay, half-life*

Considering Table 3.4, we obtain:

t **(h)**	0	3	4	5	7	9	11	15	20
$-\ln [N(t)/N_0]$	0	0.30	0.39	0.49	0.68	0.88	1.07	1.47	1.97

Table 3.6. *Variation as a function of time of the opposite of the Napierian logarithm in the ratio of the number, N(t), of non-disintegrated nuclei to the initial number, N_0, of nuclei of an α emitter*

The resulting curve is shown in Figure 3.12.

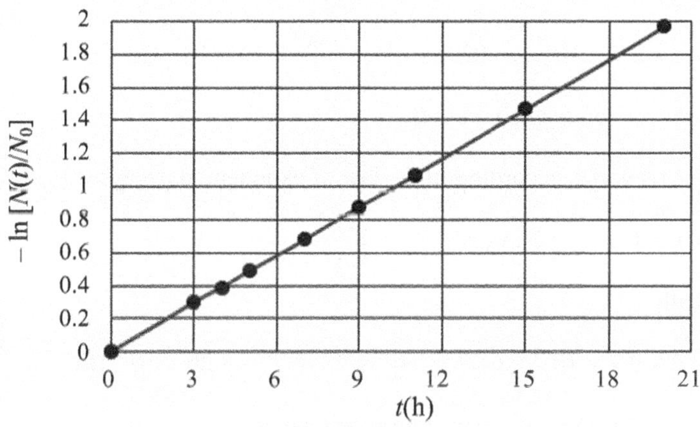

Figure 3.12. *Curve of variation as a function of time of the opposite of the Napierian logarithm in the ratio of the number, N(t), of non-disintegrated nuclei to the initial number, N_0, of nuclei of an α emitter*

Let us write $y = -\ln [N(t)/N_0]$. Graphically, the slope a of the line given by the last relationship [3.83] is equal to:

$$a = \frac{\Delta y}{\Delta t} = \frac{1.6 - 0.8}{16.4 - 8.2} = 0.09756\,\text{h}^{-1} \qquad [3.84]$$

The probability per unit time of decay and the half-life of the α emitter studied are equal to:

$$\lambda = 0.09756 \text{ h}^{-1} = 2.7 \times 10^{-5} \text{ s}^{-1};\ T = \ln 2/\lambda = 7.1 \text{ h}.$$

c) *Activity*

– *Usage of the experimental curve*

Let N_{14} be the number of nuclei remaining at instant $t = 14$ hrs. Graphically:

$$-\ln [N_{14}/N_0] = 1.39 \Rightarrow N_{14} = N_0 e^{-1.39} \qquad [3.85]$$

The activity, A_{14}, of the α emitter studied is then equal to:

$$A_{14} = \lambda N_{14} = \lambda N_0 e^{-1.39} = 2\lambda N_A e^{-1.39} \qquad [3.86]$$

Numerically:

$$A_{14} = 2 \times 2.7 \times 10^{-5} \times 6.02 \times 10^{23} \times e^{-1.39} = 8.10 \times 10^{19} \text{ Bq} \qquad [3.87]$$

– *Using the law of decay of activity*

Activity $A(t)$ varies according to the law of exponential decay:

$$-A(t) = A_0\, e^{-\lambda t} = 2\lambda N_A\, e^{-\lambda t} \qquad [3.88]$$

Numerically:

$$A_{14} = 2 \times 2.7 \times 10^{-5} \times 6.02 \times 10^{23}\, e^{-0.09756 \times 14} = 8.29 \times 10^{18} \text{ Bq}$$

SOLUTION 3.2.– Decay of an α emitting parent nucleus

a) *Identification of the α emitter*

Thorium-229 is the daughter nucleus. Under the law of conservation of mass number, we obtain, for the parent nucleus: $A = 229 + 4 = 233$: the parent nucleus is uranium-233.

b) *Calculation of released energy*

The energy released during α decay of uranium-233 is given by the relationship:

$$Q_\alpha = \Delta mc^2 = [m_U - (m_{Th} + m_\alpha)]c^2 \qquad [3.89a]$$

By subtracting the mass of Z electrons from the atomic masses recorded in Table 3.5, we obtain:

$$Q_\alpha = [232.9889 - (228.9821 + 4.0015)] \times 931.5 = 4.937 \text{ MeV} \quad [3.89b]$$

c) *Calculation of kinetic energies*

By definition:

$$Q_\alpha = E_{c\alpha} + E_{cTh} + E_\gamma \quad [3.90a]$$

For a direct transition from the ground state of ^{233}U to the ground state of ^{229}Th, we obtain:

$$Q_\alpha = E_{c\alpha} + E_{cTh} + E_\gamma \quad [3.90b]$$

Under the momentum conservation law, we obtain:

$$E_{crY} = \frac{m_\alpha}{m_{Th}} E_{c\alpha} \quad [3.90c]$$

Using [3.90a] and [3.90c], after arrangement we find:

$$E_{c\alpha} = \frac{m_{Th}}{m_{Th} + m_\alpha} Q_\alpha \Rightarrow E_{c\alpha} - 4.852 \text{ MeV} \quad [3.91]$$

For thorium, we obtain, according to [3.90c]:

$$E_{cTh} = 0.085 \text{ MeV} \quad [3.92]$$

d) *Analysis of nuclear radiation*

$5/2^+ \rightarrow 5/2^+$ *nuclear transition*

$5/2^+ \rightarrow 5/2^+$ transition occurs without γ photon emission. In this case, the particles resulting from the decay of uranium-233 have a kinetic energy equal to 4.852 MeV [3.91]. However, the spectral analysis reveals the presence of only two groups of α particles of kinetic energies:

$$E_{c\alpha 1} = 4.1 \text{ MeV}; E_{c\alpha 2} = 4.7 \text{ MeV} \quad [3.93]$$

The $5/2^+ \rightarrow 5/2^+$ nuclear transition is therefore not observed.

e) *Energies of the gamma photons emitted*

For a group of particles (i) of kinetic energy $E_{c\alpha I}$, we obtain, according to [3.90b]:

$$E_{\gamma i} = Q_\alpha - (E_{c\alpha i} + E_{cThi}) \quad [3.94]$$

Using [3.90c], we obtain:

$$E_{cTh1} = (4.0015/228.9821) \times 4.1 = 0.072 \text{ MeV} \quad [3.95a]$$

$$E_{cTh1} = (4.0015/228.9821) \times 4.7 = 0.082 \text{ MeV} \quad [3.95b]$$

Considering [3.89], [3.93] and [3.94], the gamma photon energy is equal to:

$$E_{\gamma 1} = 4.937 - (4.1 + 0.072) = 0.765 \text{ MeV} \quad [3.96a]$$

$$E_{\gamma 2} = 4.937 - (4.7 + 0.082) = 0.155 \text{ MeV} \quad [3.96b]$$

f) *Mapping of the decay energy diagram*

The mapping of the uranium-233 α decay energy diagram is shown in Figure 3.13.

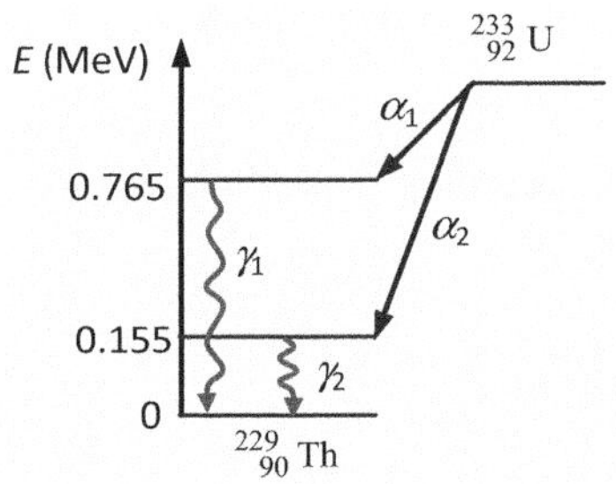

Figure 3.13. *Uranium-233 α decay energy diagram*

SOLUTION 3.3.– Nuclear power produced by the decay of ^{239}Pu.

a) *Calculation of kinetic energies*

The particles produced by the decay of plutonium-239 are the α particles and the daughter nucleus. The transition occurs without photon emission. The decay energy,

Q_α, is then equal to (the mass of Z electrons will be subtracted from the atomic masses used):

$$Q_\alpha = [m_{\mathrm{Pu}} - (m_{\mathrm{U}} + m_\alpha)]c^2 = 5.248 \text{ MeV} \quad [3.97]$$

Moreover, according to [3.90c] and [3.91]:

$$\begin{cases} E_{c\alpha} = \dfrac{m_U}{m_U + m_\alpha} Q_\alpha \\ E_{cU} = \dfrac{m_\alpha}{m_U + m_\alpha} Q_\alpha \end{cases} \quad [3.98]$$

Using [3.97], we obtain, according to [3.98]:

$$E_{c\alpha} = (234.9933299/238.9948299) \times 5.248 = 5.160 \text{ MeV} \quad [3.99a]$$

$$E_{cU} = (4.0015/238.9948299) \times 5.248 = 0.088 \text{ MeV} \quad [3.99b]$$

b) *Activity of plutonium*-239

The radioactive half-time, $T = 24{,}410$ years ($= 7.7 \times 10^{11}$ s) of plutonium-239 is long enough to consider its activity $A(t)$ virtually constant and equal to its initial activity, A_0.

c) *Expression of nuclear power*

The energy released by the decay of a ^{239}Pu nucleus is given by the relationship [3.97]. A mass m of plutonium contains $N_0 = (m \times N_A)/M$ nuclei. The decay number per second is equal to $A_0 = \lambda N_0$. Each decay produces an energy Q_α. The nuclear power produced by the decay of a mass, m, of ^{239}Pu is therefore equal to (M: molar mass in g · mol^{-1}, therefore the mass, m, is expressed in g, i.e. $m = 1{,}000$ g):

$$P = A_0 Q_\alpha = \lambda N_0 Q_\alpha = \ln 2 \frac{Q_\alpha}{T} \times \frac{m \times N_A}{M} \quad [3.100]$$

NOTE.–

$$P = \frac{0.693 \times 5.248 \times 10^6 \times 1.602 \times 10^{-19}}{7.7 \times 10^{11}} \times \frac{1000 \times 6.02 \times 10^{23}}{239.0521634} = 1.91 \text{ W}$$

4

Beta Radioactivity, Radioactive Family Tree

Overall objective	
To know the general properties of β^- and β^+ radioactivities, electron capture and radioactive family trees	
Specific objectives	
To know the experimental facts that led to the discovery of the positron	To define the β decay energy
To know the experimental facts that led to the discovery of artificial radioactivity	To explain the difference between the spectra of β^- and β^+ emissions
To describe the experiment of Frédéric and Irène Joliot-Curie that led to the discovery of artificial radioactivity	To determine the maximum energy, E_{max}, of the β spectrum
To describe the experiment of Frédéric and Irène Joliot-Curie that led to the identification of radiophosphorous-30	To know Sargent's law
To describe the Geiger-Müller counter	To interpret the Sargent diagram
To know the principle of a Geiger-Müller counter	To distinguish between the Gamow-Teller rules and the Fermi rules governing β transitions
To know the origin of the Townsend avalanche	To define the β decay energy diagram
To differentiate between the β^- and β^+ decay equations	To know the condition of the β^- emission
To make the link between the Segrè diagram and β^- or β^+	To know the condition of the β^+ emission

For a color version of all of the figures in this chapter, see www.iste.co.uk/sakho/nuclear1.zip.

To define electron capture	To use the Bateman equations to establish the law of accumulation of a daughter product
To define double β decay [$\beta\beta(2\nu)$]	To define a radioactive or secular equilibrium or a state equilibrium
To know the correspondence between Siegbahn notation and that of the IUPAC	To know the experimental facts that led to the conducting of the nuclear transmutation reaction
To use the classification of energy levels and X-rays in Siegbahn notation	To write the general equation for nuclear transmutation, $X(a, b)\,Y$
To express the energy of the X-photon emitted from a given electronic shell	To define the production rate of a radionuclide
To illustrate Auger deexcitation schematically	To define the production yield of a radionuclide
To schematically illustrate X-deexcitation	To know the unit of the yield of a nuclear reaction
To define a radioactive series or family tree	To make the analogy between the time constant of a circuit (R, C) and the "nuclear time constant"
To write the decay chain of a given family tree	To know the four radioactive families
To define the simple family tree	To know the three natural radioactive families
To establish the law of accumulation of the daughter product	To know the characteristics relative to each radioactive family
To define the Bateman equations	
Prerequisites	
Radioactive decay law	Angular momenta and parity of a system
Angular momentum and parity conservation laws	Nuclear deexcitation
Properties of a relativistic particle	

4.1. Beta radioactivity

4.1.1. *Experiment of Frédéric and Irène Joliot-Curie: discovery of artificial radioactivity*

As stated in section 3.1.3, the β-radioactivity mode consists of the β^- decay mode (negative electron) and the β^+ decay mode (positive electron or positron). The *experiment of Frédéric and Irène Joliot-Curie* carried out in 1934 allowed the discovery of artificial radioactivity [PER 82, BIM 06, SAK 16]. This discovery made it possible to highlight β^+ radiation, similar to β^- radiation.

In studying the effects of α radiation on matter, in 1934, Frédéric and Irène Joliot-Curie found that an inactive boron, aluminum or magnesium foil becomes radioactive when placed in front of a polonium source emitting α particles. Moving away from the polonium source, they noted that the radioactivity from the activated foil persisted with a β^+-positron emission. The number of β^+ particles emitted decreased exponentially with time when the source was far away. Each of the foils of activated B, Al, or Mg had a different radioactive half-life. This proves that the observed radioactive decays are not due to a nucleus common to B, Al and Mg. In the case of activated aluminum, for example, the Joliots implemented chemical processes to identify the new β^+ emitting element. The experimental set-up used by the Joliot couple is schematically presented in Figure 4.1a.

The set-up consists of a chamber filled with gaseous carbon dioxide in which a source of polonium-210 (^{210}Po) emitting α particles is placed. A valve (not shown) connects the chamber to the gas pump. For sufficient gas pressure, the α particles can be stopped before reaching the target consisting of an aluminum foil, Al. The detection of particles that may be emitted following irradiation of the aluminum foil by α particles is carried out by means of a Geiger-Müller counter coupled to an amplifier and a numerator.

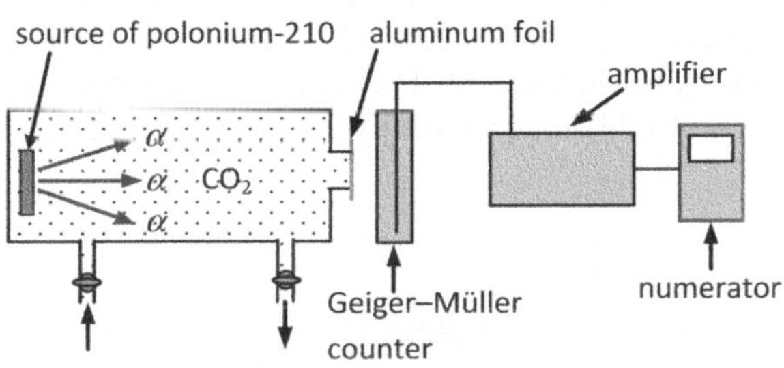

Figure 4.1a. *Experimental set-up of Frédéric and Irène Joliot-Curie that led to the discovery of artificial radioactivity*

On January 11, 1934, Frédéric Joliot, alone in the nuclear physics laboratory of Orsay, powered up the Geiger-Müller counter. In this state, the counter only detected background noise due to cosmic radiation and ambient radioactivity. Frédéric then operated the valve so as to empty the chamber. This allowed the α particles to bombard the aluminum foil. He then noted that the counter recorded an increase in the number of pulses. He thus determined the number of strokes per minute corresponding to the maximum energy of the α particles. Subsequently, he filled the chamber with carbon dioxide under sufficient pressure to stop the α particles. He then observed that the counter continued to record pulses and that the number of strokes per minute determined using the numerator decreased according to an exponential decay law. To interpret this surprising observation, Frédéric Joliot hypothesized that the aluminum foil had become radioactive under irradiation by α particles: the aluminum was activated.

Intuitively, Frédéric Joliot simplified the experimental set-up by placing the source of polonium directly in contact with the aluminum foil for several minutes. He then placed the activated sheet in contact with the counter. Minute by minute, Frédéric noted, with the help of the numerator, the radioactivity decay that he had just highlighted. This decay follows the empirical law of Rutherford and Soddy with a half-life of 3 min 15 sec: he had just discovered artificial radioactivity.

Back in 1933, the Joliots had already used an intense source of polonium to bombard an aluminum foil. They identified the particles from aluminum irradiated to be β^+ particles thanks to the curvature of their traces through a Wilson chamber. Thus, Frédéric knew that the pulses recorded by the Geiger-Müller counter were due to the β^+ particles passing through the detector. But an enigma remained following Frédéric's experimental observations of 1934: what was the chemical nature of the radioelement responsible for the observed radioactivity?

To identify the radionuclide responsible for β^+ emission, Frédéric and Irène Joliot-Curie developed a chemical process. They then placed the activated aluminum foil in a solution of hydrochloric acid (H_3O^+ Cl^-). The chemical reaction occurring led to the formation of dihydrogen mixed with traces of hydrogen phosphide, H_3P (Figure 4.1b).

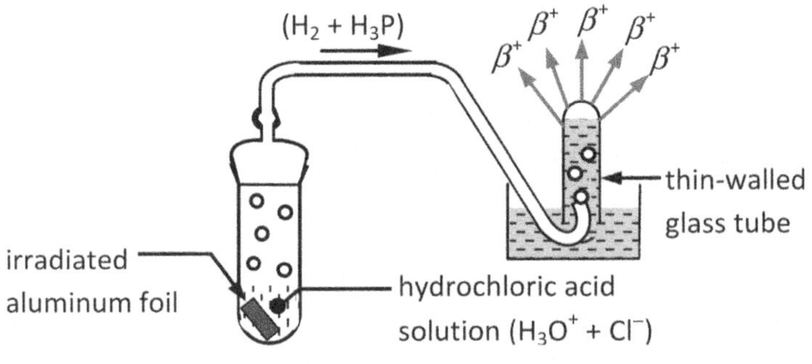

Figure 4.1b. *Experimental set-up of Frédéric and Irène Joliot-Curie that led to the identification of radiophosphorous-30 by chemical means*

The bombardment of aluminum by α particles therefore induced *transmutation of aluminum* into phosphorus according to the equation:

$$ {}^{4}_{2}He + {}^{27}_{13}Al \rightarrow {}^{1}_{0}n + {}^{30}_{15}P \qquad [4.1]$$

The phosphorus-30 thus created is β^{+} radioactive and decays according to the equation:

$$ {}^{30}_{15}P \rightarrow {}^{30}_{14}Si + {}^{0}_{1}e + \nu \qquad [4.2]$$

NOTE.– *Principle of a Geiger-Müller counter.*

The Geiger-Müller counter is a capacitor whose plates consist of a detector tube (T) filled with gas (He, Ne, Ar, etc.) under low pressure and a metal wire (W). The cylindrical tube acts as a cathode and the wire plays the role of an anode. The capacitor thus constituted is assembled in series with a generator (G) and an ohmic conductor of resistance R. The generator imposes a voltage U_{FT} of the order of 1 kV. The voltage at the terminals of the ohmic conductor is applied to a pulse counter (Figure 4.2a).

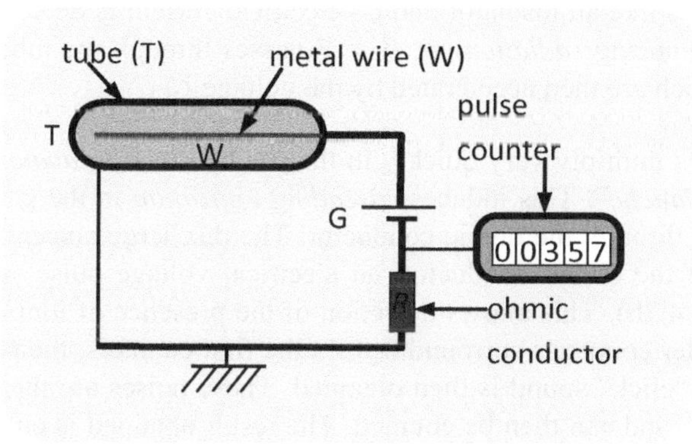

Figure 4.2a. *Description of a Geiger-Müller counter*

The inlet face of the tube (the one that is directed toward the point where the measurement is to be made) is closed by a lightweight material (generally mica leaf, or silicon or beryllium), permeable to ionizing radiation. The principle of measuring using a Geiger-Müller counter is shown in Figure 4.2b.

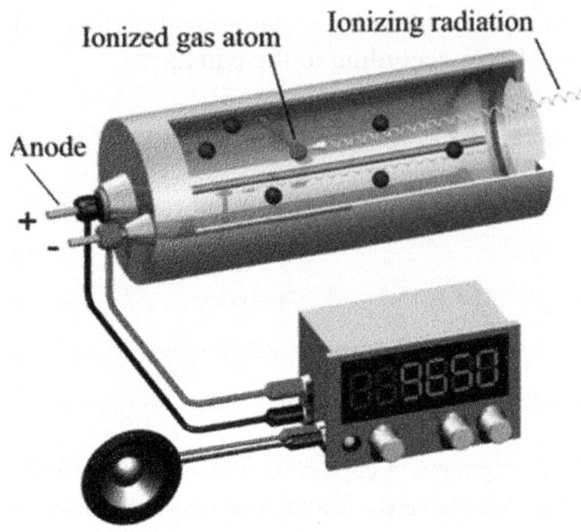

Figure 4.2b. *Measuring principle using a Geiger-Müller counter. Source: https://fr.qwe.wiki/wiki/Geiger_counter*

When no ionizing particle interacts with the gas contained in the tube, the capacitor behaves like an insulator and no electrical current is detected in the circuit formed. When *ionizing radiation* (α, β or γ) passes through the tube, electrons and ions appear, which are then accelerated by the voltage U_{FT}.

The electrons multiply very quickly in the gas by *electron avalanche*, known as "*Townsend avalanche*". This induces *cascading ionization* in the gas and discharge of the capacitor through the ohmic conductor. The discharge current thus creates, at the terminals of the ohmic conductor, an electrical voltage pulse sent to the *pulse counter* (Figure 4.2b). This allows detection of the presence of ionizing radiation in the Geiger-Müller counter's surroundings. In the first counters, the pulse is sent to a loudspeaker. A "click" sound is then obtained. These pulses are then shaped by the electronics used, and can then be counted. The result obtained is either converted to sound, or displayed by a galvanometer or a digital display, or both (Figure 4.2b).

Jean Frédéric Joliot, known as Frédéric Joliot-Curie was a French physicist and chemist. His wife Irène Joliot-Curie was also a French physicist and chemist. This couple stand out particularly for their discovery of artificial radioactivity in 1934. They were winners of the Nobel Prize in Chemistry 1935 in recognition of this discovery. Unfortunately, Irène died of acute leukemia caused by multiple exposure to polonium and X-rays, the same disease that killed her mother, Marie Curie, in 1934 (Box 3.2).

Sir John Sealy Edward Townsend was a British physicist-mathematician. He conducted numerous works on the electrical conduction of gases. Thus in 1897, he discovered the phenomenon of electron avalanche in ionized gases, called Townsend avalanche in his honor.

Box 4.1. *Irène Joliot (1897–1956); Frédéric Joliot (1900–1958); Townsend (1868–1967)*

4.1.2. *Balanced equation, β decay energy*

During β radioactivity, the mass number, A, does not vary. Thus, the parent nucleus and the daughter nucleus have the same mass number. It is therefore an isobaric transition during which the atomic number of the daughter nucleus increases (for β^- mode) or decreases (for the β^+ mode) by one unit relative to the atomic number, Z, of the parent nucleus. An electron and an electron antineutrino are emitted in the case of the β^- mode, while a positron and an electron neutrino are emitted in the case of the β^+ mode. The β^- and β^+ decay equations are written, respectively:

$$ {}^{A}_{Z}X \rightarrow {}^{A}_{Z+1}Y + {}^{0}_{-1}e + {}^{0}_{0}\bar{\nu}_e + \gamma \qquad [4.3a] $$

$$ {}^{A}_{Z}X \rightarrow {}^{A}_{Z-1}Y + {}^{0}_{1}e + {}^{0}_{0}\nu_e + \gamma \qquad [4.3b] $$

As shown in the Segrè diagram (Figure 1.18), β^- decay concerns unstable nuclei located above the nuclear energy surface grouping together stable nuclei for which $Z = N$. This radioactivity occurs for nuclei with excess neutrons. Thus, during β^- decay, a neutron disintegrates into a proton, an electron and an electron antineutrino:

$$ {}^{1}_{0}n \rightarrow {}^{1}_{1}H + {}^{0}_{-1}e + {}^{0}_{0}\bar{\nu}_e \qquad [4.4] $$

Equation [4.4] is often simplified as indicated in the first equation [3.4], which we recall here:

$$ n \rightarrow p + e^- + \bar{\nu}_e $$

Reaction [4.4] corresponds to the *free neutron decay* reaction with a period of 925 sec. In the nucleus, the stability of the neutron is ensured by attractive nuclear forces. As explained in section 3.1.4, the neutron decays by weak interaction (Figure 3.3). The transformation of neutron to proton is interpreted by the decay of the W^- boson into an electron and into an electron antineutrino.

β+ decay, by contrast, concerns unstable nuclei located below the surface of nuclear energy. It occurs for nuclei with excessive protons. Unlike β^- radioactivity, which concerns natural radionuclides, β^+ radioactivity is observed almost only for artificial radionuclides. In this decay mode, a proton decays into a neutron, a positron and an electron neutrino according to the second simplified equation [3.4] as recalled here:

$$p \rightarrow n + e^+ + \nu_e$$

Similarly, as Figure 3.4 shows, the proton decays by weak interaction. The transformation of proton to neutron is interpreted by the decay of the W^+ boson into a positron and into an electron neutrino.

Unlike the unstable free neutron, the *lifetime of the free proton* is estimated at 10^{32} years compared to the lifetime of the free neutron, equaling 925 sec. The proton is absolutely stable.

Let us now give examples of β decay. Sodium-24 and krypton-91 are β^- emitters, while iron-53 and molybdenum-91 are β^+ emitters. Decay equations are written (specific cases of isobaric transitions without photon emission):

– for β^- decay:

$$^{24}_{11}Na \rightarrow {}^{24}_{12}Mg + {}^{0}_{-1}e + {}^{0}_{0}\bar{\nu}_e \qquad [4.5a]$$

$$^{91}_{36}Kr \rightarrow {}^{91}_{37}Rb + {}^{0}_{-1}e + {}^{0}_{0}\bar{\nu}_e \qquad [4.5b]$$

– for β^+ decay:

$$^{53}_{26}Fe \rightarrow {}^{53}_{25}Mn + {}^{0}_{1}e + {}^{0}_{0}\nu_e \qquad [4.6a]$$

$$^{91}_{42}Mo \rightarrow {}^{91}_{41}Nb + {}^{0}_{1}e + {}^{0}_{0}\nu_e \qquad [4.6b]$$

The daughter nuclei then formed are magnesium(Mg)-24, rubidium(Rb)-91, manganese(Mn)-53, and niobium(Nb)-91.

Table 4.1 shows several (β, γ) emitters. The radioactive half-life, or half-life period, T, is indicated for each radionuclide. Some of these (marked with asterisks) exhibit double or triple decay.

(β^-, γ) emitter			(β^+, γ) emitter		
Z	Nucleus	Half-life T	Z	Nucleus	Half-life T
3	^{8}Li	840.3 ms	6	^{11}C	20.334 min
6	^{14}C	5,730 years	9	^{18}F	109.771 min
15	^{32}P	14.263 days	11	^{22}Na	2.6027 years
19	^{40}K*	1.248×10^9 years	12	^{23}Mg	11.317 sec
23	^{52}V	3.743 min	14	^{27}Si	4.16 sec
27	^{60}Co	5.2713 years	20	^{39}Ca	859.6 ms
53	^{131}I	8.02 days	27	^{53}Co	242 ms
55	^{137}Cs	30.1671 years	35	^{80}Br**	17.68 min

* ^{40}K: β^- (88.8%); β^+ (0.001%); *EC* (11.2%).

** ^{80}Br: β^- (91.7%); β^+ (8.3%).

Table 4.1. *Several (β, γ) emitters. The (β^+, γ) emitters are all artificial*

As in the case of α radioactivity, the *energy released during β decay*, also called *β decay energy*, noted Q_β, is equal to the sum of the kinetic energies of the β particle and of the recoiled daughter nucleus and the quantum energies of the γ photon and the neutrino or the antineutrino. Let us use E_μ to designate the energy of the neutrino or antineutrino. We obtain:

$$Q_\beta = E_{c\beta} + E_{cr}(\mathrm{Y}) + E_\gamma + E_\mu \qquad [4.7]$$

Analogously to [3.35], the relationship for calculating β decay energy is written (the β^- and β^+ particles have the same mass, noted m_β):

$$Q_\beta = [m_\mathrm{X} - (m_\mathrm{Y} + m_\beta)]c^2 \qquad [4.8]$$

The mass loss during β decay is then written according to [4.8]:

$$\Delta m = [m_\mathrm{X} - (m_\mathrm{Y} + m_\beta)] \qquad [4.9]$$

APPLICATION 4.1.– Francium (Fr) with atomic number 87 has 23 isotopes, the most well-known of which is isotope-223 with a radioactive half-life of 22 min. Almost all ^{223}Fr nuclides undergo β^- decay. Some of these nuclides can undergo α decay. Write the β^- and α decay equations for francium-223, taking into account the emission of photons. Calculate the probability per unit time for a nucleus of francium-223 to decay.

ANSWER.– The respective β^- and α decay equations for francium-223 are written:

$$^{223}_{87}Fr \rightarrow ^{223}_{88}Ra + ^{0}_{-1}e + ^{0}_{0}\bar{\nu}_e + \gamma \quad [4.10a]$$

$$^{223}_{87}\text{Fr} \rightarrow ^{219}_{85}\text{At} + ^{4}_{2}\text{He} + \gamma \quad [4.10b]$$

The probability per unit time for a nucleus of francium-223 to decay corresponds to its λ decay constant. I.e.

$$\lambda = \ln 2/T = 5.25 \times 10^{-4}\ \text{s}^{-1} \quad [4.10c]$$

APPLICATION 4.2.– Calculate the decay energy of radiosodium-24.

Given data:

– *mass of nucleus* ^{24}Na: 23.98422254 u;

– *mass of nucleus* ^{24}Mg: 23.9784568 u;

– *mass of β particle*: 5.486×10^{-4} u;

– 1 u = 931.5 MeV.

ANSWER.– Using [4.8], we obtain:

$$Q_\beta = [m_{\text{Na}} - (m_{\text{Mg}} + m_\beta)]c^2 \quad [4.11]$$

Numerically this gives:

$$Q_\beta = 0.00521714\ \text{u} = 4.86\ \text{MeV}$$

4.1.3. *Continuous β emission spectrum*

Unlike the α emission spectrum, which is often discrete (Figure 3.9), the β emission spectrum is continuous. Experimentally, the absorption curve of a β-ray decreases exponentially when the thickness of the absorbent increases. It is then canceled out when the thickness of the absorbent is equal to the *maximum course of the β-rays* in the matter that is passed through. As stated in section 3.1.5, β particles can pass through 7 mm of aluminum before being stopped. In 1914, Chadwick (Box 1.6) and other researchers showed for the first time that the *β emission spectrum* is continuous. Figures 4.3 and 4.4 show the β^- emission spectrum of phosphorus-32 (Figure 4.3) and the β^+ emission spectrum of phosphorus-30 (Figure 4.4). The two spectra, although continuous, present a slight difference.

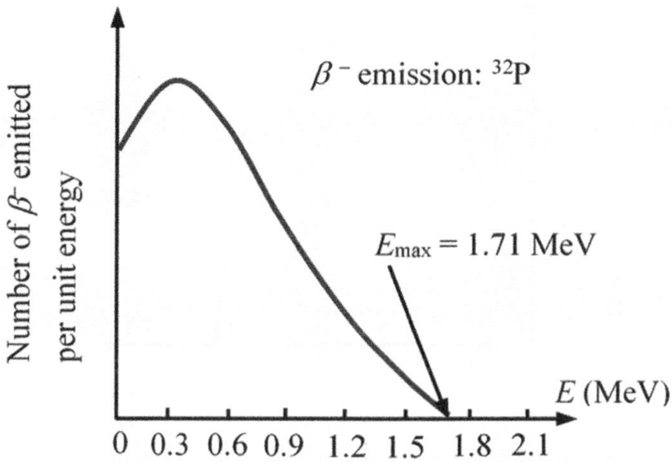

Figure 4.3. *β^- emission spectrum of phosphorus-32*

For β^- emission, the number of particles emitted is not zero when the total energy, E, is zero. It all occurs as if, during β^- emission, certain electrons emitted have just enough energy to leave the nucleus without being able to move away from it because of the Coulomb attraction. They thus remain trapped in the electric field of the nucleus for $E = 0$. When $E > 0$, those with sufficient kinetic energy can overcome the Coulomb attraction. They thus move away from the nucleus after emission. Then, the number of β^- particles emitted increases until it reaches a peak and then decreases exponentially. The curve is canceled out when the energy, E, of the β^- spectrum is equal to E_{max}. By definition, E_{max} is called *maximum energy of the β spectrum*, or *extreme energy of the β spectrum*. For phosphorus-32, $E_{max} = 1.71$ MeV (Figure 4.3).

By contrast, for β^+ emission, there is no positron emitted with zero energy. Since these particles are positive, they easily move away from the nucleus after emission due to Coulomb repulsion. Subsequently, when $E > 0$, the number of β^+ particles emitted increases until it reaches a peak and then decreases exponentially until it is canceled out when the energy, E, is equal to the extreme energy of the β spectrum. For phosphorus-30, $E_{max} = 3.20$ MeV (Figure 4.4).

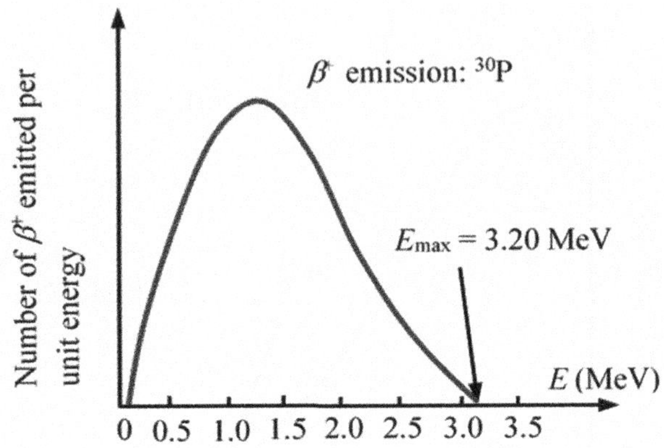

Figure 4.4. *β^+ emission spectrum of phosphorus-30*

Note that the β emission is accompanied by the emission of an electron antineutrino (case of β^- emission) or an electron neutrino (case of β^+ emission). If E_μ is used to designate the electron antineutrino or electron neutrino and $E_{c\beta}$ for the relativistic kinetic energy of the β particle, then:

$$E_{max} = E_{c\beta} + E_\mu \quad [4.12]$$

4.1.4. *Sargent diagram, β transition selection rules*

Studying the β^- emitters of the natural radioactive families of uranium (U), thorium (Th) and actinium (Ac), in 1933 Bernice Sargent established an empirical relationship between the λ decay constant and the extreme energy, E_{max}, of the β spectrum. This relationship, called *Sargent's law*, is written:

$$\log \lambda = a + b \log E_{max} \quad [4.13]$$

In Sargent's law, a and b are empirically-determined constants.

In its original form, the representation of empirical relationship [4.13] is known as the *Sargent diagram*. This diagram was established for natural β^- emitters corresponding to radium (Ra), lead (Pb), bismuth (Bi), thallium (Tl), thorium (Th), Actinium (Ac), and protactinium (Pa). The Sargent diagram is shown in Figure 4.5 for the few β^- emitters:

– RaB ($^{214}_{82}Pb$), RaC ($^{214}_{83}$Bi), RaD ($^{210}_{82}$Pb), RaE ($^{210}_{83}Bi$);

– ThB ($^{212}_{82}Pt$), ThC ($^{212}_{83}Bi$), ThC″ ($^{208}_{81}Tl$);

– AcB ($^{211}_{82}$Pt), AcC″ ($^{207}_{81}Tl$);

– UX_1 ($^{234}_{90}$Th), UX_2 ($^{234}_{91}$Pa);

– $MsTh_2$ ($^{228}_{89}$Ac) ($MsTh_2$: mesothorium-2).

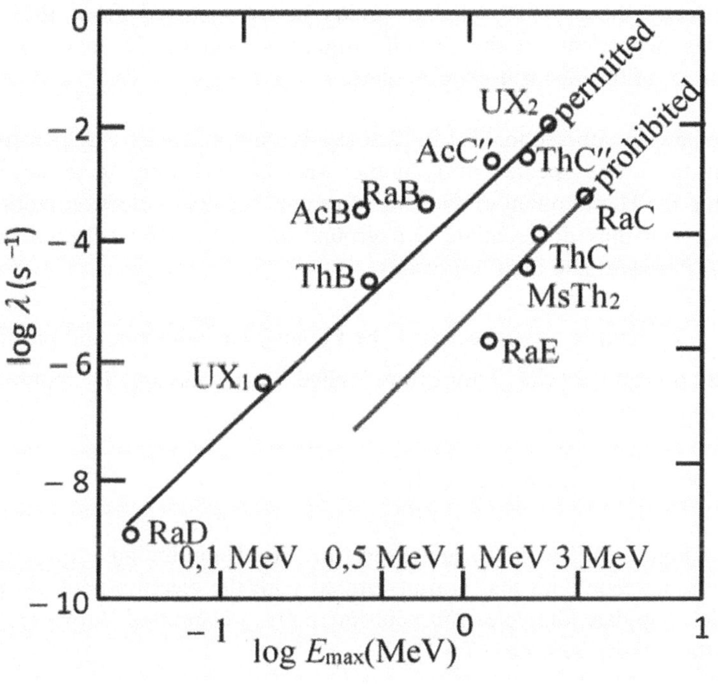

Figure 4.5. *Sargent diagram in its original form according to [SAR 33]*

In Figure 4.5, we have reproduced the Sargent diagram, respecting the writing of nuclide symbols according to the *old nucleus notations*. These notations are often used in certain nuclear physics works, especially when quoting verbatim from old works on radioactivity. This is particularly the case with Sargent's work [SAR 33]. It is important for the reader to be able to make the correspondence between the old nucleus notations and their conventional notations. These correspondences are presented for all three natural radioactive families (see section 4.4).

As shown in Figure 4.5, the Sargent diagram shows that the natural β^- emitting nuclides are approximately distributed around two lines. For the same value of energy, E_{max}, the radioactive decay constants of the nuclides distributed on either side of the upper line are 100 to 1,000 times higher than the radioactive decay constants of the nuclides distributed on either side of the lower line. To explain this difference, Sargent suggested that the group of nuclides distributed around the upper line consists of nuclides of "permitted" β transitions, while the group of nuclides distributed around the lower line consists of nuclides of "forbidden" β transitions.

In β decay theory, the selection rules are determined from the transition probabilities dependent on the orbital angular momentum, L, and spin angular momentum, S, of the electron-neutrino pair.

As we saw in sub-section 3.4.2, β decay is interpreted by a transformation of a nucleon from one species into another species according to equations [3.4]. Considering the Hamiltonian of the {transforming nucleon – electron-neutrino pair} system, we can thus differentiate two groups of β *transition selection rules*: the *Gamow–Teller rules* and the *Fermi rules*.

Let $\vec{J}_i$, $\vec{J}_f$, and π_i, π_f respectively, be the angular momenta and parities of the initial and end states of the β transition studied. Under the angular momentum and parity conservation law:

$$\vec{J}_i = \vec{J}_f + \vec{L} + \vec{S} \; ; \; \pi_i = \pi_f (-1)^L \qquad [4.14a]$$

– the *Gamow–Teller rules* are used for the Hamiltonian expression when coupling the nucleon spin that is transformed with the electron-neutrino pair spin. These rules stipulate that electron-neutrino pairs are emitted with parallel spins corresponding to *triplet states* ($S = 1$);

– not knowing the nucleon spin that is transformed in the Hamiltonian, the *Fermi rules* are used. These rules express the fact that the electron-neutrino pairs are emitted with antiparallel spins corresponding to *singlet states* ($S = 0$).

The Gamow–Teller and Fermi rules governing permitted and forbidden β transitions can be summarized as follows:

– Gamow–Teller rules

$$\Delta J = 0, \pm 1; \; \pi_i = \pi_f \qquad [4.14b]$$

– Fermi rules

$$\Delta J = 0;\ \pi_i = \pi_f \qquad [4.14c]$$

Thus,

– if the orbital angular momentum $L = 0$, then β transitions are permitted. For $S = 0$ (singlet state), we then have $\Delta J = 0$; for $S = 1$ (triplet state), then $\Delta J = \pm 1$;

– if the orbital angular momentum $L \neq 0$, then β transitions are forbidden.

Forbidden transitions are classified in terms of forbidden 1st-order transitions when $L = 1$, 2nd-order for $L = 2$, and so on.

Bernice Weldon Sargent was a Canadian physicist. He distinguished himself by his work in nuclear physics. He is especially famous for having established in 1932, in his doctoral thesis, the empirical relationship between the radioactive decay constants of radioisotopes emitting β^- particles and the corresponding Napierian logarithms of their maximum energies of β^- particles. This relationship and the corresponding curves are called Sargent's law and the Sargent diagram, respectively, in his honor.

Edward Teller was an American-Hungarian physicist. He distinguished himself by his many contributions to nuclear physics, molecular physics and spectroscopy. Gamow–Teller's selection rules developed in collaboration with George Gamow (Box 3.5) attest to his valuable contribution to nuclear physics.

Box 4.2. *Sargent (1906–1993); Teller (1908–2003)*

4.1.5. *Decay energy diagram*

As for α decay, the β *decay energy diagram* gives an overall mapping of the decay process of a β^- or β^+ emitting nucleus. It notably includes the level of the parent nucleus, the excited levels and the ground level of the daughter nucleus, the β transitions between the energy level of the parent nucleus and those of the daughter nucleus, and the γ decay transitions of the daughter nucleus.

Figures 4.6 and 4.7 show the respective β^- and β^+ *decay energy diagrams* of cesium-137 (Figure 4.6) and sodium-22 (Figure 4.7). Note that sodium-22 can decay by electron capture toward the excited level of neon-22. We will come back to this in section 4.1.7.

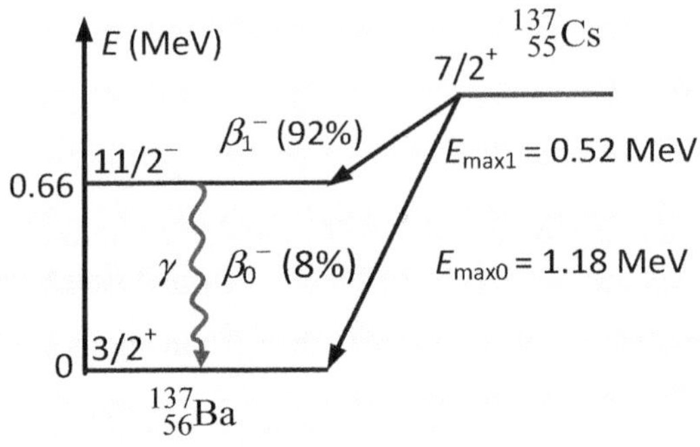

Figure 4.6. *β^- decay energy diagram of cesium-137 (T = 32 years). The diagram shows a direct β_0^- transition to the $3/2^+$ ground level of barium-137 without photon emission and a β_1^- transition supplying the $11/2^-$ excited level of barium-137. $11/2^- \rightarrow 3/2^+$ deexcitation is accompanied by the emission of a photon of energy 0.66 MeV. The extreme energy of the β spectrum is equal to 0.52 MeV for the β_1^- transition and 1.18 MeV for the β_0^- transition*

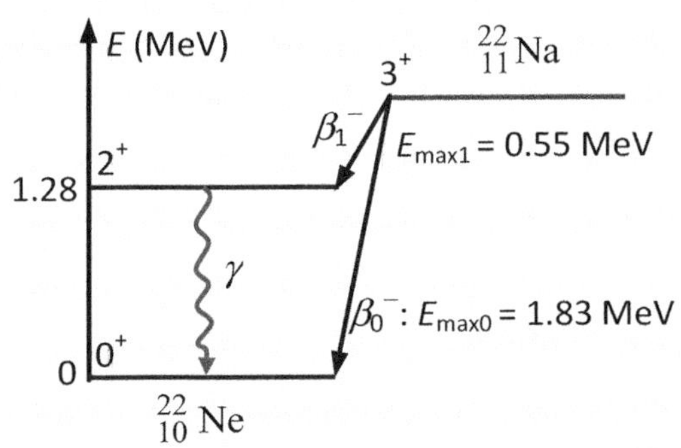

Figure 4.7. *β^+ decay energy diagram of sodium-22 (T = 2.58 years). The diagram shows a direct β_0^- transition to the 0^+ ground level of neon-22 without photon emission and a β_1^- transition feeding the 2^+ excited level of neon-22. The $2^+ \rightarrow 0^+$ deexcitation is accompanied by the emission of a 1.28 MeV energy photon. The extreme energy of the β spectrum is equal to 0.55 MeV for the β_1^-transition and 1.83 MeV for the β_0^- transition. Note that an EC transition is observed between the 3^+ levels of sodium and the 2^+ level of neon corresponding to an electron capture process (11%)*

APPLICATION 4.3.– Using the data in Figures 4.6 and 4.7, show that the energies of the emitted gamma photons are equal to 0.66 MeV (Figure 4.6) and 1.28 MeV (Figure 4.7), respectively.

ANSWER.– Considering [4.7] and [4.12], the β decay energy is equal to:

$$Q_\beta = E_{c\beta} + E_\mu + E_\gamma = E_{\max} + E_\gamma \quad [4.15]$$

According to [4.8]:

$$Q_{\beta 0} = Q_{\beta 1} = [m_X - (m_Y + m_\beta)]c^2 \quad [4.16]$$

– For cesium-137:

$$Q_{\beta 0} = E_{\max 0};\ Q_{\beta 1} = E_{\max 1} + E_\gamma \quad [4.17]$$

This then gives:

$$E_\gamma = E_{\max 0} - E_{\max 1} = 1.18 - 0.52 = 0.66 \text{ MeV}$$

– The same applies to sodium-22:

$$E_\gamma = E_{\max 0} - E_{\max 1} = 1.83 - 0.55 = 1.28 \text{ MeV}$$

4.1.6. *Condition of β^+ emission*

Let us use the following designations:

– m_X and m_Y for the respective masses of the parent nucleus, *X*, and the daughter nucleus, *Y*;

– M_X and M_Y for the respective atomic masses of the *X* and *Y* elements;

– $m_0 = m_\beta$ for the resting mass of the electron or positron.

Using [4.8] and [4.12], we obtain:

$$Q_\beta = [m_X - (m_Y + m_0)]c^2 = E_{\max} + E_\gamma \quad [4.18]$$

This then gives:

$$(m_X - m_Y)c^2 = m_0c^2 + E_{\max} + E_\gamma \quad [4.19]$$

The mass of an atom, M (X), is less than the sum of the masses of the nucleus, m (X) and the Z electrons, Zm_0. The difference is equal to the binding energy, W, of the system:

$$W = m(\mathrm{X}) + Zm_0 - M(\mathrm{X}) \qquad [4.20]$$

Note that for a bound system, $W > 0$ and for an unbound system, $W < 0$.

By overlooking the binding energy of the {nucleus-electrons} system, the mass of an atom is then equal to the sum of the masses of the Z-electrons and that of the nucleus.

– For β^- emission, the atomic numbers of the parent and daughter nucleus are Z and $(Z + 1)$, respectively. We thus obtain:

$$M_\mathrm{X} = m_\mathrm{X} + Zm_0;\ M_\mathrm{Y} = m_\mathrm{Y} + (Z+1)m_0 = m_\mathrm{Y} + Zm_0 + m_0 \qquad [4.21]$$

The difference in atomic mass $(M_\mathrm{X} - M_\mathrm{Y})$ is then written according to [4.21],

$$(M_\mathrm{X} - M_\mathrm{Y})c^2 = (m_\mathrm{X} - m_\mathrm{Y})c^2 - m_0c^2$$

This then gives:

$$(m_\mathrm{X} - m_\mathrm{Y})c^2 = (M_\mathrm{X} - M_\mathrm{Y})c^2 + m_0c^2 \qquad [4.22]$$

Equalizing [4.19] and [4.22] gives:

$$(M_\mathrm{X} - M_\mathrm{Y})c^2 + m_0c^2 = m_0c^2 + E_{\max} + E_\gamma$$

That is, ultimately:

$$(M_\mathrm{X} - M_\mathrm{Y})c^2 = E_{\max} + E_\gamma > 0 \qquad [4.23]$$

The result [4.23] reflects the β^- *emission condition*: the mass difference between the atomic masses of the X (parent nucleus) and Y (daughter nucleus) elements must be positive. This is verified for the β^- emitters.

– For β^+ emission, the atomic numbers of the parent and daughter nucleus are Z and $(Z - 1)$, respectively. We thus obtain:

$$M_\mathrm{X} = m_\mathrm{X} + Zm_0;\ M_\mathrm{Y} = m_\mathrm{Y} + (Z-1)m_0 = m_\mathrm{Y} + Zm_0 - m_0 \qquad [4.24]$$

The difference in atomic mass ($M_X - M_Y$) is then written according to [4.24]:

$$(M_X - M_Y)c^2 = (m_X - m_Y)c^2 + m_0c^2$$

This then gives:

$$(m_X - m_Y)c^2 = (M_X - M_Y)c^2 - m_0c^2 \quad [4.25]$$

Equalizing [4.19] and [4.25] gives:

$$(M_X - M_Y)c^2 - m_0c^2 = m_0c^2 + E_{max} + E_\gamma$$

That is, ultimately:

$$(M_X - M_Y)c^2 = 2m_0c^2 + E_{max} + E_\gamma \quad [4.26]$$

Using [4.26], let us express the β^+ *emission condition*: the mass difference between the atomic masses of the X (parent nucleus) and Y (daughter nucleus) elements must be strictly greater than twice the resting energy of the electron, i.e.:

$$(M_X - M_Y)c^2 > 2m_0c^2 = 1.022 \text{ MeV} \quad [4.27]$$

Note that emission is impossible if $(M_X - M_Y)c^2 = 2m_0c^2$ since if E_γ can be zero in the case of decay without photon emission, the extreme energy, E_{max}, of the β spectrum is never zero.

4.1.7. *Decay by electron capture*

When the mass difference between neighboring isobars allows, in terms of energy, a nucleus with charge number Z can capture one of its internal electrons to transform into the neighboring isobar with charge number $Z - 1$, with emission of a ν neutrino. The phenomenon of *electron capture* generally occurs in the K-shell since, as it is the deepest, its electrons have the greatest probability of being captured by the nucleus. Electron capture is a phenomenon in competition with β^+ decay.

In the general case, the emission of X-photon by a multi-electron atom occurs as a result of an internal conversion phenomenon or as a result of an electron capture process, thus creating an *electronic gap* or a *hole* in one of the inner shells of the atom.

From a chronological point of view, electron capture was first theorized by Gian-Carlo Wick in 1934. In this work, Wick drew on Fermi's β decay theory. He proposed a possible type of decay in addition to β^+ radioactivity, notably calculating the probability of decaying by positron emission and electron capture.

In 1937, Luis Alvarez discovered the electron capture of the *K*-shell by studying the radioactive decay of vanadium-48. In his experiment, vanadium-48 was produced by titanium bombardment by deuterons. Detection of the X-photons characteristic of titanium occurred using an aluminum foil.

In the general case, during the electron capture process, a proton of the nucleus changes into a neutron, with the decay energy being carried away by the neutrino according to the equation:

$$p + e^- \rightarrow n + \nu \qquad [4.28]$$

Electronic capture is an isobaric transition with general equation:

$${}^{A}_{Z}X + e^- \rightarrow {}^{A}_{Z-1}X + \nu_e + RX \qquad [4.29]$$

The neutrino (electron neutrino) formed via reaction [4.29] has no interaction with the matter it passes through because it is neutral like the photon. Therefore, the detection of an electron capture process cannot occur directly from the characteristics of the neutrino emitted via [4.29]. Nevertheless, as we explained above, during an electron capture, the occupation of the *K*-shell hole by one of the *L*-, *M*-, *N*-, etc. electrons results from an electron rearrangement accompanied by the emission of characteristic *X*-photons. These easily detectable photons help to highlight the electron capture process. Note that the rearrangement of the electronic cloud may be accompanied by the emission of *Auger electrons* by a process called *Auger effect* (see section 4.19). As a result, electron capture is accompanied by a completion of *X*-photon and Auger electron emissions.

Equation [2.163b] reflects cesium-131 decay by electron capture. Figure 4.8 shows the decay energy diagram of sodium highlighting the competition between β^+ decay and electron capture.

As noted at the beginning of this section, the phenomenon of electron capture generally occurs in the deepest *K*-shell. It follows that, the higher the number of *Z*-protons, the closer the electrons are to the nucleus, the greater the capture probabilities of their electron are. As a result, the β^+ decay/electron capture ratio increases with atomic number *Z*.

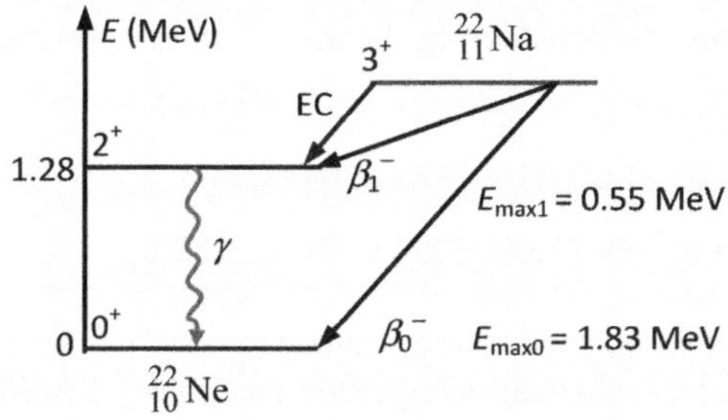

Figure 4.8. *Decay energy diagram of sodium-22 by electron capture (EC) and by β^+ electron capture*

By overlooking the differences in binding energy between {nucleus (X)- electrons} and {nucleus (Y)- electrons} systems, the *decay energy by electron capture*, Q_{EC}, is written according to [4.29]:

$$Q_{EC} = [M_X - M_Y]c^2 = E_\nu + E_X > 0 \qquad [4.30]$$

Equation [4.30] expresses the *electron capture condition.*

Using β^+ emission condition [4.27], we obtain:

$$M_X > M_Y + 2m_0 \qquad [4.31]$$

Thus, according to [4.30] and [4.31], decay by electron capture is always possible when $M_X > M_Y$. On the other hand, the competitive β^+ decay process is only possible if $M_X > M_Y + 2m_0$. Therefore, an electron capture not accompanied by β^+ emission may occur in the case of isobars of low mass difference ($M_X - M_Y$).

NOTE.–

All of the known chemical elements in the universe were created from successive nuclear reactions at the heart of the stars. Today, it is recognized that all known elements are the result of a stellar explosion around 14 billion years ago, known as the Big Bang. This explosion marked the birth of the Universe.

Practically speaking, all of the β^+ emitters formed at the time of the Big Bang no longer meet in nature. Since their decay periods were relatively short, these emitters could not remain, except in trace states, in the atmosphere and the Earth's crust. Today, the only β^+ emitter existing in its natural state is potassium-40 (see exercise 4.6).

APPLICATION 4.4.– We will study the competition between β^+ emission and electron capture. We will do this by considering the atomic masses of the following isobaric pairs:

– ^{7}Li: 7.01600455 u; ^{7}Be: 7.01692983 u; 0.00093 u.

– ^{11}C :11.0114336 u; ^{11}B: 11.0093054 u. 0.0021282 u.

For each pair of isobars, specify the spontaneous transformation likely to occur. In each case, write the transformation equation (without γ emission) and determine, if applicable, the maximum energy of the spectrum of the emitted β particles.

ANSWER.– Let us determine the differences in resting energies $(M_X - M_Y)c^2$ for each pair of isobars.

– For ^{7}Li and ^{7}Be:

$$(M_{Be} - M_{Li})c^2 = 0.00092528 \times 931.5 = 0.862 \text{ MeV} < 1.022 \text{ MeV}$$

$(M_{Be} - M_{Li})c^2 < 1.022$ MeV: β^+ emission is impossible according to [3.105]. Only electronic capture is possible in accordance with condition [4.30]. The decay equation is written, in this case:

$$^{7}_{4}\text{Be} + \text{e}^- \rightarrow {}^{7}_{3}\text{Li} + \nu_e + \text{RX} \quad [4.32a]$$

– For ^{11}C and ^{11} B:

$$(M_C - M_B)c^2 = 0.0021282 \times 931.5 = 1.982 \text{ MeV} > 1.022 \text{ MeV}$$

Note that $(M_C - M_B)c^2 > 1.022$ MeV : β^+ emission is therefore possible according to [4.27], accompanied by the competitive electron capture process. The β^+ decay and electron capture equations are written:

$$^{11}_{6}\text{C} \rightarrow {}^{11}_{5}\text{B} + {}^{0}_{1}\text{e} + \nu_e \quad [4.32b]$$

$$^{11}_{6}\text{C} + \text{e}^- \rightarrow {}^{11}_{5}\text{B} + \nu_e + \text{RX} \quad [4.32c]$$

The maximum β^+ spectrum energy is given by equation [4.26]. We obtain:

$$E_{max} = (M_X - M_Y)c^2 - 2m_0c^2 - E_\gamma \quad [4.32d]$$

Decay occurs without γ-photon emission. Giving:

$$E_{max} = (M_C - M_B)c^2 - 2m_0c^2 \quad [4.32e]$$

Numerically this gives:

$$E_{max} = 1.982 - 1.022 = 0.96 \text{ MeV}$$

For information. The decay period of ^{7}Be is T = 53.22 days and that of ^{11}C is T = 20.334 min. Moreover, ^{11}C decays by β^+ emission for 99.79% and by electron capture for 0.21%.

4.1.8. *Double β decay, branching ratio*

Certain nuclides ($T \sim 10^{19}$ years) can decay by *simultaneous emission* of two β^- particles and two electron neutrinos. This rare natural radioactivity, noted [$\beta\beta(2\nu)$], is called *double β decay*. As an example, we can cite the specific cases of [$\beta\beta(2\nu)$] radioactivity of molybdenum 100 ($T = 8.5 \times 10^{18}$ years; daughter nucleus ruthenium (Ru)) and tellurium 130 ($T = 7.9 \times 10^{20}$ years; daughter nucleus xenon (Xe)) of the respective decay equations:

$$^{100}_{42}Mo \rightarrow {}^{100}_{44}Ru + \beta^- + \beta^- + \bar{\nu}_e + \bar{\nu}_e \quad [4.33a]$$

$$^{130}_{52}Te \rightarrow {}^{130}_{54}Xe + \beta^- + \beta^- + \bar{\nu}_e + \bar{\nu}_e \quad [4.33b]$$

As equations [4.33] show, double β decay requires the emission of two *Dirac antineutrinos*. Note that in the discussion on β radioactivity, the term "electron-neutrino pair" is commonly used instead of "electron-antineutrino pair". The word "neutrino" then refers to one or other of the forms of neutrino. In Majorana's theory, for example, no distinction is made between neutrino and antineutrino (see note at the end of this section).

It should be noted that certain nuclei may decay by *simultaneous emission* of β^- and β^+ (and/or electron capture). The most widely-studied cases are of copper-64 (see exercise 4.10), potassium-40 (see exercise 4.6) and vanadium-50. For example, the decay of vanadium-50 with simultaneous emission of β^- and β^+ occurs according to the respective balanced equations:

$$^{50}_{23}V \rightarrow {}^{50}_{24}Cr + \beta^- + \bar{\nu}_e + 1.2 \text{ MeV} \quad [4.34a]$$

$$^{50}_{23}V \rightarrow {}^{50}_{22}Ti + \beta^+ + \nu_e + 2.4 \text{ MeV} \quad [4.34b]$$

In addition, certain radioactive nuclei may decay by different modes. Each of the i modes corresponds to a *partial decay constant*, λ_1. The *total decay constant*, λ_{total}, is equal to the sum of the partial decay constants. By definition, the term *branching ratio* is used to refer to the fraction, R_i, given by the relationship:

$$R_i = \frac{\lambda_i}{\lambda_{total}} \quad [4.35]$$

APPLICATION 4.5.– Radon-221 can decay either by the α mode leading to polonium-217, or by the β^- mode leading to francium-221. The decay diagram is shown in Figure 4.9.

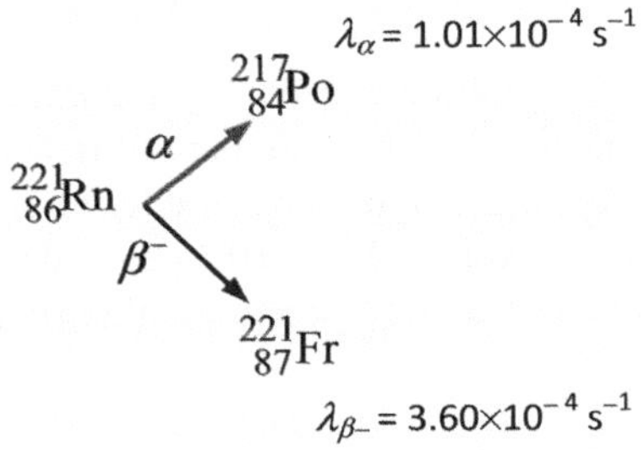

Figure 4.9. *α and β^- decay modes of radon-221*

Write the balanced equations (without γ-photon emission) of the various decay modes of radon-221, then calculate the branching ratios, R_α and R_{β^-}. The α and β^- particles will be replaced by their symbols.

ANSWER.–

– Balanced equations:

$$^{221}_{86}Rn \rightarrow \, ^{217}_{84}Po + \, ^{4}_{2}He \quad [4.36a]$$

$$^{221}_{86}Rn \rightarrow \, ^{221}_{87}Fr + \, ^{0}_{-1}e + \overline{\nu}_e \quad [4.36b]$$

– Branching ratios:

$$\lambda_{total} = \lambda_\alpha + \lambda_{\beta-} = 1.01\times10^{-4} + 3.60\times10^{-4} = 4.62\times10^{-4}\ s^{-1}$$

$$R_{\alpha} = \frac{\lambda_{\alpha}}{\lambda_{\alpha} + \lambda_{\beta-}} = 0.22 \ ; \ R_{\alpha} = 22\% \qquad [4.37a]$$

$$R_{\beta-} = \frac{\lambda_{\beta-}}{\lambda_{\alpha} + \lambda_{\beta-}} = 0.78 \ ; R_{\beta}^{-} = 78\% \qquad [4.37b]$$

Decay of radon-221 occurs at 78% according to the β^{-} mode and at 22% according to the α mode.

NOTE.– *Dirac neutrino and Majorana neutrino*

In particle physics, we can identify what are known as "Majorana" particles from what are known as "Dirac" particles.

– A *Majorana particle* or *Majorana fermion* is a particle that is its own antiparticle. The absence of an electrical charge is therefore a necessary condition to consider that an elementary particle possesses this property. The neutrino without an antiparticle is the typical example of a Majorana fermion.

– A *Dirac particle* or *Dirac fermion* is a particle that has an antiparticle. Examples of Dirac fermions are the electron (antiparticle: positive electron or positron) and proton (antiparticle: negative proton).

Experiments are currently underway, for example in the Modane Underground Laboratory (LSM) to identify whether neutrinos that are the only elementary fermions with zero electrical charge are ordinary or Majorana fermions [MAR 11]. This laboratory is located in the middle of the Fréjus tunnel connecting France and Italy and is topped by the Pointe du Fréjus located at an altitude of 2,932 m. The LSM is thus covered with 1,700 m of rock, protecting it from cosmic radiation, which cannot entirely pass through the thickness of rock [DEG 17]. The success of these experiments will then determine whether the neutrino is a Dirac particle (the neutrino and antineutrino are two distinct states) or a Majorana particle (neutrino = antineutrino).

The Majorana particles, fermions and neutrinos are named in tribute to Italian physicist Ettore Majorana (1906-presumed to have died after 1959), who is best known for his work in particle physics.

The Dirac particles, fermions and neutrinos are named in honor of British physicist Paul Adrien Maurice Dirac (1902-1984), who is famous in particular for his work in particle physics, quantum mechanics, and statistical physics.

4.1.9. *Atomic deexcitation, Auger effect*

The emission of X-photon by a multi-electron atom occurs as a result of an internal conversion phenomenon or as a result of an electron capture process, thus creating an *electronic gap* or a *hole* in one of the inner shells of the atom.

Let us assume that an electron from the *K*-shell is expelled by internal conversion or captured by a nucleus. The electronic gap thus created can be occupied by one of the electrons originating from the *L*-, *M*-, *N*-, etc. upper shells. There follows an electron rearrangement with emission of *X*-photons characteristic of the emitting atom by a process called *electron-hole recombination*. This process is in competition with the emission of an Auger electron (e^A) by a process called *Auger effect*.

Depending on the upper level where the transition occurs to the shell where the gap is located, we can identify several *spectral series* called *K*-, *L*-, *M*-, *N*-, etc. series. The *X*-photons of the *K-series* correspond to the *X*-lines produced during electronic transitions from the *L*-, *M*-, *N*-, etc. upper shells to the *K*-shell. Similarly, the *L-series* corresponds to the *X*-lines produced during electronic transitions from the *M*-, *N*-, etc. upper shells to the *L*-shell, and so on. The set of photons or X-rays of a given series constitutes the *spectral lines* of that series.

In the adopted nomenclature, the *spectral lines of a series* are designated by the name of the shell toward which the electronic transition takes place, indexed with a Greek letter, α, β, γ, etc., depending on the subshells concerned by the transition. In addition, the *K*-, *L*-, *M*-, *N*-, etc. shells correspond, respectively, to the values of the principal quantum number, $n = 1, 2, 3, 4$, etc. We either use the so-called Siegbahn notation, or that of the IUPAC (International Union of Pure and Applied Chemistry). Table 4.2 shows the correspondences between *Siegbahn notation* and that of the IUPAC for the top four electronic shells, *K*, *L*, *M* and *N*.

Siegbahn	$K\alpha_1$	$K\alpha_2$	$K\beta_1$	$K\beta_2$	$K\beta_1'$	$K\beta_2''$	...
IUPAC	KL_3	KL_2	KM_3	KM_2	KN_3	KN_2	...

Table 4.2. *Correspondence between the Siegbahn notation and that of the IUPAC in X-ray nomenclature*

In Siegbahn's notation, the *X*-photon noted $K\alpha_1$ corresponds to the photon noted KL_3 in IUPAC notation, and so on.

For a given shell, there are several subshells resulting from the spin-orbit interaction. By adopting the spectroscopic notation $^{2S+1}\mathrm{L}_J$ of the spectral terms, the total quantum number takes the values $|L + S|$ ……. $|L - S|$. As an example, let us explain the terms relating to the K- and L-shells:

– K-shell: $1s^2$, $n = 1$, $\ell = 0$; $L = 0$ which corresponds to the term S. In this orbital, the two electrons have antiparallel spin orientations. Hence the total spin, $S = 0$; the multiplicity of degeneracy $(2S + 1) = 1$. Moreover, J takes a single value, $J = 0$. We obtain the term $1\ {}^1\mathrm{S}_0$;

– L-shell: $2s^2 2p^6$, $n = 2$, $\ell = 0, 1$. For the configuration $2s^2$, $L = 0, J = S = 0$. This corresponds to the term $2\ {}^1\mathrm{S}_0$. For the configuration $2p^6$, we consider an electron (example: L-electron filling the hole created following a K-shell capture). Thus, for a 2p-electron, $\ell = L = 1$, $s = S = 1/2$, $J = 3/2\ (1+1/2)$ or $J = 1/2\ (1 - 1/2)$. Moreover, the multiplicity of degeneracy $(2S + 1) = 2$. The two terms $2\ {}^2\mathrm{P}_{3/2}$ and $2\ {}^2\mathrm{P}_{1/2}$ are obtained.

Table 4.3 presents the classification of energy levels and X-rays in Siegbahn notation.

Low-energy level	High-energy level	Line symbol
	$L_3\ ({}^2\mathrm{P}_{3/2})$	$K\alpha_1$
$K\ ({}^1\mathrm{S}_0)$	$L_2\ ({}^2\mathrm{P}_{1/2})$	$K\alpha_2$
	$M_3\ ({}^3\mathrm{P}_{3/2})$	$K\beta_1$
$L_3\ ({}^2\mathrm{P}_{3/2})$	$M_5\ ({}^3\mathrm{D}_{5/2})$	$L\alpha_1$
$L_2\ ({}^2\mathrm{P}_{1/2})$	$M_4\ ({}^3\mathrm{D}_{3/2})$	$L\beta_1$

Table 4.3. *Classification of energy levels and X-rays in Siegbahn notation*

Figure 4.10 shows atomic deexcitation by emission of X-rays or of an Auger electron following a K-shell electron capture.

We will use W_K and W_i to designate the respective K-shell and C_i-shell (L_i, M_i, N_i, etc.) binding energies. The energy, E_X, of the emitted X-photon is given by the equation:

$$E_X = W_K - W_i \qquad [4.38]$$

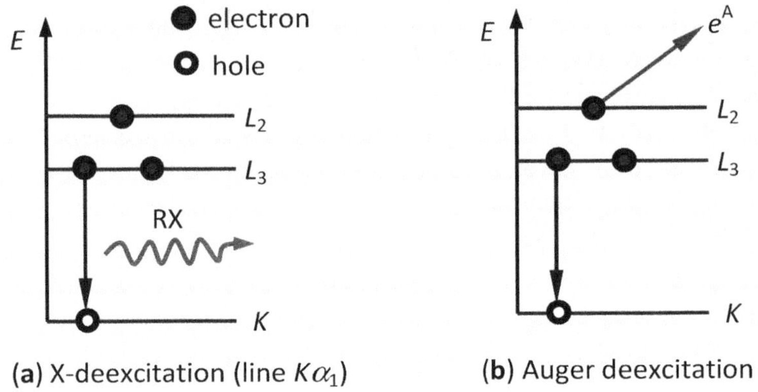

Figure 4.10. *Process of atomic deexcitation by emission of X-rays or of an Auger electron by rearrangement of the electronic cloud following a K-shell electron capture*

In Figure 4.10a, we have assumed that the hole is filled by an electron from the L_3-subshell. This corresponds to the emission of a XKL_3- or $K\alpha_1$-photon. This hole can also be filled by an L_2-electron (XKL_2-photon or $K\alpha_2$-line) or an electron of the *M*-, *N*-, etc. upper shells (not shown). For a deexcitation involving one of the *L*-electrons (an L_2-electron is indiscernible from an L_3-electron), two photons, XKL_2 and XKL_3, can then be emitted with respective energies:

$$E_{\mathrm{XKL2}} = E\,(K\alpha_1) = W_{\mathrm{K}} - W_{\mathrm{L2}} \qquad [4.39a]$$

$$E_{\mathrm{XKL2}} = E\,(K\alpha_2) = W_{\mathrm{K}} - W_{\mathrm{L3}} \qquad [4.39b]$$

APPLICATION 4.6.– The energy gap between the *L*- and *M*-shell sub-levels is unknown. In the case of the *L*-shell, this means interchanging the terms ${}^{3/2}P_2$ and ${}^2P_{1/2}$ (Table 4.4) or the L_2- and L_3-sub-levels (Figure 4.10). In this approximation, the relative arrangement of the *K*-, *L*- and *M*-shells of the copper atoms can be adopted, as shown in Figure 4.11 (scale not respected). Two lines, K_α and K_β can be detected in the X-ray spectrum. The hole on the *K*-shell is due to an electron capture. For the sake of simplicity, let us indicate one electron for each of the *L*- and *M*-shells.

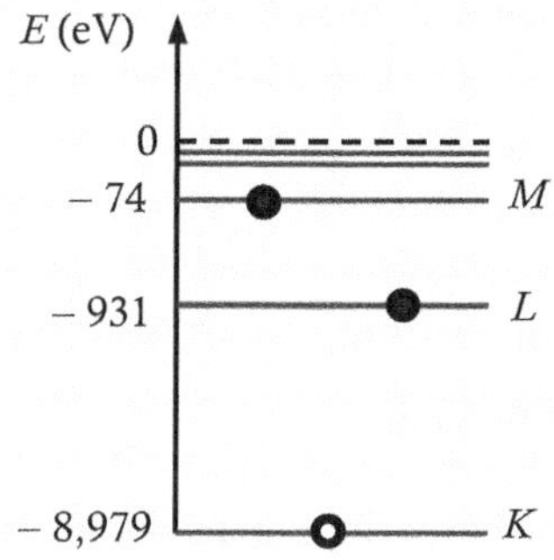

Figure 4.11. *Relative arrangement of the K-, L- and M-shells of copper atoms*

Reproduce the figure then indicate the two envisaged electron-hole recombination processes leading to the emission of lines K_α and K_β. Calculate the corresponding wavelengths.

Given data: $h = 6.63 \times 10^{-34}$ J · s; $c = 3.0 \times 10^8$ m · s^{-1}; 1 eV = 1.6×10^{-19} J.

ANSWER.– The electron-hole recombination processes are shown in Figure 4.12.

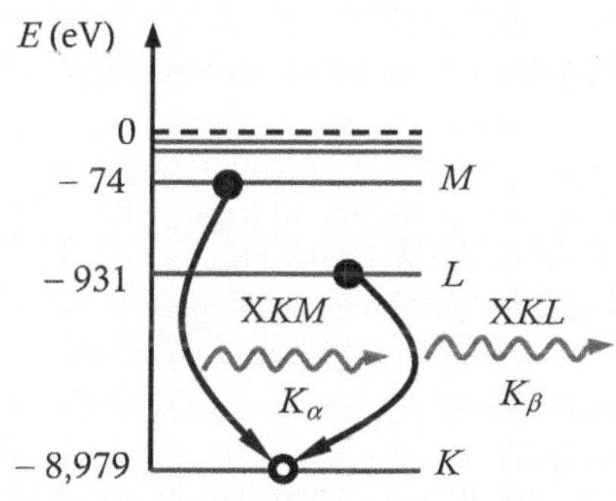

Figure 4.12. *Electron-hole recombination processes*

The variation in energy between any two shells A (upper shell) and B is given by the relationship:

$$\Delta E = E_A - E_B = \frac{hc}{\lambda_i} \qquad [4.40a]$$

The wavelengths associated with lines K_α and K_β are then written:

$$\lambda_\alpha = \frac{hc}{E_L - E_K} ; \lambda_\beta = \frac{hc}{E_M - E_K} \quad [4.40b]$$

Numerically, we find:

$$\lambda_\alpha = 154.1 \text{ pm}; \lambda_\beta = 139.2 \text{ pm}$$

The Auger deexcitation shown in Figure 4.10b occurs by Auger effect, the principle of which is as follows. The hole created in the *K*-shell is filled by an electron of a C_i-upper shell (here L_3-shell). The excess energy is then transferred to another electron of a C_j-shell (here L_2-shell) located above the C_i-shell. The electron of the C_j-shell is then expelled from the electronic cloud: this is the *Auger effect.*

The Auger electron carries a kinetic energy satisfying the energy conservation law:

$$E_c (e^A) = W_K - W_i - W_j \quad [4.41a]$$

In the case of Auger deexcitation described in Figure 3.22b, the kinetic energy of the Auger electron is equal to:

$$E_c (e^A) = W_K - W_3 - W_2 \quad [4.41b]$$

As a numerical illustration, let us consider the process of deexciting copper atoms according to the diagram described in Figure 4.12. The electron of the *L*-shell is recombined with the hole in the *K*-shell. The excess energy is then transferred to the electron of the *M*-shell. This produces an Auger electron of kinetic energy:

$$E_c (e^A) = W_K - W_L - W_M \quad [4.42]$$

Equation [3.127], for an Auger deexcitation process involving the *L*- and *M*-shells, the Auger effect is only observed if:

$$W_K > (W_L + W_M) \quad [4.43]$$

Numerically, using [3.127] we obtain (for a given shell *A*, the positive binding energy is given by the relationship: $W_A = - E_A$):

$$E_c (e^A) = 8979 - (931 + 74) = 7.974 \text{ keV}$$

Enrico Fermi was an American-naturalized Italian physicist. He made numerous contributions, notably to statistical physics (Fermi-Dirac statistics) and nuclear physics. Fermions (half-integer spin particles), the nuclear ray unit, the Fermi, Fermi gas (consisting of free fermions) and many other names are given in his honor. He won the Nobel Prize in Physics 1938 "for his demonstrations of the existence of new radioactive elements produced by neutron bombardments, and for his related discovery of nuclear reactions brought about by slow neutrons".

Karl Manne Georg Siegbahn was a Swedish physicist. He distinguished himself by his work in nuclear physics. He is especially famous for what is known as the Siegbahn notation, used to designate the lines of the X-ray spectra. He won the Nobel Prize in Physics 1924 for his research and discoveries in X-ray spectroscopy.

Pierre Auger was a French physicist. He distinguished himself by his contributions to atomic physics, nuclear physics and cosmic rays. The Auger effect named in his honor is testament to his fame.

Box 4.3. *Fermi (1901–1954); Siegbahn (1886–1978); Auger (1899–1993)*

4.2. Radioactive family trees

4.2.1. *Definition*

Certain unstable nuclei disintegrate into other nuclei, themselves radioactive.

The process gives rise to a decay sequence called *radioactive series* or *family tree*.

Let there be a sample of radioactive parent nuclei, X_1, of decay constant, λ_1. The decay produces radioactive intermediate parent nuclei, X_m, of decay constant λ_m, and leads to the stable daughter nucleus, X_n. The *decay chain* is written:

$$X_1 \xrightarrow{\lambda_1} X_2 \xrightarrow{\lambda_2} ....X_{m-1} \xrightarrow{\lambda_{m-1}} X_m \xrightarrow{\lambda_m} X_{m+1} \cdots$$

$$\ldots X_{n-1} \xrightarrow{\lambda_{n-1}} X_n \text{ (stable)} \qquad [4.44]$$

In equation [4.44], λ_m is the decay constant of the parent nucleus, X_m.

4.2.2. *Simple two-body family tree*

The simple family tree corresponds to the simple case of X_1 (m = 1) and X_2 (m = 2) nuclei. The decay chain is written according to [4.44]:

$$X_1 \xrightarrow{\lambda_1} X_2 \xrightarrow{\lambda_2} X_3 \text{ (stable)} \quad [4.45]$$

During decay, the quantity, dN_2, of nuclei X_2 decreases but also increases by the quantity $-dN_1$ originating from the decay of the parent nuclei, X_1:

$$dN_2 = -\lambda_2 N_2 dt - dN_1 \quad [4.46a]$$

For the parent nucleus:

$$dN_1 = -\lambda_1 N_1 dt \quad [4.46b]$$

Using [4.46b], equation [4.46a] is written:

$$dN_2 = -\lambda_2 N_2 dt + \lambda_1 N_1 dt \quad [4.46c]$$

If N_{01} designates the initial number of the parent nuclei, X_1, the decay law of the number N_1 (t) of these parent nuclei is written, according to Rutherford and Soddy's empirical law [3.13]:

$$N_1(t) = N_{01} e^{-\lambda_1 t} \quad [4.46d]$$

By applying solution [4.46d] to relationship [4.46c], we obtain:

$$\frac{dN_2(t)}{dt} + \lambda_2 N_2(t) = \lambda_1 N_{01} e^{-\lambda_1 t} \quad [4.47]$$

The solution to differential equation [4.47] is of the form:

$$N_2(t) = K_1 e^{-\lambda_1 t} + K_2 e^{-\lambda_2 t} \quad [4.48a]$$

In equation [4.48a], K_1 and K_2 are constants to be determined.

A specific solution to differential equation [4.47] is obtained for K_2 = 0. This then leads to the relationship:

$$N_2(t) = K_1 e^{-\lambda_1 t} \Rightarrow \frac{dN_2(t)}{dt} = -\lambda_1 K_1 e^{-\lambda_1 t} \quad [4.48b]$$

Using the last equation [4.48.], we obtain, according to [4.47]:

$$-\lambda_1 K_1 e^{-\lambda_1 t} + \lambda_2 K_1 e^{-\lambda_1 t} = \lambda_1 N_{01} e^{-\lambda_1 t} \quad [4.48c]$$

We deduce from equation [4.48c] the expression of K_1, i.e.:

$$K_1 = \frac{\lambda_1}{\lambda_2 - \lambda_1} N_{01} \quad [4.49]$$

Replacing K_1 with its expression [4.49] in [4.48a] results in:

$$N_2(t) = \frac{\lambda_1}{\lambda_2 - \lambda_1} N_{01} e^{-\lambda_1 t} + K_2 e^{-\lambda_2 t} \quad [4.50]$$

Using the initial conditions for which at $t = 0$, N_2 $(t = 0) = 0$, the expression of K_2 is determined from [4.50]. We thus obtain:

$$K_2 = -\frac{\lambda_1}{\lambda_2 - \lambda_1} N_{01} \quad [4.51]$$

Using results [4.49] and [4.51], the *law of accumulation of the daughter product*, X_2, is written according to [4.48a]:

$$N_2(t) = \frac{\lambda_1}{\lambda_2 - \lambda_1} N_{01} \left(e^{-\lambda_1 t} - e^{-\lambda_2 t}\right) \quad [4.52]$$

Let us generalize equations [4.45] and [4.52] in the case of a parent nucleus, N_p, and a daughter nucleus, N_f, of any kind. The simple two-body family-tree equation is written:

$$X_p \xrightarrow{\lambda_p} X_f \xrightarrow{\lambda_f} X_n \text{ (stable)} \quad [4.53]$$

For the number $N_f(t)$ of daughter nuclei, we obtain:

$$N_f(t) = \frac{\lambda_p}{\lambda_f - \lambda_p} N_{0p} \left(e^{-\lambda_p t} - e^{-\lambda_f t}\right) \quad [4.54]$$

Using [4.54], the activity of the daughter nucleus, $A_f(t)$, is deduced:

$$A_f(t) = \frac{dN_f(t)}{dt} = \lambda_p N_p(t) - \lambda_f N_f(t) \quad [4.55]$$

4.2.3. *Multi-body family tree, Bateman equations*

The *Bateman equations* express the *radioactive family-tree general equations*. They are a set of equations describing the abundances and activities in a decay chain as a function of time. They thus allow determination of the evolution in the activity of a radioactive source.

Let there be a sample containing N_{01} parent nuclei at initial instant $t = 0$. Let N_m (t) be the number of intermediate parent nuclei and N_n the number of stable daughter nuclei. For $m < n$ [4.44], the relationship is written:

$$X_{m-1} \xrightarrow{\lambda_{m-1}} X_m \xrightarrow{\lambda_m} X_{m+1} \qquad [4.56]$$

The variation, dN_m, in the number of daughter nuclei ($m = f$) is given by the equation:

$$dN_m(t) = -\lambda_m N_m(t)dt + \lambda_{m-1} N_{m-1}(t)dt \qquad [4.57]$$

By recurrence, we obtain the number N_m of daughter nuclei:

$$N_m(t) = \frac{N_{01}}{\lambda_m} \sum_{i=1}^{m} C_i^m e^{-\lambda_i t} \qquad [4.58]$$

In relation [4.58], the coefficients C_i^m verify the equation:

$$C_i^m = \frac{\prod_{j=1}^{m} \lambda_j}{\prod_{j=1, j\neq i}^{m} (\lambda_j - \lambda_i)} \qquad [4.59]$$

Relations [4.58] and [4.59] express the Bateman equations.

APPLICATION 4.7.– Find the law of accumulation [4.50] resulting from the simple two-body family tree based on the Bateman equations.

ANSWER.– For the simple two-body family tree, $m = 2$ and $i = 1$. Using general equation [4.58], we obtain:

$$N_2(t) = \frac{N_{01}}{\lambda_2} \sum_{i=1}^{2} C_i^2 e^{-\lambda_i t} = \frac{N_{01}}{\lambda_2} \left(C_1^2 e^{-\lambda_1 t} + C_2^2 e^{-\lambda_1 t} \right) \qquad [4.60]$$

According to [4.59]:

$$C_1^2 = \frac{\prod\limits_{j=1}^{2} \lambda_j}{\prod\limits_{j \neq 1}^{2} (\lambda_j - \lambda_1)} = \frac{\lambda_1 \times \lambda_2}{(\lambda_2 - \lambda_1)} \quad [4.61a]$$

$$C_2^2 = \frac{\prod\limits_{j=1}^{2} \lambda_j}{\prod\limits_{j=1, j \neq 2}^{2} (\lambda_j - \lambda_2)} = \frac{\lambda_1 \times \lambda_2}{(\lambda_1 - \lambda_2)} \quad [4.61b]$$

By applying [4.61] to [4.60], we obtain:

$$N_2(t) = \frac{N_{01}}{\lambda_2}\left(\frac{\lambda_1\lambda_2}{(\lambda_2 - \lambda_1)} e^{-\lambda_1 t} + \frac{\lambda_1\lambda_2}{(\lambda_1 - \lambda_2)} e^{-\lambda_1 t}\right) \quad [4.62]$$

By arranging [4.62], we indeed find the law of accumulation [4.50].

APPLICATION 4.8.– Write the decay chain for a three-body problem and then deduce from the Bateman equations, the corresponding law of accumulation of the daughter product. Show that under certain conditions that will be specified, we can find from the three-body problem, the law of accumulation [4.50] relating to the simple two-body family tree.

ANSWER.– For the three-body family tree, the decay chain is written according to [4.44]:

$$X_1 \xrightarrow{\lambda_1} X_2 \xrightarrow{\lambda_2} X_3 \xrightarrow{\lambda_3} X_4 \text{ (stable)} \quad [4.63]$$

Using [4.58], we obtain, for the parent nucleus, X_3:

$$N_3(t) = \frac{N_{01}}{\lambda_3}\sum_{i=1}^{3} C_i^3 e^{-\lambda_i t} = \frac{N_{01}}{\lambda_3}\left(C_1^3 e^{-\lambda_1 t} + C_2^3 e^{-\lambda_2 t} + C_3^3 e^{-\lambda_3 t}\right) \quad [4.64]$$

According to [4.59]:

$$C_1^3 = \frac{\prod_{j=1}^{3} \lambda_j}{\prod_{j\neq 1}^{3} (\lambda_j - \lambda_1)} = \frac{\lambda_1 \times \lambda_2 \times \lambda_3}{(\lambda_2 - \lambda_1) \times (\lambda_3 - \lambda_1)} \quad [4.\ 65a]$$

$$C_2^3 = \frac{\prod_{j=1}^{3} \lambda_j}{\prod_{j=1, j\neq 2}^{3} (\lambda_j - \lambda_2)} = \frac{\lambda_1 \times \lambda_2 \times \lambda_3}{(\lambda_1 - \lambda_2) \times (\lambda_3 - \lambda_2)} \quad [4.65b]$$

$$C_3^3 = \frac{\prod_{j=1}^{3} \lambda_j}{\prod_{j=1, j\neq 3}^{3} (\lambda_j - \lambda_3)} = \frac{\lambda_1 \times \lambda_2 \times \lambda_3}{(\lambda_1 - \lambda_3) \times (\lambda_2 - \lambda_3)} \quad [4.65c]$$

Using [4.65], the accumulation equation [4.64] of the parent nucleus, X_3, is written:

$$N_3(t) = \frac{N_{01}}{\lambda_3} \left(\frac{\lambda_1 \times \lambda_2 \times \lambda_3}{(\lambda_2 - \lambda_1) \times (\lambda_3 - \lambda_1)} e^{-\lambda_1 t} + \frac{\lambda_1 \times \lambda_2 \times \lambda_3}{(\lambda_1 - \lambda_2) \times (\lambda_3 - \lambda_2)} e^{-\lambda_2 t} + \frac{\lambda_1 \times \lambda_2 \times \lambda_3}{(\lambda_1 - \lambda_3) \times (\lambda_2 - \lambda_3)} e^{-\lambda_3 t} \right)$$

Ultimately, this gives:

$$N_3(t) = \frac{\lambda_1 \times \lambda_2}{(\lambda_2 - \lambda_1)} N_{01} \left(\frac{e^{-\lambda_1 t}}{(\lambda_3 - \lambda_1)} - \frac{e^{-\lambda_2 t}}{(\lambda_3 - \lambda_2)} + \frac{(\lambda_2 - \lambda_1) e^{-\lambda_3 t}}{(\lambda_1 - \lambda_3) \times (\lambda_2 - \lambda_3)} \right) \quad [4.66]$$

Relative law of accumulation of the simple two-body family tree

In the case where the daughter product, X_2, has a very short radioactive half-life compared to the long decay half-life of the parent nucleus, X_1 ($T_1 >> T_2$), the X_2 daughter product disappears instantly so that the X_2 nucleus practically transforms into the stable X_3 nucleus. In this case, the law of accumulation of X_3 is written according to [4.66] ($T_2 \to 0$; $\lambda_2 \to \infty$):

$$N_3(t) = \frac{\lambda_1 \times \lambda_2}{\lambda_2} N_{01}\left(\frac{e^{-\lambda_1 t}}{(\lambda_3 - \lambda_1)} + \frac{\lambda_2 e^{-\lambda_3 t}}{(\lambda_1 - \lambda_3)\times\lambda_2}\right) \quad [4.67]$$

i.e. after simplification and arrangement:

$$N_3(t) = \frac{\lambda_1}{(\lambda_3 - \lambda_1)} N_{01}\left(e^{-\lambda_1 t} - e^{-\lambda_3 t}\right)$$

We indeed find the law of accumulation [4.50] relating to the simple two-body family tree. The index "3" simply needs to be changed to the index "2".

4.2.4. *Secular equilibrium*

In the specific case where the period, T_1, of the parent nucleus, X_1, is very large in relation to the half-lives T_m ($m \geq 2$) of all the daughter nuclei, a *radioactive equilibrium* called *secular equilibrium* or *state equilibrium* is produced.

When the condition $T_1 >> T_m$ is satisfied, the exponential factors in general equation [4.58] are negligible in relation to the first factor. This then gives:

$$\lambda_m N_m(t) = N_{01} C_1^m e^{-\lambda_1 t} \quad [4.68]$$

Given that $\lambda_j >> \lambda_1$, $\lambda_j - \lambda_1 \approx \lambda_j$, general equation [4.59] gives:

$$C_1^m = \frac{\prod_{j=1}^{m} \lambda_j}{\prod_{j=2,}^{m} (\lambda_j - \lambda_1)} \approx \frac{\lambda_1 \times \prod_{j=2}^{m} \lambda_j}{\prod_{j=2,}^{m} \lambda_j} \approx \lambda_1 \quad [4.69]$$

Using [4.69], relationship [4.68] is placed in the final form:

$$\lambda_m N_m(t) = \lambda_1 N_{01} e^{-\lambda_1 t} \Rightarrow \lambda_m N_m(t) = \lambda_1 N_1(t) \quad [4.70]$$

As when secular equilibrium is established, all activities of the daughter products are equal, i.e.:

$$A_1(t) = A_2(t) = A_3(t) = \text{, etc.} = A_m(t) \quad [4.71]$$

Let us illustrate, as an example, the decay law of a radioactive source and the accumulation law of its daughter product. To achieve this, let us consider the decay of tellurium-131 ($T = 30$ hrs) and of its daughter product, iodine-131 ($T = 8.0$ days) leading to the stable daughter nucleus, xenon-131 according to decay chain [2.163a], which we recall here:

$$^{131}_{52}Te \xrightarrow{\beta^-} {}^{131}_{53}I \xrightarrow{\beta^-} {}^{131}_{54}Xe$$

Figure 4.13 shows the curves of the evolution of the activities of a pure source of ^{131}Te and of the evolution of its daughter product, ^{131}I.

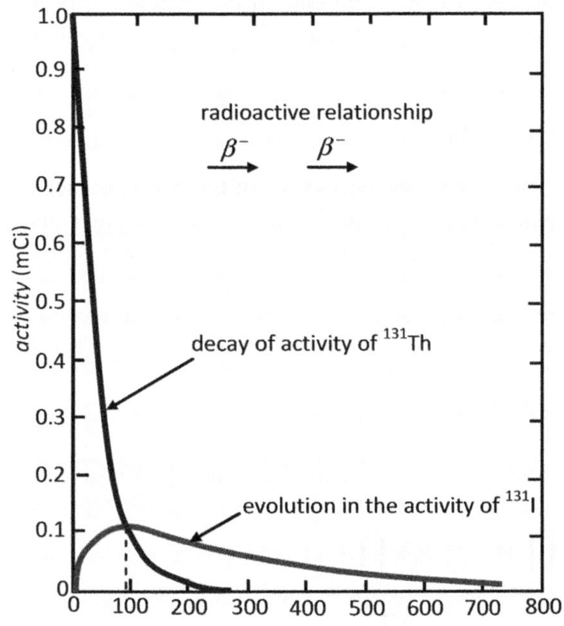

Figure 4.13. *Evolution curves of the activities of tellurium-131 and its daughter product, iodine-131. The activity of ^{131}I is maximum after 95.0 hrs. 1 Ci = 3.7×10^{10} Bq (the units of activity accepted today are the becquerel (Bq) and the curie (CI))*

As shown in Figure 4.13, for $t < 100$ hrs, the activity of tellurium-131 decays rapidly while that of its daughter product accumulates to a maximum after a period equal to 95.0 hrs, corresponding to the secular equilibrium state. Beyond 100 hours, the activities of both nuclei decay. Between the initial instant and the 700 hrs instant, the ratio of the activity of ^{131}I to that of ^{131}Te increases with time.

Using [4.71], it is shown that secular equilibrium is reached on the date t_m, given by the relationship (see application 4.9):

$$t_m = \frac{1}{\ln 2}\frac{T_1 T_2}{(T_1 - T_2)}\ln\left(\frac{T_1}{T_2}\right) \quad [4.72]$$

Numerically ($T_1 = 30$ hrs; $T_2 = 8.0$ days $= 192$ hrs), we obtain:

$$t_m = \frac{1}{\ln 2}\frac{30\times 192}{(30-192)}\ln\left(\frac{30}{192}\right) = 95.22 \text{ hrs}$$

APPLICATION 4.9.– Show that the ratio of the activity of ^{131}I to that of ^{131}Te increases continuously with time in accordance with the evolutions in the curves shown in Figure 4.13. Deduce therefrom result [4.72].

ANSWER.– The activities of ^{131}Te and ^{131}I are given, respectively, by the relationships:

$$A_1(t) = \lambda_1 N_1(t)\,;\; A_2(t) = \lambda_2 N_2(t) \quad [4.73]$$

The number of ^{131}Te parent nuclei decreases exponentially while that of the intermediate ^{131}I parent nucleus evolves according to the exponential decay law:

$$A_1(t) = \lambda_1 N_{01} e^{-\lambda_1 t} \quad [4.74a]$$

$$A_2(t) = \frac{\lambda_1 \lambda_2}{\lambda_2 - \lambda_1} N_{01}\left(e^{-\lambda_1 t} - e^{-\lambda_2 t}\right) \quad [4.74b]$$

Using [4.74], the ratio of activities $A_2(t)/A_1(t)$ is written:

$$\frac{A_2(t)}{A_1(t)} = \frac{\lambda_2}{\lambda_2 - \lambda_1} e^{\lambda_1 t}\left(e^{-\lambda_1 t} - e^{-\lambda_2 t}\right)$$

which gives:

$$\frac{A_2(t)}{A_1(t)} = \frac{\lambda_2}{\lambda_2 - \lambda_1}\left(1 - e^{(\lambda_1 - \lambda_2)t}\right) \quad [4.74c]$$

Knowing that $T_2 > T_1$ (T_1 = 30 hrs; T_2 = 8.0 days), then $\lambda_1 > \lambda_2$. Let us write $\Delta\lambda = (\lambda_1 - \lambda_2) > 0$. Equation [4.74c] ultimately gives:

$$\frac{A_2(t)}{A_1(t)} = \frac{\lambda_2}{\lambda_2 - \lambda_1} e^{\lambda_1 t}\left(e^{-\lambda_1 t} - e^{-\lambda_2 t}\right) \qquad [4.75]$$

Result [4.75] indeed shows that the ratio of activity of ^{131}I activity to ^{131}Te activity increases continuously over time since $\Delta\lambda = (\lambda_1 - \lambda_2) > 0$. Let us now deduct from [4.75] result [4.72].

When the state equilibrium is reached on the date t_m, then $A_2(t_m)/A_1(t_m) = 1$. Using [4.75], we obtain:

$$\frac{\lambda_2}{\lambda_1 - \lambda_2}\left(e^{(\lambda_1 - \lambda_2)t_m} - 1\right) = 1 \Rightarrow \frac{\lambda_1 - \lambda_2}{\lambda_2} = \left(e^{(\lambda_1 - \lambda_2)t_m} - 1\right)$$

This then gives:

$$e^{(\lambda_1 - \lambda_2)t_m} = \frac{\lambda_1}{\lambda_2} \Rightarrow (\lambda_1 - \lambda_2)t_m = \ln\left(\frac{\lambda_1}{\lambda_2}\right) \qquad [4.76a]$$

By introducing radioactive half-lives, the last equality [4.76a] gives:

$$\ln\left(\frac{T_2}{T_1}\right) = \left(\frac{\ln 2}{T_1} - \frac{\ln 2}{T_2}\right)t_m = \ln 2\left(\frac{T_2 - T_1}{T_1 T_2}\right)t_m \qquad [4.76b]$$

By arranging [4.76b], result [4.72] is found.

> Harry Bateman was a British mathematician. He is best known for his works in mathematical physics. In nuclear physics, he is famous for having established the general radioactive family-tree equations called Bateman equations.

Box 4.4. *Bateman (1882–1946)*

4.3. Radionuclide production by nuclear bombardment

4.3.1. *General aspects*

As we saw in section 1.3.1, in 1919 Rutherford performed the first artificial nuclear transmutation reaction from the bombardment of nitrogen nuclei by

α particles. The particles resulting from the reaction were identified by Blackett in 1925 as indicated in balanced equation [1.23], which we recall here:

$$ {}^{4}_{2}He + {}^{14}_{7}N \rightarrow {}^{1}_{1}H + {}^{17}_{8}O $$

In addition, nuclear transmutation reactions were carried out in 1934 by Irène and Frédéric Joliot-Curie by carrying out aluminum bombardment by α particles via equation [3.83], which we also recall here:

$$ {}^{4}_{2}\mathrm{He} + {}^{27}_{13}\mathrm{Al} \rightarrow {}^{1}_{0}\mathrm{n} + {}^{30}_{15}\mathrm{P} $$

These experiments thus enabled the creation of radiophosphorous-30, which decays according to the β^{+} mode via equation [3.84].

Thus, the production of radionuclides by nuclear bombardment has been known since 1934. Nuclear reactions [1.23] and [3.83] verify the general nuclear transmutation equation noted *X* (*a*, *b*) *Y*. The equation for these induced nuclear reactions is written:

$$ a + X \rightarrow b + Y \quad [4.77] $$

In equation [4.77]:

- *X* is the *target core*;
- *Y* represents the *residual core*;
- *a* designates the *incident particle*;
- *b* is the *emerging particle*.

APPLICATION 4.10.– Give the notation of nuclear transmutation reactions [1.23] and [3.84] according to the label *X*(*a*, *b*)*Y*.

ANSWER.– $^{14}\mathrm{N}(\alpha,p)^{17}\mathrm{O}$ and $^{27}\mathrm{Al}(\alpha,n)^{30}\mathrm{P}$. Note that in these notations, the helium-4 nucleus is noted α (not ^{4}He) and the proton *p* (not ^{1}H).

4.3.2. *Production rate of a radionuclide*

Let us consider a target nucleus, X_1, producing β-emitting radioactive nuclei, X_2, following a nuclear bombardment. The decay product of nucleus X_2 is designated by X_3. The radioactive relationship may be described by the series:

$$^{A_1}X_1 \xrightarrow[\lambda_1]{(a,b)} {}^{A_2}X_2 \xrightarrow[\lambda_2]{\beta} {}^{A_3}X_3 \qquad [4.78]$$

In equation [4.78], λ_1 is the probability per unit time of a nucleus X_1 transforming into a nucleus X_2 by bombardment using the incident particle a, and λ_2 is the probability per unit time of a nucleus X_2 decaying according to the β mode.

An illustrative example is the reaction ^{23}Na(d, p)^{24}Na of production of β radioactive sodium-24 by continuous bombardment of sodium-23 by a beam of deuterons. The radioactive relationship can be written according to [4.78]:

$$^{23}Na \xrightarrow[\lambda_1]{(d,p)} {}^{24}Na \xrightarrow[\lambda_2]{\beta^-} {}^{24}Mg \qquad [4.79]$$

If N_{01} designates the initial number of target nuclei X_1, the decay law of the number, $N_1(t)$, of nuclei X_1 is written:

$$N_1(t) = N_{01} e^{-\lambda_1 t} \qquad [4.80]$$

In the general case, if the probability per unit time λ_1 is very small ($\lambda_1 \to 0$), the number of nuclei, on the other hand, is very large. In this case, decay law [4.80] is written, approximately:

$$N_1(t) \approx N_{01}(1 - \lambda_1 t) = N_{01} - \lambda_1 N_{01} t \qquad [4.81]$$

Using [4.81], we obtain:

$$\lambda_1 N_{01} t = N_{01} - N_1(t) \qquad [4.82]$$

For a continuous bombardment time τ, equation [4.82] gives:

$$\lambda_1 N_{01} = \frac{N_{01} - N(\tau)}{\tau} \qquad [4.83]$$

By definition, the quantity $\lambda_1 N_{01}$ is referred to as the production rate of radioactive nucleus X_2. In practice, the *rate of production of a radionuclide* is low ($\lambda_1 \to 0$). For this reason, experiments of intense and prolonged irradiation of target nuclei are carried out to obtain an appreciable amount of radionuclides.

4.3.3. *Production yield of a radionuclide*

Let us determine the law of evolution of radionuclide X_2. Its activity a_2 (t) is given by equation [4.74b], which we recall here:

$$A_2(t) = \frac{\lambda_1 \lambda_2}{\lambda_2 - \lambda_1} N_{01} \left(e^{-\lambda_1 t} - e^{-\lambda_2 t} \right)$$

In addition, the term *production yield of a radionuclide* is used to refer to the *yield of a nuclear reaction, X*(*a*, *b*)*Y* producing radionuclides, *Y*. By definition, the production yield of a radionuclide, X_2, via equation [4.78] is equal to the value of the derivative of its activity [4.74b] with respect to time, i.e.:

$$r = \frac{dA_2(t)}{dt} \qquad [4.84a]$$

According to [4.84a], the yield, *r*, is the size of an activity per unit time. It is therefore expressed in becquerel per second ($Bq \cdot s^{-1}$).

When irradiation produces a single type of radionuclide, X_2, the yield, *r*, is given by the relationship:

$$r = \left. \frac{dA_2(t)}{dt} \right|_{t=0} \qquad [4.84b]$$

Considering [4.84b], we obtain:

$$\frac{dA_2(t)}{dt} = \frac{\lambda_1 \lambda_2}{\lambda_2 - \lambda_1} N_{01} \left(-\lambda_1 e^{-\lambda_1 t} + \lambda_2 e^{-\lambda_2 t} \right) \qquad [4.84c]$$

At instant $t = 0$, we then obtain:

$$r = \frac{\lambda_1 \lambda_2}{\lambda_2 - \lambda_1} N_{01} (-\lambda_1 + \lambda_2) = \lambda_1 \lambda_2 N_{01} \qquad [4.85]$$

The yield, *r*, is therefore constant.

Considering the radioactive half-life, we obtain:

$$T = \frac{\ln 2}{\lambda} \Rightarrow \frac{1}{\lambda} = \frac{T}{\ln 2} = 1.44T \qquad [4.86a]$$

Let us now introduce the time variable:

$$\tau = \frac{1}{\lambda} = 1.44T \qquad [4.86b]$$

The activity of X_2 is given by equation [4.84b]. Since $\lambda_1 \to 0$ and $\lambda_2 >> \lambda_1$, we thus obtain:

$$A_2(t) = \frac{\lambda_1 \lambda_2}{\lambda_2 - \lambda_1} N_{01} \left(e^{-\lambda_1 t} - e^{-\lambda_2 t} \right) \approx \frac{\lambda_1 \lambda_2}{\lambda_2 - \lambda_1} N_{01} \left(1 - e^{-\lambda_2 t} \right)$$

That is:

$$A_2(t) \approx \lambda_1 N_{01} \left(1 - e^{-\lambda_2 t} \right) \qquad [4.87]$$

According to [3.87] and [3.86b]:

$$\lambda_1 N_{01} = \frac{r}{\lambda_2} = r\tau_2 \qquad [4.88]$$

Considering [4.88], equation [4.87] is then written:

$$A_2(t) \approx r\tau_2 \left(1 - e^{-\lambda_2 t} \right) \qquad [4.89]$$

The activity of X_2 is maximum for $t \to \infty$. That is, according to [4.89]: $A_{2\max} = r\tau_2$.

Let us examine graphically the variation in activity A_2 (t) of product X_2. The resulting curve is shown in Figure 4.14.

It is interesting to compare the evolution in the activity, A_2 (t), shown in Figure 4.14 with that of the charge curve of a capacitor of capacity C through an ohmic conductor of resistance R. The circuit (R, C) considered of *time constant* $\tau = RC$, is fed by an ideal e.m.f. generator, E. During the experiment, the electrical charge, q (t), of the capacitor varies according to the law [SAK 16]:

$$q(t) = Q_{\max} \left(1 - e^{-t/\tau} \right) = CE \left(1 - e^{-t/\tau} \right) \qquad [4.90]$$

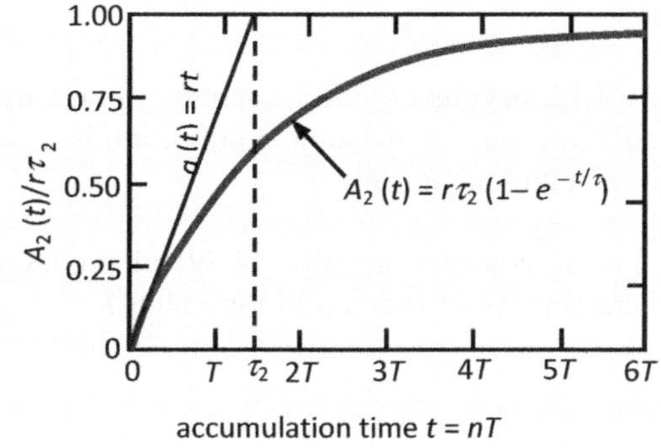

Figure 4.14. *Variation in the activity, A_2 (t), of a radionuclide, X_2, obtained by nuclear irradiation. The maximum activity is equal to $r\tau_2$. When X_2 is unique, the yield, r, is constant and equal to the slope of the tangent to curve A_2 (t) at instant t = 0. The tangent to curve A_2 (t) intersects the horizontal asymptote, $A_2(t) = r\tau_2$ at time $t = r\tau_2$, with τ_2 the nuclear time constant: $\tau_2 = 1/\lambda_2$*

The curve $q(t)$ given by [4.90] has the same shape as the curve giving the evolution, a_2, over time (Figure 4.14). Using [4.90], we obtain:

– at $t = \tau$, $q(\tau) = 0.63\ Q_{max} = 63\%\ Q_{max}$;

– at $t = 5\tau$, $q(5\tau) = 0.99\ Q_{max} = 99\%\ Q_{max}$.

Thus for circuit (R, C), the charge, $q(t)$, of the capacitor reaches 63% of its maximum charge, Q_{max}, at the instant $t = \tau$. At $t = 5\tau$, the capacitor is almost charged.

By analogy with circuit (R, C), we obtain, using the law of evolution [4.89] of activity $A_2(t)$:

– at $t = \tau_2$, $A_2(\tau) = 0.63\ A_{2max} = 63\%\ a_{2max}$;

– at $t = 5\tau$, $A_2(5\tau) = 0.99\ A_{2max} = 99\%\ a_{2max}$.

Thus for a radioactive source, X_2, produced by nuclear irradiation, the activity $A_2(t)$ of the source reaches 63% of its maximum charge A_{2max} at instant $t = \tau$. At $t = 5\tau$, the accumulation of the radionuclide virtually stops and remains constant. By analogy with the time constant of the circuit (R, C), $\tau = 1/\lambda$ [4.86b] can be called the "*nuclear time constant*" of a radioactive source produced by nuclear irradiation. It provides an estimate of the accumulation time of the radionuclide produced by nuclear bombardment.

As shown in Figure 4.14, beyond $t = 6T$, the activity of radionuclide X_2 no longer increases with accumulation time. A dynamic equilibrium is established where the production and decay velocities are equal.

APPLICATION 4.11.– Let us consider equation [4.79] of production of sodium-24. Determine the maximum activity of sodium-24 for which $T = 15.0$ hrs and $r = 4.107 \times 10^8$ Bq · h^{-1}.

ANSWER.– We have:

$$\tau_2 = 1.44\ T_2 \Rightarrow A_{2\max} = r \times 1.44\ T_2 \qquad [4.91]$$

Numerically this gives:

$$A_{2\max} = 4.107 \times 10^8 \times 1.44 \times 15.0 = 8.87 \times 10^9 \text{ Bq}$$

APPLICATION 4.12.– Show that for a single type of radionuclide, X_2, of activity $A_2\ (t)$, the yield, r, is constant and equal to the slope of the tangent to curve $A_2\ (t)$ at instant $t = 0$ (Figure 4.14).

Let us recall the equation for the tangent at point x_0:

$$f(x) = f'(x_0)(x - x_0) + f(x_0) \qquad [4.92a]$$

ANSWER.– The equation for the tangent to curve *a2 (t)* at the origin verifies the relationship:

$$a(t) = a_2'(t_0)(t - t_0) + a_2(t_0) = a_2'(0)t + a_2(0) \qquad [4.92b]$$

Using [4.90], we obtain:

$$a_2(0) = 0\,;\, a_2'(0) = \left.\frac{da_2(t)}{dt}\right|_{t=0} = r\tau_2\lambda_2 \qquad [4.92c]$$

Yet $\tau = 1/\lambda$. Taking [4.86] into account, we then obtain:

$$a_2(0) = r\tau_2\lambda_2 = r = \lambda_1\lambda_2 N_{01} \qquad [4.92d]$$

The equation for the tangent of curve a_2 (t) at the origin is then written:

$$a(t) = \lambda_1 \lambda_2 N_{01}\, t = r\, t \quad [4.93]$$

Thus, when irradiation produces a single type of radionuclide, X_2, of activity A_2 (t), the yield, r, is constant and equal to the slope of the tangent to curve A_2 (t) at instant $t = 0$ (Figure 4.14).

4.4. Natural radioactive series

4.4.1. *Presentation*

There are four *radioactive families*, three of which are natural. They are the *thorium family*, the *neptunium family,* the *uranium-238 family* and the *uranium-235 family*. The *natural radioactive families* are the thorium, uranium-238 and uranium-235 families. The member of the neptunium family with the longest half-life is neptunium-237 with radioactive half-life $T = 2.44 \times 10^6$ years. As a consequence, no single member of the neptunium family exists in nature. All members of this family are artificially created, making the neptunium family the only *artificial radioactive family*. For each of the three natural radioactive families, the family tree ends with a daughter nucleus, which is a stable isotope of lead (Pb). Inversely, the neptunium family yields stable thallium-205. All nuclei of the family tree derived from the same parent nucleus is called the *radioactive family* or *radioactive series*.

In a given series, each family member decays either by the α mode or by the β mode. For each of these members, the mass number, A, is of the type:

$$A = 4n + q,\ n > 0;\ 0 \le q \le 3 \quad [4.94]$$

In relationship [4.94], n is a real number and q is an integer.

The value of q is used to differentiate the four radioactive families:

– thorium family: $q = 0 \Rightarrow A = 4n$;

– neptunium family: $q = 1 \Rightarrow A = 4n + 1$;

– uranium-238 family: $q = 2 \Rightarrow A = 4n + 2$;

– uranium-235 family: $q = 2 \Rightarrow A = 4n + 3$.

In the following, we will present the characteristics relating to each of the radioactive families mentioned above.

4.4.2. *Thorium (4n) family*

In the thorium family, each member has a mass number, A, equal to a multiple of 4, i.e. $A = 4n$. The variations in A and n are different depending on whether the decay is either of α type or β type.

As an example:

– for the $^{232}\text{Th} \rightarrow {}^{228}\text{Ra}$ transition according to the α mode, $\Delta A = (232 - 228) = 4$; $n\,(232) = 232/4 = 58$; $n\,(228) = 228/4 = 57 \Rightarrow \Delta n = (57 - 58) = -1$;

– for the $^{212}\text{Pb} \rightarrow {}^{212}\text{Bi}$ isobaric transition according to the β mode, $\Delta A = 0$; $\Delta n = (57 - 58) = -1$.

The precursor element of the family is thorium-232 with a radioactive half-life of $T = 14.05 \times 10^9$ years. The stable end product of the family tree is lead-208.

The long half-life of thorium-232 makes it possible to find it in nature with its daughter products. The parent substance of thorium-232 is uranium-236, which is an α emitter with radioactive half-life $T \approx 5 \times 10^7$ years. Given that the Earth's age is estimated at around 4.5×10^9 years, uranium-236 is no longer found in terrestrial sources. It follows that thorium-232 has not formed in the Earth's crust since the extinction of its parent substance.

The decay chain of the thorium-232 family giving all the descendants of the family tree is written:

$$
\begin{gathered}
{}^{232}_{90}\text{Th} \xrightarrow[1.41\times10^{10}\,\text{a}]{\alpha} {}^{228}_{88}\text{Ra} \xrightarrow[5.75\,\text{a}]{\beta^-} {}^{228}_{89}\text{Ac} \xrightarrow[6.13\,\text{h}]{\beta^-} {}^{228}_{90}\text{Th} \xrightarrow[1.91\,\text{a}]{\alpha} {}^{224}_{88}\text{Ra} \\
{}^{224}_{88}\text{Ra} \xrightarrow[3.62\,\text{j}]{\alpha} {}^{220}_{86}\text{Rn} \xrightarrow[55.6\,\text{s}]{\alpha} {}^{216}_{84}\text{Po} \xrightarrow[146\,\text{ms}]{\alpha} {}^{212}_{82}\text{Pb} \xrightarrow[10.64\,\text{h}]{\beta^-} {}^{212}_{83}\text{Bi} \\
{}^{212}_{83}\text{Bi} \left\{ \begin{array}{l} \xrightarrow[60.55\,\text{min}]{64.07\%\,\beta^-} {}^{212}_{84}\text{Po} \xrightarrow[299\,\text{ns}]{\alpha} \\ \\ \xrightarrow[60.55\,\text{min}]{35.97\%\,\alpha} {}^{208}_{81}\text{Tl} \xrightarrow[3.05\,\text{min}]{\beta^-} \end{array} \right\} {}^{208}_{82}\text{Pb (stable)}
\end{gathered}
\qquad [4.95]
$$

The decay chain [3.95] of thorium-232 can be represented more simply using a diagram (Z, N). The diagram shown in Figure 4.15 is then obtained.

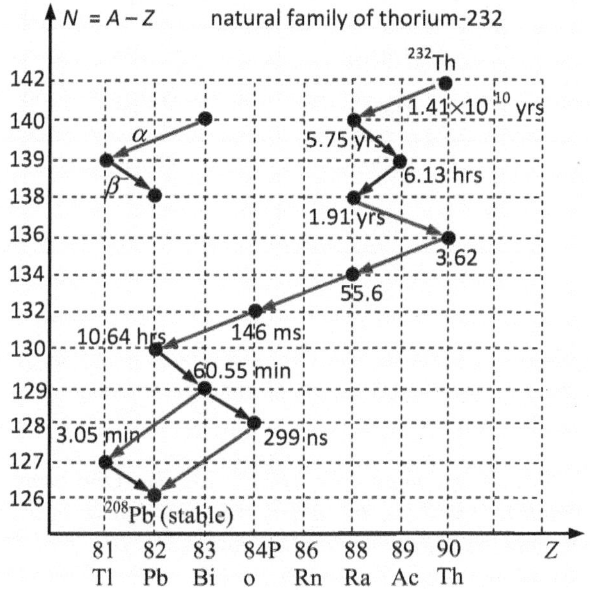

Figure 4.15. *Natural series of thorium. The stable end product of the family tree is lead-208*

Table 4.4 shows the correspondence between the old and conventional notations of nuclei belonging to the natural family of thorium.

Radionuclide	Old notation	Conventional notation	Radioactive half-time
Radium-228	MésoTh1	$^{228}_{88}$Ra	5.75 yrs
Actinium-228	MésoTh2	$^{228}_{89}$Ac	6.13 hrs
Thorium-228	RdTh*	$^{228}_{90}$Th	1.91 yrs
Radon-220	Tn**	$^{220}_{86}$Rn	55.6 sec
Polonium-216	ThA	$^{216}_{84}$Po	146 ms
Lead-212	ThB	$^{212}_{82}$Pb	10.64 hrs

Radionuclide	Old notation	Conventional notation	Radioactive half-time
Bismuth-212	ThC	${}^{212}_{83}Bi$	60.55 min
Lead-208	ThD	${}^{208}_{82}Pb$	∞
Polonium-212	ThC′	${}^{212}_{84}Po$	299 ns
Thallium-208	ThC″	${}^{208}_{81}Tl$	3.05 min
Radium-224	ThX	${}^{224}_{88}Ra$	3.62 days

* RadioThorium

** Thoron

Table 4.4. *Correspondence between old and conventional nucleus notations. The radionuclides listed belong to the natural family of thorium*

NOTE.– *Natural thorium is composed of 100% 232 Th. One gram of thorium-232 has an activity equal to 4,070 Bq. In nature, it is found in secular equilibrium with its descendant, thorium-228, at the rate of around* 1.3×10^{-10}*g of* ^{228}Th *per gram of* ^{232}Th.

4.4.3. *Neptunium (4n + 1) family*

Each member of the neptunium family with a mass number $A = 4n + 1$ decays according to the α mode or β mode. As with the thorium family:

– for the α mode: $\Delta A = 4$ or $\Delta n = -1$;

– for the β mode: $\Delta A = 0$ or $\Delta n = 0$

The first member of the neptunium family is plutonium-241. For all members of this family, neptunium-237 has the longest radioactive half-life of 2.44×10^6 years. Therefore, no member of the neptunium ($4n$ +1) family is present in nature. They have all decayed since the creation of the Earth some 4.5×10^9 years ago. The stable end product of the family tree is thallium-205. Among the descendants of the family, only bismuth-209 and thallium-205 are observed in nature. All other descendants of the ($4n$ +1) family including neptunium-237 are created artificially.

The decay chain of the neptunium family giving all the descendants of the family tree is written:

$$
\begin{array}{l}
{}^{237}_{93}Np \xrightarrow[2.144\times10^{6}y]{\alpha} {}^{233}_{91}Pa \xrightarrow[26.967\ d]{\beta^-} {}^{233}_{92}U \xrightarrow[1.592\times10^{5}y]{\alpha} {}^{229}_{90}Th \\
{}^{229}_{90}Th \xrightarrow[7340\ y]{\beta^-} {}^{225}_{88}Ra \xrightarrow[14.9\ d]{\beta^-} {}^{225}_{89}Ac \xrightarrow[10\ d]{\alpha} {}^{221}_{87}Fr \xrightarrow[4.9\ \min]{\alpha} {}^{217}_{85}At \\
{}^{217}_{85}At \xrightarrow[323\ ms]{\alpha} {}^{213}_{83}Bi \left\{ \begin{array}{l} \xrightarrow[46.59\ \min]{98\%\beta^-} {}^{213}_{84}Po \xrightarrow[4.2\ \mu s]{\alpha} \\ \\ \xrightarrow[46.59\ \min]{2\%\alpha} {}^{209}_{81}Tl \xrightarrow[2.161\ \min]{\beta^-} \end{array} \right\} {}^{209}_{82}Pb \\
{}^{209}_{82}Pb \xrightarrow[3.253\ h]{\beta^-} {}^{209}_{83}Bi \xrightarrow[1.9\times10^{19}y]{\alpha} {}^{205}_{81}Tl\ (stable)
\end{array}
\qquad [4.96]
$$

The diagram (*Z*, *N*) corresponding to the decay chain [4.96] is shown in Figure 4.16. This diagram reflecting the *neptunium series* has eleven descendants, among which the stable end product, ^{205}Tl. Bismuth-209 with radioactive half-life $T = 19$ billion billion can be considered stable.

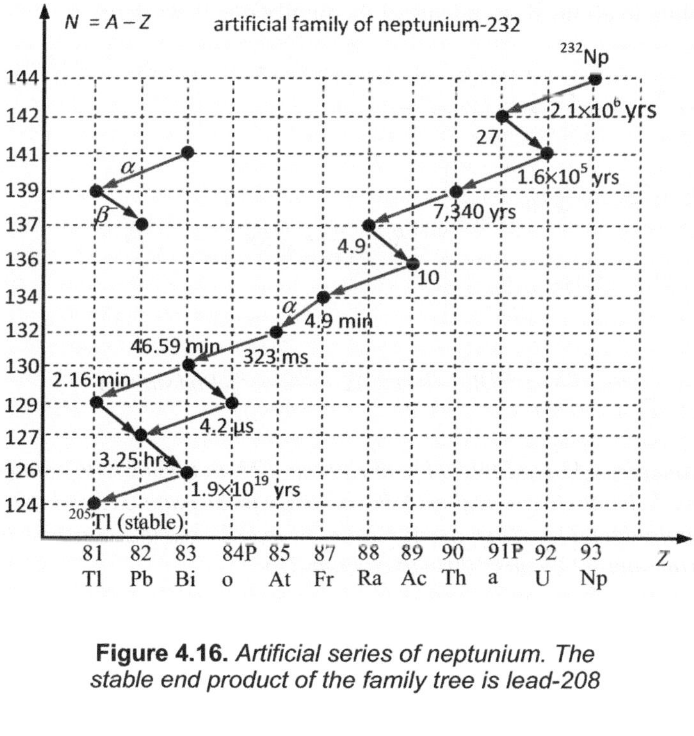

Figure 4.16. *Artificial series of neptunium. The stable end product of the family tree is lead-208*

4.4.4. *Uranium-235 (4n +2) family*

The uranium-235 family was formerly called the actinium family. Thus, ^{235}U was called Actinium Uranium, noted as AcU. Each member of the uranium-235 family has a mass number $A = 4n + 2$, and decays according to the α mode or β mode. In the uranium-235 family, the mass number, A, is odd. It follows that n is not an integer. As examples:

– for the ^{235}U → ^{231}Th transition, $\Delta A = (232 - 228) = 4$; $n\ (235) = (235 - 2)/4 = 58.25$; $n\ (231) = 57.25 \Rightarrow \Delta n = -1$;

– for the ^{211}Pb → ^{211}Bi isobaric transition; $\Delta A = 0$; $\Delta n = 0$.

In summary:

– for the α mode: $\Delta A = 4$ or $\Delta n = 1$;

– for the β mode: $\Delta A = 0$ or $\Delta n = 0$.

The first member of the uranium-235 family is plutonium-239. The very long radioactive half-life of uranium-235 is equal to 7×10^9 years. In the Earth's crust, natural uranium is composed of approximately 93% ^{238}U and 0.7% ^{235}U. The stable end product of the family tree is lead-207. The decay chain of the uranium-235 family leading to all the descendants of the family tree is written:

$$^{235}_{92}U \xrightarrow[7\times10^9 y]{\alpha} {}^{231}_{90}Th \xrightarrow[25.2\ d]{\beta^-} {}^{231}_{91}Pa \xrightarrow[32700\ y]{\alpha} {}^{227}_{89}Ac$$

$$^{227}_{89}Ac \begin{cases} \xrightarrow[21.8\ y]{98.62\%\beta^-} {}^{227}_{90}Th \xrightarrow[18.72\ d]{\alpha} \\ \xrightarrow[21.8\ y]{1.38\%\alpha} {}^{223}_{87}Fr \xrightarrow[21\ min]{\beta^-} \end{cases} {}^{223}_{88}Ra \xrightarrow[11.43d]{\alpha} {}^{219}_{86}Rn \xrightarrow[3.96\ s]{\alpha} {}^{215}_{84}Po \qquad [4.97]$$

$$^{215}_{84}Po \xrightarrow[1.78\ ms]{\alpha} {}^{211}_{82}Pb \xrightarrow[36.1\ min]{\beta^-} {}^{211}_{83}Bi \xrightarrow[2.15\ min]{\alpha} {}^{207}_{81}Tl \xrightarrow[4.77\ min]{\beta^-} {}^{207}_{82}Pb\ (stable)$$

NOTE.– *Francium-223 lead 99.994% to radium-223 and 0.006% to astate-219, itself radioactive. It leads 3% to radon-219 and 97% to bismuth-215, which transforms 100% to polonium-215. This decay chain is overlooked by considering that the decay of francium-223 leads 100% to radium-223.*

The diagram (*Z*, *N*) corresponding to the decay chain [4.97] is shown in Figure 4.17.

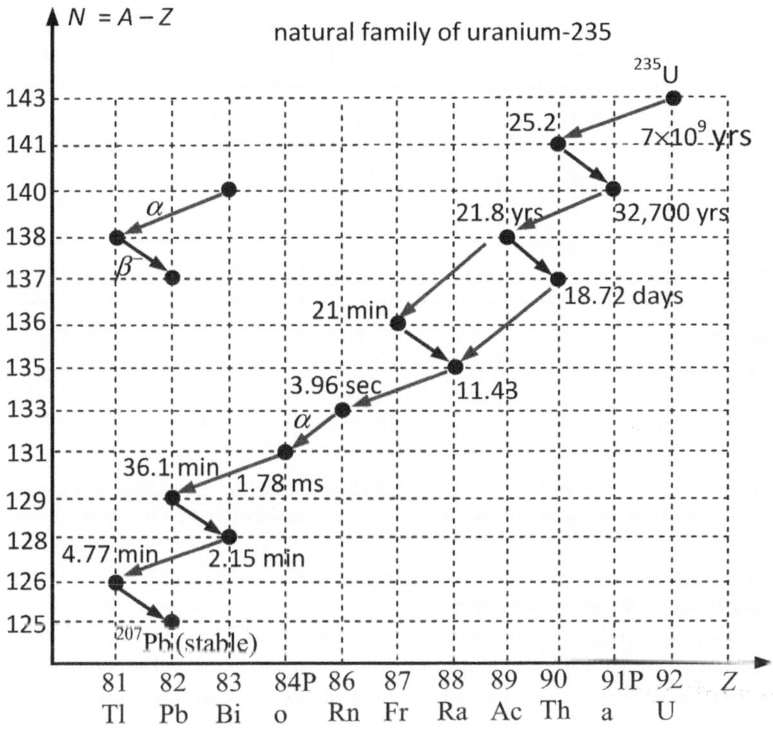

Figure 4.17. *Natural series of uranium-235. The stable end product of the family tree is lead-207*

Table 4.5 shows the correspondence between the old and conventional notations of nuclei belonging to the natural family of uranium-235.

Radionuclide	Old notation	Conventional notation	Radioactive half-time
Polonium-215	AcA	$^{215}_{84}Po$	1.78 ms
Lead-211	AcB	$^{211}_{82}Pb$	36.1 min
Bismuth-211	AcC	$^{211}_{83}Bi$	2.15 min

Radionuclide	Old notation	Conventional notation	Radioactive half-time
Lead-207	AcD	${}^{207}_{82}\mathrm{Pb}$	∞
Francium-223	AcK	${}^{223}_{87}\mathrm{Fr}$	21 min
Uranium-235	AcU	${}^{235}_{92}\mathrm{U}$	7×10^{9} yrs
Radium-223	AcX	${}^{223}_{88}\mathrm{Ra}$	11.43 days
Thallium-207	AcC″	${}^{207}_{81}\mathrm{Tl}$	1.3 ms
Radon-219	An	${}^{219}_{86}\mathrm{Rn}$	7×10^{9} yrs
Thorium-227	RdAc	${}^{231}_{90}\mathrm{Th}$	24.1 days

Table 4.5. *Correspondence between old and conventional nucleus notations. The radionuclides listed belong to the natural family of uranium-235*

4.4.5. *Uranium-238 (4n + 3) family*

The uranium family refers to radionuclides with a mass number, $A = 4n + 3$. The decay chain of uranium-238 giving all the descendants of the family tree is written:

$$
\begin{gathered}
{}^{238}_{92}U \xrightarrow[4.468\times10^{9}y]{\alpha} {}^{234}_{90}Th \xrightarrow[24.1\,d]{\beta^-} {}^{234}_{91}Pa \xrightarrow[1.17\text{ min}]{\beta^-} {}^{234}_{92}U \xrightarrow[2.445\times10^{5}y]{\alpha} {}^{230}_{90}Th \\
{}^{230}_{90}Th \xrightarrow[7.7\times10^{4}y]{\alpha} {}^{226}_{88}Ra \xrightarrow[1600\,y]{\alpha} {}^{222}_{86}Rn \xrightarrow[3.8235\,d]{\alpha} {}^{218}_{84}Po \xrightarrow[3.05\text{ min}]{\alpha} {}^{214}_{82}Pb \\
{}^{214}_{82}Pb \xrightarrow[26.8\text{ min}]{\beta^-} {}^{214}_{83}Bi \left\{ \begin{array}{l} \xrightarrow[19.9\text{ min}]{99.96\,\%\beta^-} {}^{214}_{84}Po \xrightarrow[163.7\,\mu s]{\alpha} \\ \xrightarrow[19.9\text{ min}]{0.04\,\%\alpha} {}^{210}_{81}Tl \xrightarrow[1.3\,ms]{\beta^-} \end{array} \right\} {}^{210}_{82}Pb \xrightarrow[22.26\,y]{\beta^-} {}^{210}_{83}Bi \\
{}^{210}_{83}Bi \xrightarrow[5.013\,d]{\beta^-} {}^{210}_{84}Po \xrightarrow[138.38\,d]{\alpha} {}^{206}_{82}Pb\,(stable)
\end{gathered}
\tag{4.98}
$$

The decay chain [4.98] is often translated into a (Z, N) diagram to represent the natural uranium series (Figure 4.18).

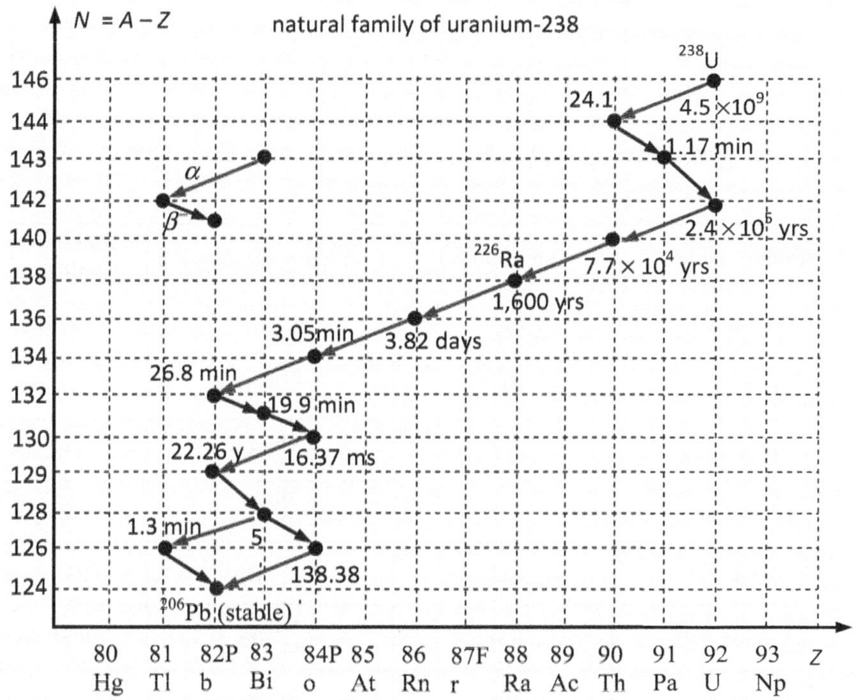

Figure 4.18. *Natural series of uranium-238 (often called uranium-radium series). The stable end product of the family tree is lead-206*

Table 4.6 shows the correspondence between the old and conventional notations of nuclei belonging to the natural family of uranium-238.

NOTE.– *Nuclear fuels*

Natural thorium 100% composed of ^{232}Th is around three times more abundant than natural uranium consisting 93% of ^{238}U and 0.7% of ^{235}U. Thorium-232 and uranium-238 are not fissile. In nuclear power plants, uranium-235 is used as a fuel. Its fission under the impact of a thermal neutron produces nuclear energy that can be converted into electrical energy. However, thorium is a fertile nucleus in the same way as uranium 238. Under the impact of a slow or thermal neutron, a fertile nucleus produces a fissile nucleus. Uranium-233 and plutonium-239 are fissile nuclei obtained by neutron bombardment of uranium-238 and thorium-232, respectively, according to the equations:

$$ {}_{0}^{1}n + {}_{92}^{238}U \xrightarrow{capture} {}_{92}^{239}U \xrightarrow[23\,\text{min}]{\beta^-} {}_{93}^{239}Np \xrightarrow[2.35\,d]{\beta^-} {}_{94}^{239}Pu \xrightarrow[24410\,y]{\alpha} {}_{92}^{235}U \qquad [4.99] $$

$$ {}_{0}^{1}n + {}_{90}^{232}Th \xrightarrow{capture} {}_{90}^{233}Th \xrightarrow[22\,\text{min}]{\beta^-} {}_{91}^{233}Pa \xrightarrow[27\,d]{\beta^-} {}_{92}^{233}U \xrightarrow[159200\,y]{\alpha} {}_{90}^{219}Th \qquad [4.100] $$

Radionuclide	**Old notation**	**Conventional notation**	**Radioactive half-time**
Polonium-218	RaA	${}_{84}^{218}\text{Po}$	3.05 min
Lead-214	RaB	${}_{82}^{214}\text{Pb}$	26.8 min
Bismuth-214	RaC	${}_{83}^{214}\text{Bi}$	19.9 min
Lead-210	RaD	${}_{82}^{210}\text{Pb}$	22.26 yrs
Bismuth-210	RaE	${}_{83}^{210}\text{Bi}$	5.013 days
Polonium-210	RaF	${}_{84}^{210}\text{Po}$	138.38 days
Lead-206	RaG	${}_{82}^{206}\text{Pb}$	∞
Polonium-214	RaC′	${}_{84}^{214}\text{Po}$	163.7 μs
Thallium-210	RaC″	${}_{81}^{210}\text{Tl}$	1.3 ms
Uranium-238	UI	${}_{92}^{238}\text{U}$	4.5×10^9 yrs
Uranium-234	UII	${}_{92}^{234}\text{U}$	2.4×10^5 yrs
Thorium-234	UX1	${}_{90}^{234}\text{Th}$	24.1 days
Protactinium-234	UX2	${}_{91}^{234}\text{Pa}$	1.17 hrs

Table 4.6. *Correspondence between old and conventional nucleus notations. The radionuclides listed belong to the natural family of uranium-238*

Mixed with depleted uranium to form MOX (mixed oxide) nuclear fuel consisting of PuO_2 and UO_2, the fission of plutonium-239 can be exploited for the manufacture of nuclear weapons such as atomic bombs. For this reason, its use is particularly controlled. As for uranium-233, its use as a nuclear fuel is one of the challenges of the future.

Today, the most widely used nuclear fuel is composed of pellets of uranium oxide, UO_2, enriched by 3% uranium-235. To demonstrate the importance of this fuel, let us calculate the energy released by the fission of 1 g of uranium-235 via the balanced equation [1.65], which we recall here:

$$ {}_{0}^{1}\mathrm{n} + {}_{92}^{235}\mathrm{U} \rightarrow {}_{38}^{94}\mathrm{Sr} + {}_{54}^{140}\mathrm{Xe} + 2\,{}_{0}^{1}\mathrm{n} + \gamma $$

The atomic masses and that of the neutron are equal to:

m (U) = 235.04392 u; m (Sr) = 93.91536 u;

m (Xe) = 139.91879 u; m_n = 1.00866 u.

The mass loss in reaction [1.65] has the value (overlooking the mass of the electrons):

$$ \Delta m = [m\ (\mathrm{U}) + m_n] - [m\ (\mathrm{Sr}) + m\ (\mathrm{Xe}) + 2\ m_n] = 0.2011\ \mathrm{u} \qquad [4.101] $$

The energy released by the fission of a nucleus of uranium-235 is then equal to:

$$ Q_l = \Delta mc^2 = 0.2011 \times 931.5 = 218708.6\ \mathrm{MeV} \qquad [4.102] $$

A mass m of nuclei contains a number N of nuclei equal to (with N_A Avogadro's number):

$$ \frac{m}{A} = \frac{N}{N_A} \Rightarrow N = \frac{mN_A}{A} \qquad [4.103] $$

The number of nuclei contained in 1 g of uranium-235 is equal to N_A/A. The energy released by the fission of a gram of uranium-235 is then equal to:

$$ Q = \frac{N_A}{235} \times Q_l \qquad [4.104] $$

Numerically, we obtain ($N_A = 6.02 \times 10^{23}\ \mathrm{mol}^{-1}$):

$$ Q = \frac{6.02 \times 10^{23}}{235} \times 187.33 \times 10^6 \times 1.602 \times 10^{-19} = 7.69 \times 10^{10}\ \mathrm{J} \qquad [4.105] $$

Let us compare the energy released by fission of one gram of uranium-235 with the heat value, $Q_p = 42$ MJ · kg $^{-1}$. The mass of oil producing the same amount of energy [4.105] by combustion is equal to:

$$m = 7.69 \times 10^{10}/4.2 \times 10^7 = 1.830.95 \text{ kg} \qquad [4.106]$$

Result [4.106] shows that in order to obtain the same amount of energy produced by the fission of **one gram** of uranium-235, approximately **2 tons of oil** needs to be burned. The stakes for nuclear energy are therefore high.

4.5. Exercises

For certain exercises, the following data will be used:

$M_0c^2 = 0.511$ MeV; 1 u = 931.5 MeV; 1 eV = 1.602×10^{-19} J; $N_A = 6.02 \times 10^{23}$ mol $^{-1}$.

EXERCISE 4.1.– Determination of the maximum energy of the β^+ spectrum from the semi-empirical Bethe–Weizsäcker formula

The semi-empirical Bethe–Weizsäcker formula [2.119] is recalled here:

$$M(A,Z)c^2 = ZM_Hc^2 + (A-Z)M_nc^2 - a_vA + a_sA^{2/3} + a_c\frac{Z^2}{A^{1/3}} + a_a\frac{(A-2Z)^2}{A} \pm |\delta|$$

Let us consider a radionuclide, ${}^AX$, of mass number $A = 2Z - 1$. Its β^+ decay from its ground level leads to stable isobar ${}^AY$.

a) Show that ${}^AX$ is a mirror nucleus.

b) Let m_X and m_Y be the respective masses of nuclei ${}^AX$ and ${}^AY$. Express the mass difference $(m_X - m_Y)$ as a function of m_p, m_n, A and a_c.

c) Then establish, as a function of A and a_c, the expression of the maximum energy, $E_{\max\beta+}$ of the β^+ spectrum, valid for all mirror nuclei of mass number $A = 2Z - 1$.

d) Table 4.7 shows several mirror nuclei ($A = 2Z - 1$), all β^+ emitters, without γ-photon emission. Reproduce and then complete the table, presenting the theoretical values of $E_{\max\beta+}$. Then verify the accuracy of the nuclear model with constant nucleon density.

Nucleus	$E_{max\beta+}$ (MeV) experiment	$E_{max\beta+}$ (MeV) theory
^{11}C	0.99	
^{13}N	1.24	
^{15}O	1.68	
^{17}F	1.72	
^{19}Ne	2.18	
^{21}Na	2.50	
^{23}Mg	2.99	
^{27}Si	3.48	
^{29}P	3.94	
^{31}S	3.9	
^{33}Cl	4.1	
^{35}Ar	4.4	
^{37}K	4.6	
^{39}Ca	5.1	
^{41}Sc	4.94	

Table 4.7. *Experimental values of maximum energy E_{max} of the β^+ spectra of several mirror nuclei for which A = 2Z – 1. The experimental values are taken from [EVA 61]*

Given data: $A_c = 0.585$ MeV; $(m_n - m_p) = 1.293$ MeV.

EXERCISE 4.2.– Decay of indium-114: maximum energies of β^- and β^+ spectra

Analysis of the emission spectra of particles resulting from the decay of a mass of 10 g of indium-114 (Z = 49) shows that the latter decays 99.5% according to the β^- mode and only 0.5% according to the β^+ mode. For each of these decay modes, isobaric transition leads directly to the ground level of the daughter nucleus. The radioactive half-life of indium-114 is $T = 71.9$ sec. The decay of indium-114 occurs from its ground state. In this exercise, we propose to determine the maximum energies of the β^- and β^+ spectra observed.

Given data (atomic masses): indium-114 (Z = 49): 113.904914 u; tin-114 (Z = 50): 113.902779 u; cadmium-114 (Z = 48): 113.9033585 u.

a) Justify the relative positions of the ground levels of the daughter nuclei relative to that of the parent nucleus. Then represent the decay diagram of indium-114. Include on the diagram the ground levels of the isobars involved in decay, the

J^{π} values corresponding to these levels, and the isobaric transitions observed including their percentages.

b) Write the β decay equations involved. Deduce therefrom the maximum energies of the β^- and β^+ spectra observed.

c) Calculate the masses (in grams) of daughter products collected after 143.8 sec of indium-114 decay. What masses do we obtain to infinity?

EXERCISE 4.3.– Decay of vanadium-48: energies of photons and Auger electrons

Studying the decay of vanadium-48 enabled Luis Alvarez to discover the phenomenon of electron capture in 1937. This radionuclide decays by β^+ emission and by electron capture in the *K*- or *L*-shell. This decay feeds approximately 100% of the 4^+ excited level of titanium-48. The *decay energy diagram of vanadium-48* is mapped out in Figure 4.19.

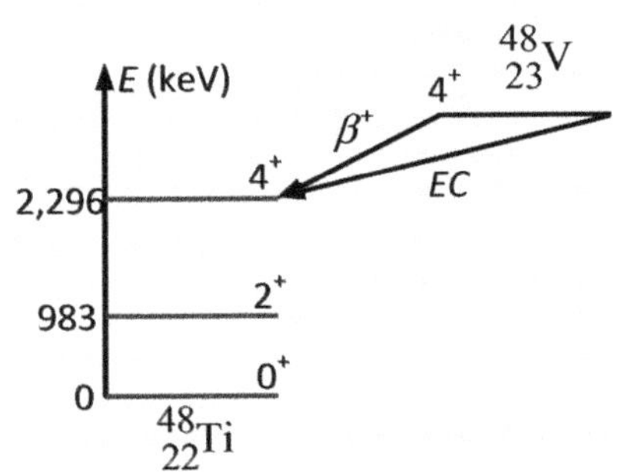

Figure 4.19. *Decay energy diagram of vanadium-48*

In this exercise, we propose to use the decay energy diagram of radiovanadium-48 (Figure 4.19) to determine the energies of X- and γ-photons as well as the energies of Auger electrons.

Given data:

– *Theoretical branching ratios*:

$$R_{\beta} = BR(\beta^+) = 45.8\%;\ R_{ECK} = BR^K(EC) = 48.2\%;\ R_{ECl} = BR^L(EC) = 6\%$$

– *Internal conversion coefficients*:

- $4^+ \rightarrow 0^+$ transition: $\alpha_K = 0.359$; $\alpha_L = 0.0385$;

- $4^+ \rightarrow 2^+$ transition: $\alpha_K = 5.88 \times 10^{-5}$; $\alpha_L = 5.25 \times 10^{-6}$;

- $2^+ \rightarrow 0^+$ transition: $\alpha_K = 1.15 \times 10^{-4}$; $\alpha_L = 1.03 \times 10^{-5}$.

– *C-shell binding energy for titanium 48*:

$W_K = 4.9664$ eV; $W_{L1} = 0.5637$ *eV*; $W_{L2} = 0.4615$ eV; $W_{L3} = 0.4555$ eV

– *Intensity, I_γ, of emitted photons*:

$$I_\gamma = \frac{I_{tot}}{\alpha_{tot} + 1} \quad [4.107]$$

In relationships [3.199]: $\alpha_{tot} = \alpha_K + \alpha_L +$, etc.; I_{tot}: total intensity of the observed γ transitions.

a) Indicate in the form of diagrams, the distributions of the vanadium and titanium nucleons according to the shell structure derived from a harmonic potential (a) and according to the distribution derived from the Woods–Saxon potential with spin-orbit coupling (b). Then justify the values of J^π shown in Figure 4.19. Figure 2.7 will be used as a basis.

b) Determine the types of multipole transitions that can be observed when titanium deexcites to its ground level. Specify, with supporting justification, the most probable transitions.

c) Reproduce Figure 4.19 and then represent the most probable multipole transitions and the corresponding γ-photons.

d) List all observed photons and calculate their intensities and energies.

e) Determine the energies of the *X*-photons and Auger electrons observed.

EXERCISE 4.4.– Decay of arsenic-74: masses of products formed and maximum energies of the β^- and β+ spectra

Let us consider a radioactive sample containing 80 mg of radioarsenic-74 (Z = 33) of radioactive half-life T = 17.77 days. This radioelement decays according to the β^- (34%) and β^+ (29%) modes and by electron capture (37%).

Given data (*atomic masses*): arsenic-74: 73.9239287 u; germanium-74: 73.9211778 u; selenium-74: 73.9224764 u.

a) Write the β^-, β^+ decay equations and *EC* of radioarsenic-74.

b) Determine the mass percentages of the elements present in the sample after 53.31 days of decay of ^{74}As.

c) Calculate the maximum energy of the β^+ spectrum.

d) The β^- decay mode of ^{74}As reveals the presence of two groups of β^- particles: a group noted β_0^- and consisting of 53% of the transitions leading to the ground state of the daughter nucleus and a group noted β_1^- consisting of 47% of the transitions leading to the excited state of the daughter nucleus, whose energy measured in relation to its ground level is equal to 596 keV. Map out the complete decay energy diagram of vanadium-74. The percentages of the observed transitions will be shown on the diagram.

e) Calculate the maximum energy of the β^- spectrum for each group of particles.

EXERCISE 4.5.– Decay of copper-64: determination of partial decay probabilities per unit time

Certain radioelements such as copper-64 have the particular feature of presenting three decay pathways: β^-, β^+ and EC. Figure 4.20 shows the energy spectra of β^- (39%) and β^+ (18%) particles emitted by copper-64. This radionuclide also decays by electron capture, EC (43%) in the K-shell.

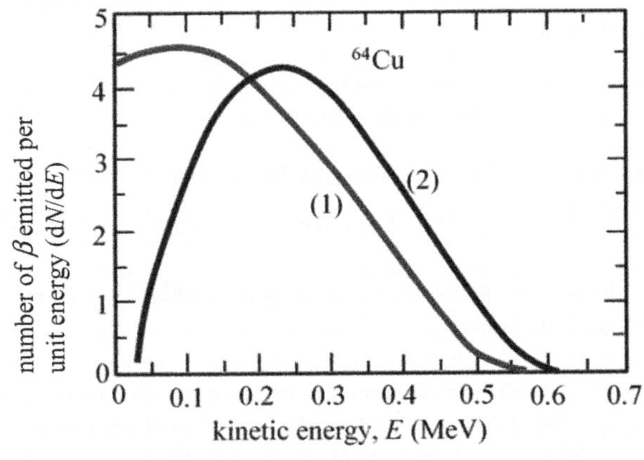

Figure 4.20. *Energy spectra of β particles emitted by copper-64*

In this exercise we propose to identify the β spectrum corresponding to each decay pathway and to determine the partial decay constants relative to the three decay pathways of copper-64.

Given data:

– *atomic masses:* copper-64: 63.9297642 u; zinc-64: 63.9291422 u;

– nickel-64: 63.927966 u;

– *branching ratios*: $\lambda_{EC}/\lambda_{\beta}^{+} = 2.3$; $\lambda_{\beta}^{-}/\lambda_{\beta}^{+} = 2$.

a) By comparing the shapes of the spectra shown in Figure 4.20, identify, with supporting justification, the spectrum corresponding to the β^- particles and that corresponding to the β^+ particles.

b) Write the β^- and β^+ decay equations.

c) Graphically determine the maximum energies, $E_{\max\beta^-}$ and $E_{\max\beta^+}$, of the β^- and β^+ spectra, respectively. Justify the results obtained by the calculation.

d) Calculate the partial decay probabilities per unit time and the partial periods relative to the three decay pathways of copper-64. What can be noted? Briefly justify this remark.

EXERCISE 4.6.– Study of the various decay pathways of potassium-40

Potassium-40 of half-life $T = 1.248 \times 10^9$ years decays:

– 88.8% to the ground level of calcium-40;

– 11% to the ground level of argon-40;

– 0.001% to the excited level (at 1,460.8 keV) of argon-40.

Deexcitation of the excited level of argon to its ground level occurs by emission of γ-photons.

Given data:

– *isotopic abundance of* ^{40}*K*: 0.01167%;

– *atomic masses*: ^{40}K: 39.96399848 u; ^{40}Ar: 39.962383122 u;

– *maximum observed spectrum energy*: 1,311.6 keV;

– *molar mass* of *potassium nitrate*: 101 g · mol^{-1}.

a) Identify, with supporting justification, the various decay pathways of ^{40}K.

b) Theoretically justify the decay path leading to the ground level of argon-40.

c) Map out the decay energy diagram of potassium-40, showing the various decay pathways and the observable gamma transition.

d) Determine the atomic mass of calcium-40.

e) Let us consider a 50-kg bag of potassium nitrate, KNO_3. Calculate its intimate activity as well as the number of photons emitted per hour by this bag.

EXERCISE 4.7.– Accumulation of xenon-131 by irradiation of tellurium-130

In this exercise, we propose to study the family tree leading to xenon-131 from deuteron irradiation of tellurium-130. The nuclear reaction ^{130}Te (d, p) ^{131}Te generates the decay chain β^- leading to the accumulation of ^{131}Xe:

$$ {}^{130}_{52}Te \xrightarrow[\lambda_1]{(d,\,p)} {}^{131}_{52}Te \xrightarrow[\lambda_2]{\beta^-} {}^{131}_{53}I \xrightarrow[\lambda_3]{\beta^-} {}^{131}_{54}Xe \text{ (stable)} $$

In this decay chain, the nuclear reaction ^{130}Te (d, p) ^{131}Te plays the role of a parent substance with a long half-life for tellurium ^{131}Te. This enables the approximation $\lambda_1 \to 0$. The radioactive half-lives of tellurium-130 and iodine-131 are equal to 30 hrs and 8 days, respectively.

Let N_{01} be the initial number of ^{130}Te nuclei. Under experimental conditions, the parent substance is equivalent to a radioactive source of decay constant λ_1 and initial activity $A_{01} = 7.4 \times 10^7$ Bq.

a) Explain the nuclear reaction ^{130}Te (d, p) ^{131}Te.

b) Determine the activity of iodine-131 after 12 hrs of irradiation.

c) Using the Bateman equations, show that the number of ^{131}Xe nuclei accumulating over time is given by the equation (taking account of the approximation: $\lambda_1 \to 0$).

$$ N_4(t) = \lambda_1 N_{01} \left(t - \frac{\lambda_3}{\lambda_2 \times (\lambda_2 - \lambda_3)} e^{-\lambda_2 t} + \frac{\lambda_2}{\lambda_3 \times (\lambda_2 - \lambda_3)} e^{-\lambda_3 t} \right) $$

d) Calculate the number of xenon-131 nuclei accumulated in 12 hrs of irradiation.

4.6. Solutions to exercises

SOLUTION 4.1.– Determination of the maximum energy of the β^+ spectrum from the semi-empirical Bethe–Weizsäcker formula

a) Nature of the ^AX nucleus

The mass number $A = Z + N$. The mirror nuclei verify the relationship:

$$Z - N = \pm 1 \Rightarrow 2Z - A = \pm 1 \Rightarrow A = 2Z \pm 1.$$

Radionuclide ^AX with mass number $A = 2Z - 1$ is indeed a mirror nucleus.

b) Expression of mass difference

Let us write the decay equation for the β^+-emitting ^AX nucleus. We obtain:

$$_Z^AX \rightarrow {}_{Z-1}^{\;\;A}Y + {}_1^0e + \nu_e \qquad [4.108a]$$

In semi-empirical Bethe–Weizsäcker formula [2.119], let us replace the mass of the hydrogen atom, M_H, with the mass of the proton and use a lowercase letter to designate the mass, M_n, of the neutron. Furthermore, knowing that the mass number is odd ($A = 2Z - 1$), then the δ pair energy is zero [2.133a]. Taking [2.119] into account we obtain:

$$\begin{aligned} m_X c^2 &= Zm_p c^2 + (A-Z)m_n c^2 \\ &\quad - a_v A + a_s A^{2/3} + a_c \frac{Z^2}{A^{1/3}} + a_a \frac{(A-2Z)^2}{A} \end{aligned} \qquad [4.108b]$$

$$\begin{aligned} m_Y c^2 &= (Z-1)m_p c^2 + (A-Z+1)m_n c^2 \\ &\quad - a_v A + a_s A^{2/3} + a_c \frac{(Z-1)^2}{A^{1/3}} + a_a \frac{(A-2Z+2)^2}{A} \end{aligned} \qquad [4.108c]$$

Let us expand equation [4.108c]. This gives:

$$\begin{aligned} m_Y c^2 &= Zm_p c^2 - m_p c^2 + (A-Z)m_n c^2 + m_n c^2 - a_v A + a_s A^{2/3} \\ &\quad + \frac{a_c Z^2}{A^{1/3}} - \frac{2a_c Z}{A^{1/3}} + \frac{a_c}{A^{1/3}} + \frac{a_a (A-2Z)^2}{A} + \frac{4a_a (A-2Z)}{A} + \frac{4a_a}{A} \end{aligned} \qquad [4.108d]$$

By subtracting, member-to-member, equations [4.108b] and [4.108d], we find:

$$(m_X - m_Y)c^2 = -(m_n - m_p)c^2 + \frac{2a_c Z}{A^{1/3}} - \frac{a_c}{A^{1/3}} - \frac{2a_a(A-2Z)}{A} - \frac{4a_a}{A} \quad [4.108e]$$

Yet $A = 2Z - \Rightarrow 2Z = A + 1$. Equation [4.108e] is then written:

$$(m_X - m_Y)c^2 = -(m_n - m_p)c^2 + \frac{a_c(A+1)}{A^{1/3}} - \frac{a_c}{A^{1/3}} + \frac{4a_a}{A} - \frac{4a_a}{A}$$

Ultimately, this gives:

$$(m_X - m_Y)c^2 = a_c A^{2/3} - (m_n - m_p)c^2 \quad [4.109]$$

c) Expression of the maximum energy of the β^+ spectrum

The decay energy via equation [4.108a] is given by relationship [4.18]:

$$Q_\beta = [m_X - (m_Y + m_0)]c^2 = E_{\max\beta+} + E_\gamma$$

This then gives:

$$E_{\max\beta+} = (m_X - m_Y)c^2 - m_0c^2 \quad [4.110]$$

Using [4.109], equation [4.110] is then written:

$$E_{\max\beta+} = a_c A^{2/3} - (m_n - m_p)c^2 - m_0c^2 \quad [4.111]$$

As a function of A, equation [4.111] is ultimately written (in MeV):

$$E_{\max\beta+} = 0.595\, A^{2/3} - 1.804 \quad [4.112]$$

d) Reproduction, accuracy of the nuclear model with constant nucleon density

Using [4.112], we obtain the theoretical values of $E_{\max\beta+}$ gathered in Table 4.8.

Reading the results gathered in Table 4.8 we can see sound agreement between theory and experiment. This justifies the validity of the nuclear model with constant density, $R = r_0 A^{1/3}$. Nevertheless, this model based on the spherical nucleus remains a simplistic model. This also justifies the small differences observed between the experimental and theoretical results listed in Table 4.8. Indeed, as we saw in Chapter 1, the nucleus has a diffuse zone of around 2.2 fm corresponding to the skin thickness in which the nuclear charge distribution density $\rho(r)$ gradually decreases (Figure 1.17). In addition, the value of the Coulomb coefficient varies from author to author (Table 2.8). For example, using the most recent value, $a_c = 0.695$ MeV obtained by Roy Chowdhury and Basu [ROY 06], for ^{11}C we find: $E_{\max\beta+} = 1.63$ MeV (experimental value: 0.99 MeV) and for ^{39}Ca: $E_{\max\beta+} = 6.2$ MeV (experimental value: 5.1 MeV). An increase in the energy gap between the theoretical and experimental values of $Emax_{\beta+}$ is clearly noted by considering the theoretical results obtained with a_c = 0.585 MeV [FEE 47, FRI 49]. It follows that the accuracy of the theoretical values of the maximum energy of the β^+ spectrum depends on the value of a_c chosen.

Nucleus	$E_{max\beta+}$ (MeV) experiment	$E_{max\beta+}$ (MeV) theory
^{11}C	0.99	1.09
^{13}N	1.24	1.43
^{15}O	1.68	1.75
^{17}F	1.72	2.06
^{19}Ne	2.18	2.36
^{21}Na	2.50	2.65
^{23}Mg	2.99	2.93
^{27}Si	3.48	3.46
^{29}P	3.94	3.72
^{31}S	3.9	3.97
^{33}Cl	4.1	4.21
^{35}Ar	4.4	4.45
^{37}K	4.6	4.69
^{39}Ca	5.1	4.92
^{41}Sc	4.94	5.15

Table 4.8. *Comparison of the experimental values of maximum energy, Emax, of the β^+ spectra of several mirror nuclei for which A = 2Z – 1*

SOLUTION 4.2.– Decay of indium-114: maximum energies of β^- and β^+ spectra

a) Relative positions, decay diagram of indium-114

The mass of cadmium-114 is greater than the mass of tin-114. It follows that the β^+ decay energy of ^{114}In is less than its β^- decay energy. The ground level of cadmium is therefore lower than that of tin relative to the ground level of indium-114. The decay diagram is shown in Figure 4.21.

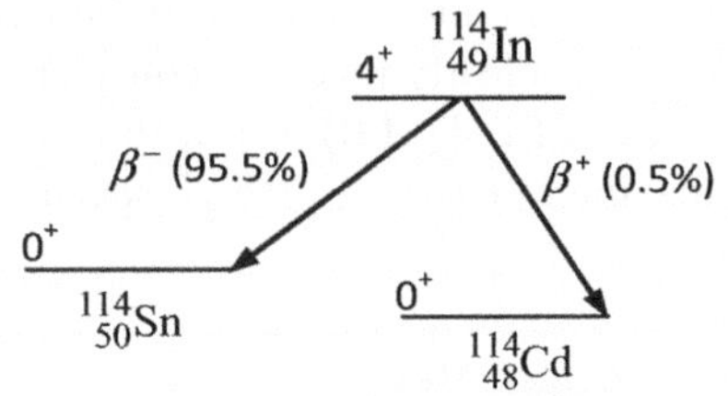

Figure 4.21. *β decay energy diagram of indium-114*

b) Decay equations, maximum energies of the β spectra

– Decay equations

For the observed β^- and β^+ modes, the following are obtained, respectively:

$$^{114}_{49}\mathrm{In} \to {}^{114}_{50}\mathrm{Sn} + {}^{0}_{-1}\mathrm{e} + \overline{\nu}_\mathrm{e} \quad [4.113a]$$

$$^{114}_{49}In \to {}^{114}_{48}Cd + {}^{0}_{1}e + \nu_e \quad [4.113b]$$

– Maximum energies of the spectra

We will use the β^- emission condition [3.106] and the β^+ emission condition [3.109]. Considering [4.113], we then obtain:

$$E_{\max\beta-} = (M_{\mathrm{In}} - M_{\mathrm{Sn}})c^2; \; E_{\max\beta+} = (M_{\mathrm{In}} - M_{\mathrm{Cd}})c^2 - 2m_0c^2 \quad [4.114a]$$

Numerically we find:

$$E_{\max\beta-} = (113.904914 - 113.902779) \times 931.5 = 1.989 \text{ MeV} \quad [4.114b]$$

$$E_{\max\beta+} = (113.904914 - 113.9033585) \times 931.5 - 1.022 = 0.427 \quad [4.114c]$$

c) Masses of daughter products collected

Let N_0 be the number of nuclei contained in $m_0 = 10$ g of indium-114 and $N(t)$ the number of nuclei contained in $m(t)$ grams of indium-114 at instant t. We obtain:

$$m_0 = \frac{N_0}{N_A} \times A; \; m(t) = \frac{N(t)}{N_A} \times A \quad [4.115a]$$

The radioactive decay law of indium-114 is written:

$$N(t) = N_0\, e^{-\lambda t} = N_0\, e^{-0.693\, t\,/\,T} \qquad [4.115b]$$

The mass of indium-114 not decayed at instant t is written according to [4.115a] and [4.115b]:

$$m\,(t) = m_0\, e^{-0.693t/T} \qquad [4.116a]$$

– At t = 143.8 sec. The decay half-life of indium-114 is T = 71.9 sec. It is then noted that the ratio t/T = 143.8/71.9 = 2 $\Rightarrow$ t = 2T. Equation [4.116a] is explicitly written:

$$m_{2T} = 10e^{-0.693\times 2} = 2.5 \text{ g} \qquad [4.116b]$$

The decayed mass, $m_{\mathrm{d}} = m_0 - m_{2T}$ = 7.5 g. The masses of products collected on the date t = 143.8 sec are then equal to:

$$m_{\mathrm{Sn}} = 99.5\%\ m_0 = 7.463 \text{ g};\ m_{\mathrm{Cd}} = 0.5\%\ m_{\mathrm{d}} = 0.0375 \text{ g} \approx 38 \text{ mg}.$$

– At $t = \infty$. The total mass, m_0, has decayed. Hence

$$m_{\mathrm{Sn}} = 99.5\%\ m_0 = 9.95 \text{ g};\ m_{\mathrm{Cd}} = 0.5\%\ m_0 = 0.05 \text{ g} = 50 \text{ mg}.$$

SOLUTION 4.3.– Decay of vanadium-48: energies of photons and Auger electrons

a) Nucleon distribution, justification of values of J^{π}

– For titanium-48: Z = 22 and N = 26. The nucleon distribution is shown in Figure 4.22. It is recalled that the number of g_j-nucleons distributed in a ℓ_j-shell is $g_j = 2j + 1$.

As shown in Figure 4.22, $j_n = j_p = j$ = 7/2. Moreover, $\pi_n = \pi_p = -\ 1$. Hence the total parity, $\pi = \pi_n \times \pi_p = +\ 1$. In addition, titanium-48 is an even-even nuclide. The angular momentum is then determined by rule [2.55], which we recall here:

$$J = 2j - 1,\ 2j - 3,\ 2j - 5,\ 2j - 7, \text{ etc.},\ 0$$

Taking parity into account, we obtain:

$$J^{\pi} = 6^+, 4^+, 2^+, 0^+$$

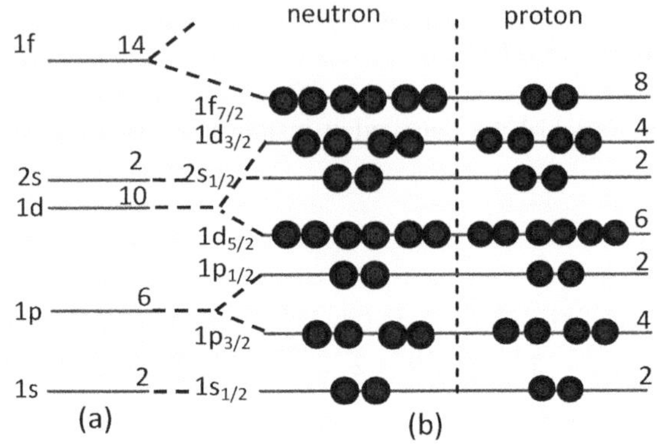

Figure 4.22. *Distribution of nucleons of titanium-48: (a) shell structure derived from a harmonic potential; (b) distribution derived from the Woods–Saxon potential with spin-orbit coupling*

The ground state corresponds to $J^{\pi} = 4^{+}$ according to the indication in Figure 4.19. The other J^{π} values refer to the excited states of titanium-48.

– For vanadium-48: $Z = 23$ and $N = 25$. The nucleon distribution is shown in Figure 4.23.

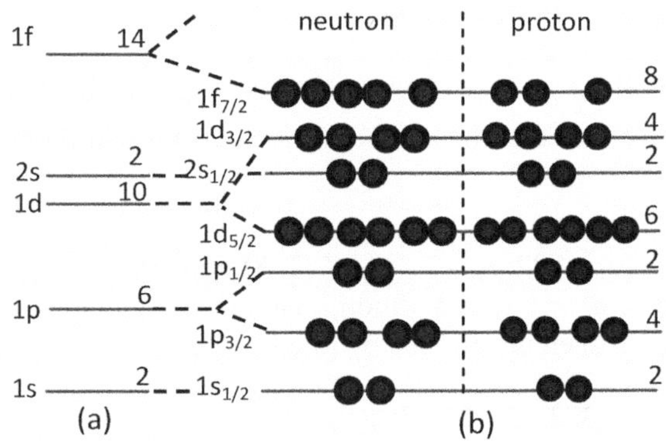

Figure 4.23. *Distribution of nucleons of vanadium-48: (a) shell structure derived from a harmonic potential; (b) distribution derived from the Woods–Saxon potential with spin-orbit coupling*

According to Figure 4.23, $j_n = j_p = j = 7/2$; $\pi = +1$. The coupling of the individual angular momenta of the unpaired nucleons gives the different values of the total angular momentum, J. Thus:

$$J^\pi = (7/2 + 7/2)^\pi, \text{ etc., } (7/2 - 7/2)^\pi = 7^+, 6^+, 5^+, 4^+, 3^+, 2^+, 1^+, 0^+$$

We find $J^\pi = 0^+$ for the ground state and $J^\pi = 4^+$ and $J^\pi = 2^+$, in accordance with the indications in Figure 4.19. The other J^π values refer to the excited states of vanadium-48, which are not shown.

b) Multipole transition types, the most probable transitions

According to Figure 4.19, three transitions are probable:

$$4^+ \to 0^+; 4^+ \to 2^+; 2^+ \to 0^+$$

These transitions occur without a change in parity. They are therefore electric multipole transitions:

– for the $4^+ \to 0^+$ transition, $\ell = 4$: E_4 transition;

– for the $4^+ \to 2^+$ transition, $\ell = 6, 5, 4, 2$: transition E_6, ... E_2;

– for the $2^+ \to 0^+$ transition, $\ell = 2$: E_2 transition.

The most probable transitions correspond to the smallest values of ℓ. They are therefore the two electric multipole transitions, E_2: $4^+ \to 2^+$ and $2^+ \to 0^+$.

c) Representation of multipole transitions and γ-photons

The representation of electric quadrupole transitions and photons observed is shown in Figure 4.24.

d) Enumeration and energies of observed photons

– Photon types: five types of photons are observed: the γ_1- and γ_2-photons resulting from the quadrupole transitions, E_2, between the excited levels of titanium, and a third photon, γ_3, due to the annihilation of the β^+-positron with an electron of matter. In addition, there are the two X-photons resulting from the electron (from the L-shell) - hole (in the K-shell) recombinations.

– Energy: taking into account Figure 4.24 and the annihilation process, we obtain:

$$E_{\gamma 1} = 1313 \text{ keV}; E_{\gamma 2} = 983 \text{ keV}; E_{\gamma 3} = m_0c^2 = 511 \text{ keV}.$$

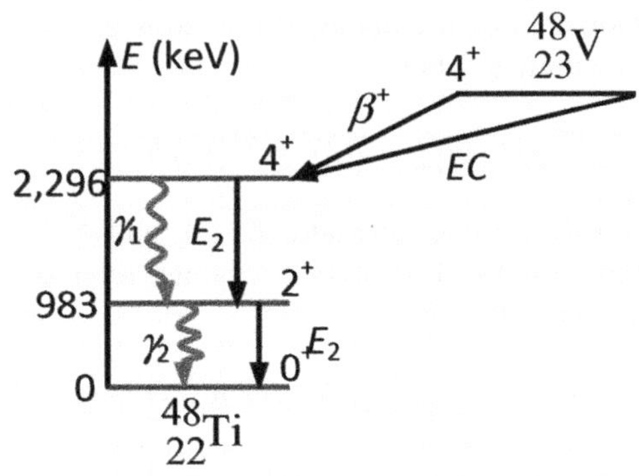

Figure 4.24. *The most probable electric quadrupole transitions (E_2) between excited levels of titanium-48*

Intensity:

– $4^+ \to 2^+$ transition: $\alpha_K + \alpha_L + 1 \approx 1 \Rightarrow I_{\gamma 1} = I_{tot} = 100\%$;

– $2^+ \to 0^+$ transition: $\alpha_K + \alpha_L + 1 \approx 1 \Rightarrow I_{\gamma 2} = I_{tot} = 100\%$;

– annihilation process: $I_{\gamma 3} = BR(\beta^+) = 45.8\%$.

e) Determination of energies of the X-photons and of the Auger electrons

– *X-photon energies*

For *X*-photons emitted by processes of electronic rearrangement between the *K*- and L_2- and *K*- and L_3-shells of titanium, we obtain:

$$E_{X1} = W_K - W_{L2} = 4.9664 - 0.4615 = 4.5049 \text{ keV}$$

$$E_{X2} = W_K - W_{L3} = 4.9664 - 0.4555 = 4.5109 \text{ keV}$$

– *Auger electron energies*

The Auger electron carries a kinetic energy satisfying energy conservation law [4.41], which we recall here:

$$E_c\,(e^A) = W_K - W_i - W_j$$

Similarly, let us recall, due to the Auger effect, a gap created in the K-shell is filled by an electron of an upper shell (i). The excess energy is then transmitted to an electron of a shell (j), causing its expulsion from the electronic cloud.

In this exercise, the Auger effect is involved in the K- and L (L_1, L_2, L_3)-shells. Let L_i and L_j designate one of the three L-subshells with L_j as the upper subshell. The binding energies of the three subshells are known. By considering the total energies, $E = -W$, we obtain:

$$E(L_1) = -0.5637 \text{ keV}; \quad E(L_2) = -0.4615 \text{ keV}; \quad E(L_3) = -0.4555 \text{ keV}.$$

Six energies can then be identified for the Auger electrons (considering the increasing arrangement order of the subshells: $L_1 \rightarrow L_2 \rightarrow L_3$):

– a L_1-electron fills the K-shell gap and another electron of the same L_1-subshell is expelled by Auger effect with a kinetic energy, E_{c11} (e^A);

– a L_2-electron fills the K-shell gap and another electron of the same L_2-subshell is expelled by Auger effect with a kinetic energy, E_{c22} (e^A);

– a L_3-electron fills the K-shell gap and another electron of the same L_3-subshell is expelled by Auger effect with a kinetic energy, E_{c33} (e^A);

– a L_1-electron fills the K-shell gap and another electron of the same L_2-subshell is expelled by Auger effect with a kinetic energy, E_{c12} (e^A);

– a L_1-electron fills the K-shell gap and another electron of the same L_3-subshell is expelled by Auger effect with a kinetic energy, E_{c13} (e^A);

– a L_2-electron fills the K-shell gap and another electron of the same L_3-subshell is expelled by Auger effect with a kinetic energy, E_{c33} (e^A).

Using [4.41], we then obtain, respectively:

$$E_{c11}(e^A) = W_K - W_1 - W_1 = W_K - W(L_1) - W(L_1)$$

$$E_{c22}(e^A) = W_K - W_2 - W_2 = W_K - W(L_2) - W(L_2)$$

$$E_{33}(e^A) = W_K - W_3 - W_3 = W_K - W(L_3) - W(L_3) \qquad [4.117]$$

$$E_{c12}(e^A) = W_K - W_1 - W_2 = W_K - W(L_1) - W(L_2)$$

$$E_{c13}(e^A) = W_K - W_1 - W_3 = W_K - W(L_1) - W(L_3)$$

$$E_{c23}(e^A) = W_K - W_2 - W_3 = W_K - W(L_2) - W(L_3)$$

Using [4.117], the Auger kinetic energies are thus theoretically obtained:

$$E_{c11}(e^A) = W_K - 2W(L_1)$$

$$E_{c22}(e^A) = W_K - 2W(L_2) \qquad [4.118a]$$

$$E_{33}(e^A) = W_K - 2W(L_3)$$

$$E_{c12}(e^A) = W_K - W(L_1) - W(L_2)$$

$$E_{c13}(e^A) = W_K - W(L_1) - W(L_3) \qquad [4.118b]$$

$$E_{c23}(e^A) = W_K - W(L_2) - W(L_3)$$

Numerically we find, according to [4.118]:

$$E_{c11}(e^A) = 4.9664 - 2 \times 0.5637 = 3.8390 \text{ keV}$$

$$E_{c22}(e^A) = 4.9664 - 2 \times 0.4615 = 4.0434 \text{ keV} \qquad [4.119a]$$

$$E_{33}(e^A) = 4.9664 - 2 \times 0.4555 = 4.0554 \text{ keV}$$

$$E_{c12}(e^A) = 4.9664 - 0.5637 - 0.4615 = 3.9412 \text{ keV}$$

$$E_{c13}(e^A) = 4.9664 - 0.5637 - 0.4555 = 3.9472 \text{ keV} \qquad [4.119b]$$

$$E_{c23}(e^A) = 4.9664 - 0.4615 - 0.4555 = 4.0494 \text{ keV}$$

NOTE.– *Experimentally, the six kinetic energies of the Auger electrons [4.118] are not solved. Thus, only the number of Auger K-electrons expelled within the energy range comprised between* $E_{cmin}(e^A)$ *and* $E_{cmax}(e_A)$ *is given. As indicated by results [4.119], the minimum kinetic energy of the Auger electrons corresponds to the energy* E_{c11}= *3.8390 keV and the maximum kinetic energy at* E_{c33} = *4.0554 keV. We thus obtain the energy range*:

$$E_{cmin}(e^A) = W_K - 2W(L_1); \; E_{cmax}(e^A) = W_K - 2W(L_3) \qquad [4.120]$$

SOLUTION 4.4.– Decay of arsenic-74: masses of products formed and maximum energies of the $\beta-$ and $\beta+$ spectra

a) Decay equations of radioarsenic-74

The equations for the β^-, β^+ and electronic capture (*EC*) decay of arsenic-74 are written, respectively:

$${}^{74}_{33}\text{As} \rightarrow {}^{74}_{34}\text{Se} + {}^{0}_{-1}\text{e} + \bar{\nu}_e \qquad [4.121a]$$

$$ {}^{74}_{33}\text{As} \rightarrow {}^{74}_{32}\text{Ge} + {}^{0}_{1}\text{e} + \nu_{\text{e}} \quad [4.121\text{b}] $$

$$ {}^{74}_{33}\text{As} + \text{e}^- \rightarrow {}^{74}_{32}\text{Ge} + \text{RX} + \nu_{\text{e}} \quad [4.121\text{c}] $$

b) Determination of mass percentages

The decay law of the radioarsenic mass is:

$$ m(t) = m_0\, e^{-0.693\, t/T} \quad [4.122] $$

The radioactive half-life of ^{74}As is T = 17.77 days. For t = 53.31 days = $3T$, equation [4.121] gives:

$$ m_{3T} = 80 \times e^{-0.693 \times 3} = 10.0 \text{ mg} \quad [4.123] $$

Arsenic-74 decays 34% by β^-, producing selenium-74, and 29% by β^+ and 37% by electron capture, producing germanium-74 according to [4.121]. The mass of decayed arsenic after 53.31 days is m_{d} = 70 mg. We then obtain a mass:

– m_{As} = 10.0 mg;

– m_{Se} = 70 × 34% = 23.8 mg;

– m_{Ge} – 70 × 66% = 46.2 mg.

c) Calculation of the maximum energy of the β^+ spectrum

The maximum energy of the β^+ spectrum is:

$$ E_{\max\beta+} = (M_{\text{As}} - M_{\text{Ge}})c^2 - 2m_0c^2 \quad [4.124] $$

Numerically:

$$ E_{\max\beta+} = (73.9239287 - 73.9211778) \times 931.5 - 1.022 = 1.540 \text{ MeV} $$

d) Decay energy diagram

Of the 34% decay according to the β^- mode, 53% of transitions, i.e. 18.02%, lead to the ground state of selenium-74 (β_0^- group) and 47% of transitions, i.e. 15.98%, lead to the excited state at 596 keV of selenium-74 (β_1^- group). The decay diagram of vanadium-74 is shown in Figure 4.25.

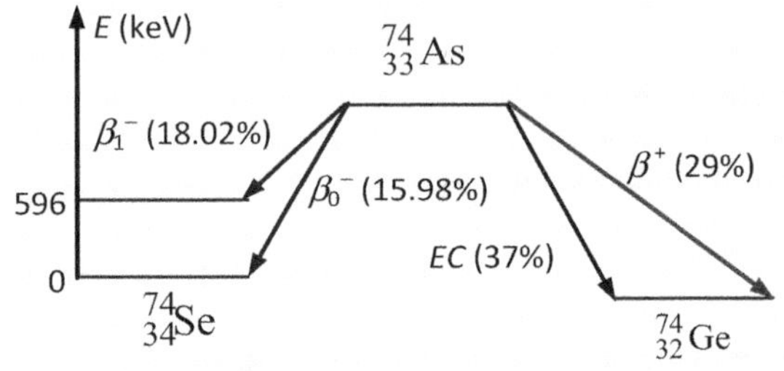

Figure 4.25. *Decay energy diagram of vanadium-74*

e) Maximum energy of the β^- spectra

For the β_0^- group, we obtain:

$$E_{\max\beta 0}^{\;-} = (M_{\text{As}} - M_{\text{Se}})c^2 \qquad [4.125a]$$

Numerically:

$$E_{\max\beta 0}^{\;-} = (73.9239287 - 73.9224764) \times 931.5 = 1.353 \text{ MeV}$$

For the β_1^- group, the energy of the emitted γ-photon ($E_\gamma = 596$ keV) during the deexcitation of selenium to its ground state is taken into account. Thus:

$$E_{\max\beta 1}^{\;-} = (M_{\text{As}} - M_{\text{Se}})c^2 - E_\gamma \qquad [4.125b]$$

Numerically:

$$E_{\max\beta 1}^{\;-} = 1.353 - 0.596 = 0.757 \text{ MeV}$$

SOLUTION 4.5.– Decay of copper-64: determination of partial decay probabilities per unit time

a) Identification of spectra

For a β^- spectrum, there is a number of particles emitted and retained in the nucleus environment by the electron-nucleus coulomb attraction for $E = 0$.

Inversely, for a β^+ spectrum, there are no particles emitted for $E = 0$. Spectrum (1) corresponds to the β^- spectrum and spectrum (2) corresponds to the β^+ spectrum.

b) Decay equations

The β^- and β^+ decay equations are written, respectively:

$$^{64}_{29}\text{Cu} \rightarrow {}^{64}_{30}\text{Zn} + {}^{0}_{-1}\text{e} + \bar{\nu}_e \qquad [4.126a]$$

$$^{64}_{29}\text{Cu} \rightarrow {}^{64}_{28}\text{Ni} + {}^{0}_{1}\text{e} + \nu_e \qquad [4.126b]$$

c) Graphical determination of the maximum energies of the β spectra, justification

– Graphical determination

The spectra are reproduced to scale. We find, graphically (Figure 4.26):

$$E_{\max\beta+} = 0.660 \text{ MeV}; E_{\max\beta-} = 0.570 \text{ MeV} \qquad [4.127]$$

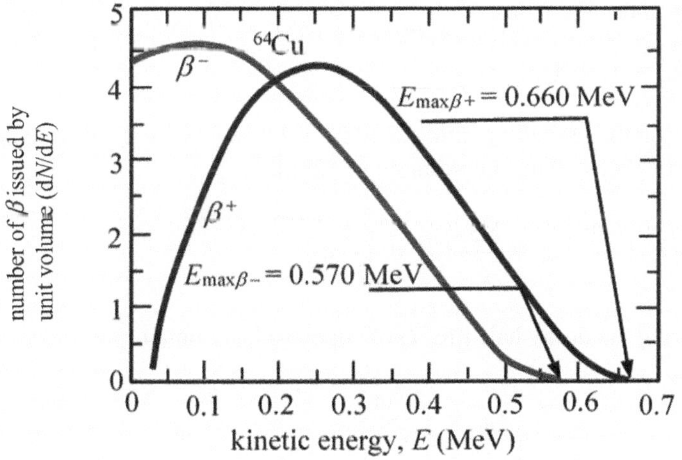

Figure 4.26. *Maximum energy of β^- and β^+ spectra of copper-64*

– Theoretical justification

Theoretically:

$$E_{\max\beta^-} = (M_{\mathrm{Cu}} - M_{\mathrm{Zn}})c^2 = 0.579 \text{ MeV} \qquad [4.128a]$$

$$E_{\max\beta+} = (M_{\mathrm{Cu}} - M_{\mathrm{Ni}})c^2 - 2m_0c^2 = 0.653 \text{ MeV} \qquad [4.128b]$$

The theoretical results [4.128] are well corroborated by the experimental observations [4.127].

Given data:

– *atomic masses*: copper-64: 63.9297642 u; zinc-64: 63.9291422 u;

– *nickel-64*: 63.927966 u;

– *branching ratios*: $\lambda_K/\lambda_\beta^+ = 2.3$; $\lambda_\beta^-/\lambda_\beta^+ = 2$.

d) Partial decay probabilities, partial half-lives

The partial decay probabilities per unit time correspond to the λ_β^-, λ_β^+ and λ_{EC} decay constants relative to the β^-, β+ and *EC* decay modes, respectively. The total decay constant, λ, of copper-64 for the three pathways is equal to the sum of the partial decay constants. Thus:

$$\lambda = \lambda_\beta^- + \lambda_\beta^+ + \lambda_{EC} \qquad [4.129a]$$

Considering the branching ratios, $\lambda_{\mathrm{EC}}/\lambda_\beta^+ = 2.3$; $\lambda_\beta^-/\lambda_\beta^+ = 2$, let us express λ as a function of λ_β^-. We obtain, according to [4.129a]:

$$\lambda = \lambda_{\beta-} + \frac{1}{2}\lambda_{\beta-} + \frac{2.3}{2}\lambda_{\beta-} = \frac{5.3}{2}\lambda_{\beta-} \qquad [4.129b]$$

Considering the decay half-life, *T*, of copper-64, we obtain:

$$\frac{\ln 2}{T} = \frac{5.3}{2}\lambda_{\beta-} \Rightarrow \begin{cases} \lambda_{\beta-} = \dfrac{2}{5.3}\dfrac{\ln 2}{T} \\ \lambda_{\beta+} = \dfrac{1}{5.3}\dfrac{\ln 2}{T} \\ \lambda_{CE} = \dfrac{2.3}{5.3}\dfrac{\ln 2}{T} \end{cases} \qquad [4.129c]$$

Given that $T = 12.8$ hrs, we obtain according to [4.129c]:

$$\lambda_\beta^- = 0.020 \text{ hrs}^{-1} = 5.7 \times 10^{-6}\text{ s}^{-1};\ T_\beta^- = 33.92 \text{ hrs}$$

$$\lambda_\beta^+ = 0.010 \text{ hrs}^{-1} = 2.8 \times 10^{-6}\text{ s}^{-1};\ T_\beta^+ = 67.84 \text{ hrs} \qquad [4.129d]$$

$$\lambda_{EC} = 0.023 \text{ hrs}^{-1} = 6.5 \times 10^{-6}\text{ s}^{-1};\ T_{EC} = 29.50 \text{ hrs}$$

Comparing the partial decay half-lives, T_i [4.129d] with the radioactive half-life $T = 12.8$ hrs of copper-64, we note that $T_i > T$. This predictable result is due to the various decay pathways of copper.

SOLUTION 4.6.– Study of the various decay pathways of potassium-40

a) *Identification of the various decay pathways of* ^{40}K

The various decay pathways of potassium-40 studied all correspond to isobaric transitions (A = constant). Let us now consider the variation in the charge number, Z, for each transition. We thus obtain:

– $_{19}K \rightarrow {}_{20}Ca$ transition (88.8%): $\Delta Z = +1$: β^- decay;

– $_{19}K \rightarrow {}_{18}Ar$ transition (11.2%): $\Delta Z = -1$: β^+ decay;

– $_{19}K \rightarrow {}_{18}Ar$ transition (0.001%): $\Delta Z = -1$: electron capture in competition with β^+ decay.

b) *Decay energy diagram of potassium-40*

The decay energy diagram of potassium-40 is mapped out below (Figure 4.27).

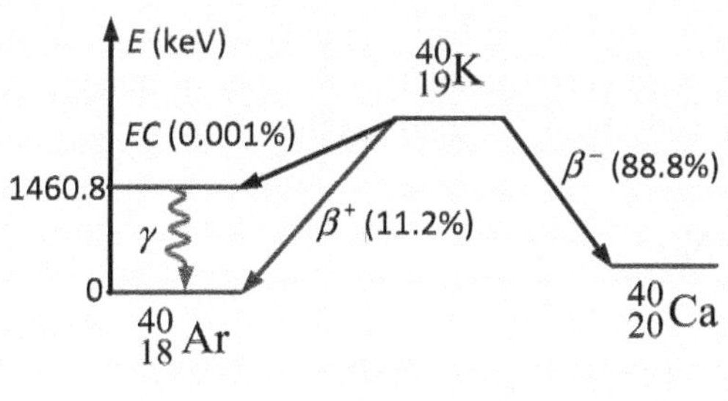

Figure 4.27. *Decay energy diagram of potassium-48. Of particular note are its three decay pathways*

c) Theoretical justification

Let us determine the mass energy difference $(M_K - M_{Ar})c^2$. We obtain:

$$(M_K - M_{Ar})c^2 = (39.96399848 - 39.962383122) \times 931.5 = 1.505 \text{ MeV}$$

The β^+ emission is clearly observed since:

$$(M_K - M_{Ar})c^2 > 1.022 \text{ MeV} \quad [4.130]$$

d) Determination of the atomic mass of calcium-40

For β^- emission:

$$(M_K - M_{Ca})c^2 = E_{\max\beta-} \Rightarrow M_{Ca}c^2 = (M_K c^2 - E_{\max\beta-}) \quad [4.131a]$$

Numerically, we obtain:

$$M_{Ca}c^2 = (39.96399848 \times 931.5 - 1.3116) = 37{,}225.153 \text{ MeV}$$

That is, in atomic units:

$$M_{Ca} = 39.9625904 \text{ u} \quad [4.131b]$$

e) Initial activity, number of photons

– Initial activity of the bag of ammonium nitrate

The molar mass of potassium, $M = 39$ g · mol $^{-1}$. One mole of potassium nitrate of mass 101 g therefore contains 39 g of potassium. The 50,000 g bag of potassium nitrate then contains 19,306.93 g of potassium. Given that the isotopic abundance of ^{40}K is equal to 0.01167%, the 50-kg bag contains a mass, *m*, of potassium-40 such that: $m = 2.25$ g. This mass corresponds to a number of nuclei, $N_0 = (m \times N_A/M) = 3.47 \times 10^{22}$. The initial activity of ^{40}K then has the value:

$$A_0 = \lambda N_0 = \frac{\ln 2}{T} N_0 \quad [4.132a]$$

Numerically ($T = 1.248 \times 10^9$ years $= 3.938 \times 10^{16}$ sec):

$$A_0 = 610{,}582.2 \text{ Bq} \approx 6.1 \times 10^5 \text{ Bq}.$$

– *Number of photons emitted per hour by this bag*

The photons are produced by the electron capture process occurring for 0.001%. Each *EC* decay produces a photon. The total number N_γ of photons is therefore equal to

$$N_\gamma = 0.001\%\, A_0 \approx 6 \text{ photons s}^{-1} = 21{,}600 \text{ photons hr}^{-1}.$$

SOLUTION 4.7.– Accumulation of xenon-131 by irradiation of tellurium-130

a) Explication of the nuclear reaction

The nuclear reaction ^{130}Te (d, p) ^{131}Te is explicitly written:

$$^{2}_{1}\text{H} + {}^{130}_{52}\text{Te} \rightarrow {}^{1}_{1}\text{H} + {}^{131}_{52}\text{Te} \qquad [4.132b]$$

b) Determination of the activity of iodine-131

Let us consider the production chain leading to stable xenon:

$$^{130}_{52}Te \xrightarrow[\lambda_1]{(d,p)} {}^{131}_{52}Te \xrightarrow[\lambda_2]{\beta^-} {}^{131}_{53}I \xrightarrow[\lambda_3]{\beta^-} {}^{131}_{54}Xe\ (stable) \qquad [4.133]$$

Let us determine the law of evolution of iodine-131 using the Bateman equations. We obtain result [4.66], which we recall here:

$$N_3(t) = \frac{\lambda_1 \times \lambda_2}{(\lambda_2 - \lambda_1)} N_{01} \left(\frac{e^{-\lambda_1 t}}{(\lambda_3 - \lambda_1)} - \frac{e^{-\lambda_2 t}}{(\lambda_3 - \lambda_2)} + \frac{(\lambda_2 - \lambda_1) e^{-\lambda_3 t}}{(\lambda_1 - \lambda_3) \times (\lambda_2 - \lambda_3)} \right)$$

The activity, $A_3(t)$, of iodine-131 then has the value ($A_0 = \lambda_1 N_{01}$):

$$A_3(t) = \frac{\lambda_2 \times \lambda_3}{(\lambda_2 - \lambda_1)} A_{01} \left(\frac{e^{-\lambda_1 t}}{(\lambda_3 - \lambda_1)} - \frac{e^{-\lambda_2 t}}{(\lambda_3 - \lambda_2)} + \frac{(\lambda_2 - \lambda_1) e^{-\lambda_3 t}}{(\lambda_1 - \lambda_3) \times (\lambda_2 - \lambda_3)} \right) \qquad [4.134a]$$

Let us now take account of the approximation $\lambda_1 \rightarrow 0$ ($e^{-\lambda 1 t} \approx 1$). We obtain:

$$A_3(t) = \frac{\lambda_2 \times \lambda_3}{\lambda_2} A_{01} \left(\frac{1}{\lambda_3} - \frac{e^{-\lambda_2 t}}{(\lambda_3 - \lambda_2)} + \frac{\lambda_2 e^{-\lambda_3 t}}{\lambda_3 \times (\lambda_2 - \lambda_3)} \right)$$

That is, ultimately:

$$A_3(t) = A_{01}\left(1 + \frac{\lambda_3}{(\lambda_2 - \lambda_3)} e^{-\lambda_2 t} + \frac{\lambda_2}{(\lambda_2 - \lambda_3)} e^{-\lambda_3 t}\right) \qquad [4.134b]$$

NOTE.– $T_2 = 30\text{ h} \Rightarrow \lambda_2 = 0.0231\text{ h}^{-1}$; $T_3 = 8\text{ d} = 192\text{ h} \Rightarrow \lambda_3 = 0.0036\text{ h}^{-1}$.

Which gives:

$$A_{3(12)} = 7.4\times10^7\left(1+\frac{0.0036}{(0.0231-0.0036)}e^{-0.0231\times12} + \frac{0.0231}{(0.0231-0.0036)}e^{-0.0036\times12}\right) = 1.68\times10^8$$

Thus: $A_{3(12)} = 1.68 \times 10^8$ Bq.

c) Demonstration

Let N_4 (t) be the number of ^{131}Xe nuclei that accumulate over time. Since xenon-131 is stable, its decay period is infinite. Consequently, its probability of decay per unit time is zero ($\lambda_4 = 0$). Using the Bateman equations, we obtain by using [4.58]:

$$N_4(t) = \frac{N_{01}}{\lambda_4}\sum_{i=1}^{4} C_i^4 e^{-\lambda_i t} = \frac{N_{01}}{\lambda_4}\left(C_1^4 e^{-\lambda_1 t} + C_2^4 e^{-\lambda_2 t} + C_3^4 e^{-\lambda_3 t} + C_4^4 e^{-\lambda_4 t}\right) \qquad [4.135]$$

According to [4.59]:

$$C_1^4 = \frac{\prod_{j=1}^{4}\lambda_j}{\prod_{j\neq1}^{4}(\lambda_j - \lambda_1)} = \frac{\lambda_1\times\lambda_2\times\lambda_3\times\lambda_4}{(\lambda_2-\lambda_1)\times(\lambda_3-\lambda_1)\times(\lambda_4-\lambda_1)} \qquad [4.136a]$$

$$C_2^4 = \frac{\prod_{j=1}^{4}\lambda_j}{\prod_{j=1, j\neq2}^{4}(\lambda_j - \lambda_2)} = \frac{\lambda_1\times\lambda_2\times\lambda_3\times\lambda_4}{(\lambda_1-\lambda_2)\times(\lambda_3-\lambda_2)\times(\lambda_4-\lambda_2)} \qquad [4.136b]$$

$$C_3^4 = \frac{\prod_{j=1}^{4} \lambda_j}{\prod_{j=1, j\neq 3}^{4} (\lambda_j - \lambda_3)} = \frac{\lambda_1 \times \lambda_2 \times \lambda_3 \times \lambda_4}{(\lambda_1 - \lambda_3)\times(\lambda_2 - \lambda_3)\times(\lambda_4 - \lambda_3)} \qquad [4.136c]$$

$$C_4^4 = \frac{\prod_{j=1}^{4} \lambda_j}{\prod_{j=1, j\neq 4}^{4} (\lambda_j - \lambda_4)} = \frac{\lambda_1 \times \lambda_2 \times \lambda_3 \times \lambda_4}{(\lambda_1 - \lambda_4)\times(\lambda_2 - \lambda_4)\times(\lambda_3 - \lambda_4)} \qquad [4.136d]$$

Using [4.136], accumulation equation [4.135] of the parent nucleus, X_3, is written:

$$N_4(t) = \lambda_1 \times \lambda_2 \times \lambda_3 N_{01}\left(\frac{e^{-\lambda_1 t}}{(\lambda_2 - \lambda_1)\times(\lambda_3 - \lambda_1)\times(\lambda_4 - \lambda_1)} + \frac{e^{-\lambda_2 t}}{(\lambda_1 - \lambda_2)\times(\lambda_3 - \lambda_2)\times(\lambda_4 - \lambda_2)} + \frac{e^{-\lambda_3 t}}{(\lambda_1 - \lambda_3)\times(\lambda_2 - \lambda_3)\times(\lambda_4 - \lambda_3)} + \frac{e^{-\lambda_4 t}}{(\lambda_1 - \lambda_4)\times(\lambda_2 - \lambda_4)\times(\lambda_3 - \lambda_4)}\right)$$

Given that $\lambda_4 = 0$, we obtain:

$$N_4(t) = \lambda_1 \times \lambda_2 \times \lambda_3 N_{01}\left(-\frac{(1-\lambda_1 t)}{\lambda_1 \times(\lambda_2 - \lambda_1)\times(\lambda_3 - \lambda_1)} - \frac{e^{-\lambda_2 t}}{\lambda_2 \times(\lambda_1 - \lambda_2)\times(\lambda_3 - \lambda_2)} - \frac{e^{-\lambda_3 t}}{\lambda_3 \times(\lambda_1 - \lambda_3)\times(\lambda_2 - \lambda_3)} + \frac{1}{\lambda_1 \times \lambda_2 \times \lambda_3}\right)$$

That is:

$$N_4(t) = \lambda_2 \times \lambda_3 N_{01}\left(\frac{\lambda_1 t}{(\lambda_2 - \lambda_1)\times(\lambda_3 - \lambda_1)} - \frac{1}{(\lambda_2 - \lambda_1)\times(\lambda_3 - \lambda_1)} - \frac{\lambda_1}{\lambda_2 \times(\lambda_1 - \lambda_2)\times(\lambda_3 - \lambda_2)} e^{-\lambda_2 t} - \frac{\lambda_1}{\lambda_3 \times(\lambda_1 - \lambda_3)\times(\lambda_2 - \lambda_3)} e^{-\lambda_3 t} + \frac{1}{\lambda_2 \times \lambda_3}\right)$$

That is, ultimately:

$$N_4(t) = \lambda_1 N_{01}\left(t - \frac{\lambda_3}{\lambda_2 \times (\lambda_2 - \lambda_3)} e^{-\lambda_2 t} + \frac{\lambda_2}{\lambda_3 \times (\lambda_2 - \lambda_3)} e^{-\lambda_3 t}\right) \qquad [4.137]$$

d) Determination of the number of nuclei of xenon-131

We have: $\lambda_1 N_{01} = A_{01} = 7.4 \times 10^7$ Bq; $\lambda_2 = 6.4 \times 10^{-6}\ \text{s}^{-1}$; $\lambda_3 = 1.0 \times 10^{-6}\ \text{s}^{-1}$.

Using [4.137], we obtain, at t = 12 hrs (t in seconds since the activity is becquerel):

$$N_{4(12)} = 7.4 \times 10^7\left(43200 - \frac{1.0 \times 10^6}{6.4 \times 5.4} e^{-\ 0.0231 \times 12} + \frac{6.4 \times 10^6}{1.0 \times 5.4} e^{-\ 0.0036 \times 12}\right)$$

$$= 8.6 \times 10^{13}$$

Appendices

Appendix 1

Quantified Energy of the Three-Dimensional Quantum Harmonic Oscillator

In this appendix, we propose to demonstrate expression [2.19] of the quantified energy of a three-dimensional quantum harmonic oscillator. To do this, we will adopt two approaches. Firstly, we will integrate the Schrödinger equation describing the state of a one-dimensional quantum harmonic oscillator. Then, we will take a much more flexible approach to finding relationship [2.19], based on the creation and annihilation operators.

A1.1. Integration of the Schrödinger equation

Let us consider a quantum harmonic oscillator with a dimension (Ox) of mass m, energy E, pulsatance ω and potential energy $U(x)$ [SAK 12]:

$$U(x) = \frac{1}{2}kx^2 = \frac{1}{2}m\omega^2 x^2 \qquad \text{[A1.1]}$$

The steady-state Schrödinger equation of the harmonic oscillator is given by the expression:

$$\left[-\frac{\hbar^2}{2m}\frac{d^2}{dx^2} + U(x)\right]\Phi(x) = E\Phi(x) \qquad \text{[A1.2]}$$

Using [A1.1], equation [A1.2] is written:

$$\left[-\frac{\hbar^2}{2m}\frac{d^2}{dx^2} + \frac{1}{2}m\omega^2 x^2\right]\Phi(x) = E\Phi(x) \qquad \text{[A1.3]}$$

Let us introduce into [A1.3] the dimensionless quantities:

$$q = \sqrt{\frac{m\omega}{\hbar}}\, x \,;\ \varepsilon = \frac{2E}{\hbar\omega} \qquad [A1.4]$$

We obtain:

$$\frac{d^2\Phi(q)}{dq^2} + \left(\varepsilon - q^2\right)\Phi(q) = 0 \qquad [A1.5]$$

The solution to equation [A1.5] is of the form:

$$\Phi(q) = Au(q)e^{-q^2/2} \qquad [A1.6]$$

In [A1.5], A is a constant and function $u\ (q)$ is an *integer series* of power q given by the expression:

$$u(q) = \sum_{k=0}^{\infty} a_k\, q^k \qquad [A1.7]$$

By deriving the wave function [A1.6] we obtain:

$$\begin{cases} \dfrac{d\Phi(q)}{dq} = A\left[\dfrac{du(q)}{dq} - qu(q)\right]e^{-q^2/2} \\ \dfrac{d^2\Phi(q)}{dq^2} = A\left[\dfrac{d^2u(q)}{dq^2} - q\dfrac{du(q)}{dq} - u(q)\right]e^{-q^2/2} - qA\left[\dfrac{du(q)}{dq} - qu(q)\right]e^{-q^2/2} \end{cases}$$

The second of these equations is written in condensed form:

$$\frac{d^2\Phi(q)}{dq^2} = \left[\frac{d^2u(q)}{dq^2} - 2q\frac{du(q)}{dq} + (q^2 - 1)u(q)\right]\Phi(q) \qquad [A1.8]$$

If we apply [A1.6] and [A1.8] in [A1.5], after arrangement we find:

$$\frac{d^2u(q)}{dq^2} - 2q\,\frac{du(q)}{dq} + (\varepsilon - 1)\,u(q) = 0 \qquad [A1.9]$$

To establish the quantified expression of the energy of the one-dimensional harmonic oscillator, we simply need to determine the recurrence relationship satisfied by the expansion coefficients, a_k, of the entire series, $u(q)$, and apply the *cutoff condition*.

Using [A1.7], we then obtain:

$$\begin{cases} \dfrac{du(q)}{dq} = \sum_{k=0}^{\infty} k a_k \, q^{k-1} \\ \dfrac{d^2u(q)}{dq^2} = \sum_{k=0}^{\infty} k(k-1) a_k \, q^{k-2} \end{cases} \quad \text{[A1.10]}$$

If we apply [A1.10] in [A1.9], we obtain:

$$\sum_{k=0}^{\infty} k(k-1) a_k \, q^{k-2} - 2q \sum_{k=0}^{\infty} k a_k \, q^{k-1} + (\varepsilon - 1) \sum_{k=0}^{\infty} a_k \, q^k = 0$$

That is:

$$\sum_{k=0}^{\infty} k(k-1) a_k \, q^{k-2} + \sum_{k=0}^{\infty} (\varepsilon - 2k - 1) a_k \, q^k = 0 \quad \text{[A1.11]}$$

By identifying the terms of the same power in q^k, we find (by simply replacing k with $k + 2$ in the first term of equation [A1.11]):

$$(k+2)(k+1) a_{k+2} + (\varepsilon - 2k - 1) a_k = 0$$

Thus:

$$a_{k+2} = \frac{(2k + 1 - \varepsilon)}{(k+2)(k+1)} a_k \quad \text{[A1.12]}$$

Now we can use the cutoff condition applied to equation [A1.12].

From a quantum mechanics perspective, wave function [A1.6] must be finite (convergent) for all values of q (including $q \to \pm \infty$). To meet this requirement, series [A1.10] must stop for a certain value, n, of the integer, k. This cutoff condition thus implies that $a_{n+2} = 0$. That is, according to [A1.12]:

$$2n + 1 - \varepsilon = 0 \Rightarrow \varepsilon = 2n + 1 \quad \text{[A1.13]}$$

Using the last quantity [A1.4] and [A1.13], we obtain:

$$\varepsilon = \frac{2E}{\hbar\omega} = 2n+1 \Rightarrow E_n = \hbar\omega\left(n+\frac{1}{2}\right) \quad [A1.14]$$

A1.2. Use of creation and annihilation operators

Let us consider a harmonic oscillator with a dimension (Ox) of mechanical energy E given by the relation:

$$E = E_c + U(x) = \frac{p^2}{2m} + \frac{1}{2}m\omega^2 x^2 \quad [A1.15]$$

The Hamiltonian H associated with mechanical energy [A1.15] is written:

$$H = \frac{P^2}{2m} + \frac{1}{2}m\omega^2 X^2 \quad [A1.16]$$

The steady-state Schrödinger equation of the harmonic oscillator is given by expression [A1.3]. Let us introduce the dimensionless operators $\hat{X}$ and $\hat{P}$ defined by the relationships [COH 77]:

$$\begin{cases} \hat{X} = \sqrt{\frac{m\omega}{\hbar}} X \\ \hat{P} = \frac{1}{\sqrt{\hbar m\omega}} P \end{cases} \Rightarrow \begin{cases} X = \sqrt{\frac{\hbar}{m\omega}} \hat{X} \\ P = \sqrt{\hbar m\omega}\, \hat{P} \end{cases} \quad [A1.17]$$

Using [A1.17], Hamiltonian [A1.16] is then written in the form:

$$H = \frac{m\hbar\omega}{2m}\hat{P}^2 + \frac{m\omega^2}{2} \times \frac{\hbar}{m\omega}\hat{X}^2 \quad [A1.18]$$

Factoring [A1.18] by the energy of the photon, we obtain:

$$H = \hbar\omega[\frac{1}{2}(\hat{P}^2 + \hat{X}^2)] \quad [A1.19]$$

The Hamiltonian H has the dimension of an energy. Let us then introduce the dimensionless operator, $\hat{H}$, given by the relationship:

$$\hat{H} = \frac{1}{2}(\hat{P}^2 + \hat{X}^2) \Rightarrow H = \hbar\omega\hat{H} \qquad \text{[A1.20]}$$

According to [A1.20], the Hamiltonian, H, is proportional to the operator, $\hat{H}$: they therefore have the same eigenvector, which we note $|\Phi_n\rangle$.

To determine the eigenvalues of H, we must now introduce three other operators noted $a^\dagger$ (read: *a dagger*) and a. By definition, the $a^\dagger$ and a operators are called (photon) creation and (photon) annihilation operators, respectively, given by the relationships:

$$\begin{cases} a = \dfrac{1}{\sqrt{2}}(\hat{X} + i\hat{P}) \\ a^\dagger = \dfrac{1}{\sqrt{2}}(\hat{X} - i\hat{P}) \end{cases} \Rightarrow \begin{cases} \hat{X} = \dfrac{1}{\sqrt{2}}(a^\dagger + a) \\ \hat{P} = \dfrac{i}{\sqrt{2}}(a^\dagger - a) \end{cases} \qquad \text{[A1.21]}$$

Knowing that the commutator $[X, P] = i\hbar$, using [A1.17], it is easy to show that $[\hat{X}, \hat{P}] = i$. Let us thus determine the product $a^\dagger a$ to introduce the operator N. Using [A1.22], we obtain:

$$a^\dagger a = \frac{1}{\sqrt{2}}(\hat{X} - i\hat{P}) \times \frac{1}{\sqrt{2}}(\hat{X} - i\hat{P}) = \frac{1}{2}(\hat{X}^2 + \hat{P}^2 + i\hat{X}\hat{P} - i\hat{P}\hat{X})$$

That is, knowing that $[\hat{X}, \hat{P}] = i$:

$$a^\dagger a = \frac{1}{2}(\hat{X}^2 + \hat{P}^2) - \frac{i}{2}[\hat{P}, \hat{X}] = \hat{H} - \frac{1}{2} \qquad \text{[A1.22]}$$

The result [A1.22] allows us to introduce the dimensionless operator noted N given by the relationship:

$$N = a^\dagger a = \hat{H} - \frac{1}{2} \qquad \text{[A1.23]}$$

Using [A1.20] and [A1.23], the Hamiltonian, H, is expressed as a function *of* N. Thus:

$$H = \hbar\omega\left(N + \frac{1}{2}\right) \quad \text{[A1.24]}$$

Let us designate $|\Phi_n\rangle$ with the eigenkets of N of eigenvalues n, $n = 0, 1, 2, 3, \ldots, \infty$. Taking into account [A1.24] we obtain:

$$\begin{cases} N|\Phi_n\rangle = n|\Phi_n\rangle \\ H|\Phi_n\rangle = E_n|\Phi_n\rangle \end{cases} \Rightarrow \hbar\omega\left(N + \frac{1}{2}\right)|\Phi_n\rangle = E_n|\Phi_n\rangle = \hbar\omega\left(n + \frac{1}{2}\right)|\Phi_n\rangle$$

Thus:

$$E_n|\Phi_n\rangle = \hbar\omega\left(n + \frac{1}{2}\right)|\Phi_n\rangle \quad \text{[A1.25]}$$

With the $|\Phi_n\rangle$ kets being orthonormal ($\langle\Phi_n|\Phi_m\rangle = \delta_{nm}$), we ultimately obtain:

$$E_n\langle\Phi_n|\Phi_n\rangle = \hbar\omega\left(n + \frac{1}{2}\right)\langle\Phi_n|\Phi_n\rangle \Rightarrow E_n = \hbar\omega\left(n + \frac{1}{2}\right)$$

Result [A1.14] is indeed found. It now remains for us to generalize it to three dimensions.

In order to generalize result [A1.14] to three dimensions, let us consider the *tensor product space*, $\mathcal{E} = \mathcal{E}_x \otimes \mathcal{E}_y \otimes \mathcal{E}_z$ (the "⊗" sign symbolizes the tensor product). We will use $|\Phi_I\rangle$ to designate the ket describing the state of the harmonic oscillator in space relative to the i dimension ($i = x, y, z$):

$$|\Phi_x\rangle \in \mathcal{E}_x;\ |\Phi_y\rangle \in \mathcal{E}_y;\ |\Phi_z\rangle \in \mathcal{E}_z$$

$$H_x \in \mathcal{E}_x;\ H_y \in \mathcal{E}_y;\ H_z \in \mathcal{E}_z$$

The global ket, $|\Phi_{xyz}\rangle$, describing the state of the harmonic oscillator in three-dimension space is then written ($|\Phi_{xyz}\rangle \in \mathcal{E}$):

$$|\Phi_{xyz}\rangle = |\Phi_x\rangle \otimes |\Phi_y\rangle \otimes |\Phi_z\rangle \quad \text{[A1.26a]}$$

The global Hamiltonian, H, is the sum of the one-dimensional Hamiltonians, H_i, i.e.:

$$H_{xyz} = H = H_x + H_y + H_z \quad \text{[A1.26b]}$$

The Hamiltonian H_x of the quantum harmonic oscillator relative to the x dimension is given by expression [A1.16], which we will rewrite as follows:

$$H_x = \frac{P_x^2}{2m} + \frac{1}{2} m\omega^2 X^2 \quad \text{[A1.27]}$$

The Schrödinger equation [A1.2] is then written:

$$\left[-\frac{\hbar^2}{2m}\frac{d^2}{dx^2} + U(x)\right]\Phi_x(x) = E_x \Phi_x(x) \quad \text{[A1.28]}$$

The solution to equation [A1.28] and that of each of the equivalent equations for the y and z coordinates is given by [A1.14]. Thus:

$$E_{nx} = \hbar\omega\left(n_x + \frac{1}{2}\right); E_{ny} = \hbar\omega\left(n_y + \frac{1}{2}\right); E_{nz} = \hbar\omega\left(n_z + \frac{1}{2}\right) \quad \text{[A1.29]}$$

Let us now determine the expression of the total energy, E_{nxnynz}, of the three-dimensional harmonic oscillator. Considering [A1.26], the eigenvalue equation for Hamiltonian H is written in the tensor product space:

$$H|\Phi_{xyz}\rangle = (H_x + H_y + H_z)(|\Phi_x\rangle \otimes |\Phi_y\rangle \otimes |\Phi_z\rangle)$$

That is:

$$E_{nxnynz}\,|\Phi_{xyz}\rangle = H_x|\Phi_x\rangle \otimes (|\Phi_y\rangle \otimes |\Phi_z\rangle) + |\Phi_x\rangle \otimes (H_y\,|\Phi_y\rangle) \otimes |\Phi_z\rangle$$

$$+ (|\Phi_x\rangle \otimes |\Phi_y\rangle) \otimes H_z\,|\Phi_z\rangle$$

i.e.:

$$E_{nxnynz}\,|\Phi_{xyz}\rangle = (E_{nx} + E_{ny} + E_{nz})|\Phi_x\rangle \otimes |\Phi_y\rangle \otimes |\Phi_z\rangle$$

$$= (E_{nx} + E_{ny} + E_{nz})\,|\Phi_{xyz}\rangle$$

Thus, ultimately:

$$E_{nxnynz} = (E_{nx} + E_{ny} + E_{nz} \qquad [A1.30]$$

Using [A1.29], solution [A1.30] is written in the form:

$$E_{nxnynz} = \hbar\omega\left(n_x + n_y + n_z + \frac{3}{2}\right) \qquad [A1.31]$$

The energy of the three-dimensional quantum harmonic oscillator is then written:

$$E_n = \hbar\omega\left(n + \frac{3}{2}\right); n = n_x + n_y + n_z \qquad [A1.32]$$

By placing $n = N$ in [A1.32], equation [2.20] is found in the case of a nucleon subject to potential [2.12].

Appendix 2

Atomic Masses of Several Nuclides

NOTE.– The mass of the nucleus corresponding to the indicated element is obtained by subtracting from the atomic mass, M, the mass of Z electrons; that is: M $({}^{A}{}_{Z}X) = M\,(X) - Zme$.

EXAMPLE.– Mass of the ${}^{4}{}_{2}He$ helium nucleus: $m\,({}^{4}{}_{2}He) = M\,(He) - 2\,me$.

NOTE.– $m\,({}^{4}{}_{2}He) = 4.00260\text{ u} - 2\ \times 5.486 \times 10^{-4}\text{ u} = 4.0015028\text{ u} = 4.00150\text{ u}$.

Element (X)	Z	${}^{A}X$	M (X) (u)	Element (X)	Z	${}^{A}X$	M (X) (u)
Hydrogen (H)	1	${}^{1}H$	1.007825	Oxygen (O)	8	${}^{16}O$	15.99491
	1	${}^{2}H$	2.01400		8	${}^{17}O$	16.99910
	1	${}^{3}H$	3.016050		8	${}^{18}O$	17.99920
Helium (He)	2	${}^{3}He$	3.01603	Fluorine (F)	9	${}^{19}F$	18,99840
	2	${}^{4}He$	4.002 60	Neon (Ne)	10	${}^{20}Ne$	19.99244
Lithium (Li)	3	${}^{6}Li$	6.01512		10	${}^{21}Ne$	20.99335
	3	${}^{7}Li$	7.01600		10	${}^{22}Ne$	21.99138
Beryllium (Be)	4	${}^{7}Be$	7.01690	Sodium (Na)	11	${}^{22}Na$	21.99440
	4	${}^{9}Be$	9.01218		11	${}^{23}Na$	22.98980
	4	${}^{10}Be$	10.01350		11	${}^{24}Na$	23.99096

Element (X)	Z	$^{A}\mathrm{X}$	**M (X) (u)**		**Element (X)**	Z	$^{A}\mathrm{X}$	**M (X) (u)**
Boron (B)	5	$^{10}\mathrm{B}$	10.01290		Magnesium (Mg)	12	$^{24}\mathrm{Mg}$	23.98504
	5	$^{11}\mathrm{B}$	11.00931		Aluminum (Al)	13	$^{26}\mathrm{Al}$	25.98689
Carbon (C)	6	$^{12}\mathrm{C}$	12.00000			13	$^{27}\mathrm{Al}$	26.98153
	6	$^{13}\mathrm{C}$	13.00335		Silicon (Si)	14	$^{28}\mathrm{Si}$	27.97693
	6	$^{14}\mathrm{C}$	14.00320			14	$^{29}\mathrm{Si}$	28.97649
Nitrogen (N)	7	$^{14}\mathrm{N}$	14.00307			14	$^{30}\mathrm{Si}$	29.97376
	7	$^{15}\mathrm{N}$	15.00031			14	$^{31}\mathrm{Si}$	30.97553

Table A2.1. *Atomic masses of nuclides with atomic numbers Z = 1-14*

Element (X)	Z	$^{A}\mathrm{X}$	**M (X) (u)**		**Element (X)**	Z	$^{A}\mathrm{X}$	**M (X) (u)**
Silicon (Si)	14	$^{32}\mathrm{Si}$	31.97400		Iron (Fe)	26	$^{57}\mathrm{Fe}$	56.93540
Phosphorus (P)	15	$^{31}\mathrm{P}$	30.99376			26	$^{58}\mathrm{Fe}$	57.93300
	15	$^{32}\mathrm{P}$	31.97390			26	$^{59}\mathrm{Fe}$	58.93500
	15	$^{33}\mathrm{P}$	32.97170		Cobalt (Co)	27	$^{56}\mathrm{Co}$	55.94000
Sulfur (S)	16	$^{32}\mathrm{S}$	31.97207		Copper (Cu)	29	$^{63}\mathrm{Cu}$	62.92980
Chlorine (Cl)	17	$^{35}\mathrm{Cl}$	34.96885			29	$^{65}\mathrm{Cu}$	64.92780
	17	$^{36}\mathrm{Cl}$	35.97970		Zinc (Zn)	30	$^{64}\mathrm{Zn}$	63.92910
	17	$^{37}\mathrm{Cl}$	36.96580		Gallium (Ga)	31	$^{69}\mathrm{Ga}$	68.92570

Element (X)	Z	^{A}X	M (X) (u)	Element (X)	Z	^{A}X	M (X) (u)
Argon (Ar)	18	^{36}Ar	35.96755	Germanium (Ge)	32	^{70}Ge	69.92430
	18	^{37}Ar	36.96670		32	^{72}Ge	71.92170
	18	^{38}Ar	37.96272		32	^{74}Ge	73.92190
	18	^{39}Ar	38.96400	Arsenic (As)	33	^{75}As	74.92160
	18	^{40}Ar	39.96240	Selenium (Se)	34	^{78}Se	77.91730
Potassium (K)	19	^{39}K	38.96371		34	^{80}Se	79.91650
	19	^{40}K	39.97400	Bromine (Br)	35	^{77}Br	76.92100
	19	^{41}K	40.95200		35	^{79}Br	78.91830
	19	^{42}K	41.96300		35	^{81}Br	80.91830
Calcium (Ca)	20	^{40}Ca	39.96259		35	^{82}Br	81.91700
Scandium (Sc)	21	^{45}Sc	44.95592	Krypton (Kr)	36	^{84}Kr	83.91200
	21	^{46}Sc	45.95500	Rubidium (Rb)	37	^{85}Rb	84.91170
Titanium (Ti)	22	^{48}Ti	47.94800		37	^{87}Rb	86.90900
Vanadium (V)	23	^{51}V	50.94400	Strontium (Sr)	38	^{84}Sr	83.91340
Chrome (Cr)	24	^{52}Cr	51.94050		38	^{85}Sr	84.91300
Manganese (Mn)	25	^{55}Mn	54.93810		38	^{86}Sr	35.90940
Iron (Fe)	26	^{54}Fe	53.96960		38	^{87}Sr	86.90890
	26	^{55}Fe	54.93800		38	^{88}Sr	87.90560
	26	^{56}Fe	55.93490		38	^{89}Sr	88.90700

Table A2.2. *Atomic masses of nuclides with atomic numbers Z = 14-38*

Element (X)	Z	^{A}X	M (X) (u)
Strontium (Sr)	38	^{90}Sr	89.90700
Yttrium (Y)	39	^{87}Y	86.91100
	39	^{88}Y	87.91000
	39	^{89}Y	88.90540
	39	^{91}Y	90.90700
Zirconium (Zr)	40	^{90}Zr	89.90430
Niobium (Nb)	41	^{93}Nb	92.90600
Molybdenum (Mo)	42	^{98}Mo	97.90551
Technetium (Tc)	43	^{98}Tc	97.90720
Ruthenium (Ru)	44	^{102}Ru	101.90370
	44	^{104}Ru	103.90550
Rhodium (Rh)	45	^{103}Rh	102.90480
Palladium (Pd)	46	^{105}Pd	104.90460
	46	^{106}Pd	105.90320
	46	^{108}Pd	107.90300
Silver (Ag)	47	^{107}Ag	106.90410
	47	^{109}Ag	108.90470

Element (X)	Z	^{A}X	M (X) (u)
Tellurium (Te)	52	^{130}Te	129.90670
Iodine (I)	53	^{127}I	126.90040
Xenon (Xe)	54	^{129}Xe	128.90480
	54	^{132}Xe	131.90420
Cesium (Ce)	55	^{133}Ce	132.90510
	55	^{137}Ce	136.90750
Barium (Ba)	56	^{132}Ba	131.90570
	56	^{134}Ba	133.90430
	56	^{135}Ba	134.90560
	56	^{136}Ba	135.90440
	56	^{137}Ba	136.90630
	56	^{138}Ba	137.90500
Lanthanum (La)	57	^{139}La	138.90610
Cerium (Ce)	58	^{138}Ce	137.90570
	58	^{140}Ce	139.90530
	58	^{142}Ce	141.90900
Praseodymium (Pr)	59	^{141}Pr	140.90740

Element (X)	Z	^{A}X	M (X) (u)		Element (X)	Z	^{A}X	M (X) (u)
Cadmium (Cd)	48	^{112}Cd	111.90280		Neodymium (Nd)	60	^{142}Nd	141.90750
	48	^{114}Cd	113.90360			60	^{144}Nd	143.90990
Indium (In)	49	^{115}In	114.90410			60	^{146}Nd	145.91720
Tin (Sn)	50	^{118}Sn	117.90180		Promethium (Pm)	61	^{143}Pm	142.91100
	50	^{122}Sn	121.90340		Samarium (Sm)	62	^{152}Sm	151.91950
	50	^{124}Sn	123.90520			62	^{154}Sm	153.92200
Antimony (SB)	51	^{121}Sb	120.90380		Europium (Eu)	63	^{151}Eu	150.91960
	51	^{123}Sb	122.90410			63	^{153}Eu	152.92090
Tellurium (Te)	52	^{122}Te	121.90300		Gadolinium (Ga)	64	^{158}Ga	157.92410
	52	^{128}Te	127.90470			64	^{160}Ga	159.90710

Table A2.3. *Atomic masses of nuclides with atomic numbers Z = 38-64*

Element (X)	Z	^{A}X	M (X) (u)		Element (X)	Z	^{A}X	M (X) (u)
Terbium (Tb)	65	^{159}Tb	158.92500		Platinum (Pt)	78	^{196}Pt	195.96500
Dysprosium (Dy)	66	^{162}Dy	161.92650			78	^{198}Pt	197.96750
	66	^{163}Dy	162.92840		Gold (Au)	79	^{197}Au	196.96600
	66	^{164}Dy	163.92880		Mercury (Hg)	80	^{196}Hg	195.96580
Holmium (Ho)	67	^{165}Ho	164.93030			80	^{198}Hg	197.96680

Element (X)	Z	^{A}X	M (X) (u)
Erbium (Er)	68	^{166}Er	165.93040
	68	^{167}Er	166.93200
	68	^{168}Er	167.93240
	68	^{170}Er	169.93550
Thulium (Tm)	69	^{169}Tm	168.93440
Ytterbium (Yb)	70	^{170}Yb	169.93490
	70	^{171}Yb	169.93490
	70	^{172}Yb	171.93660
	70	^{173}Yb	172.93830
	70	^{174}Yb	173.93900
	70	^{176}Yb	175.94270
Lutetium (Lu)	71	^{175}Lu	174.94090
Hafnium (Hf)	72	^{180}Hf	179.94680
Tantalum (Ta)	73	^{181}Ta	180.94800
Tungsten (W)	74	^{182}W	181.94830
	74	^{184}W	183.95100
	74	^{186}W	185.95430
Thallium (Tl)	81	^{203}Tl	202.97230
	81	^{205}Tl	204.97450
Lead (Pb)	82	^{204}Pb	203.97310
	82	^{206}Pb	205.97450
	82	^{207}Pb	206.97590
	82	^{208}Pb	207.97660
Bismuth (Bi)	83	^{209}Bi	208.98040
Polonium (Po)	84	^{206}Po	206.98050
	84	^{208}Po	207.98130
	84	^{209}Po	208.98250
	84	^{210}Po	209.98290
Astate (At)	85	^{211}At	210.98750
Radon (Rn)	86	^{222}Rn	222.01750
Francium (Fr)	87	^{223}Fr	223.01980
Radium (Ra)	88	^{226}Ra	226.02540
Actinium (Ac)	89	^{225}Ac	225.02310
Thorium (Th)	90	^{232}Th	232.03820

Element (X)	Z	^{A}X	M (X) (u)		Element (X)	Z	^{A}X	M (X) (u)
Rhenium (Re)	75	^{185}Re	184.95300		Protactinium (Pr)	91	^{231}Pr	231.03590
	75	^{187}Re	186.95600		Uranium (U)	92	^{234}U	234.04090
Osmium (Os)	76	^{188}Os	187.95600			92	^{235}U	235.04390
	76	^{189}Os	188.95860			92	^{238}U	238.05080
Iridium (Ir)	77	^{191}Ir	190.96090		Neptunium (Np)	93	^{236}Np	236.04660
	77	^{193}Ir	192.96330			93	^{237}Np	237.04800
Platinum (Pt)	78	^{194}Pt	193.96280					
	78	^{195}Pt	194.96480					

Table A2.4. *Atomic masses of nuclides with atomic numbers Z = 65-93*

References

[AND 33a] ANDERSON C.D., "The positive electron", *Phys. Rev.*, vol. 43, p. 491, 1933.

[AND 33b] ANDERSON C.D., NEDDERMEYER S.H., "Positrons from gamma rays", *Phys. Rev.*, vol. 43, p. 1034, 1933.

[ARN 10] ARNAUD N., DESCOTES-GENON, S., KERHOAS-CAVATA, S. *et al.*, *Passeport pour les Deux Infinis*, CNRS, CEA, Université Paris Diderot, Obs. Paris, GISP2i, Dunod, Paris, 2010.

[ASL 08] ASLANGUL C., *Mécanique quantique 2 : développements et applications à basse énergie*, Éditions De Boeck Université, Brussels, 2008.

[BES 17] BESSE G., Description théorique de la dynamique nucléaire lors de collisions d'ions lourds aux énergies de Fermi, PhD thesis, Université de Nantes, Nantes, 2017.

[BET 36] BETHE H.A., "Nuclear radius and many-body problem", *Phys. Rev.*, vol. 50, p. 977, 1936.

[BIÉ 06] BIÉMONT E., *Spectroscopie atomique*, Éditions De Boeck, Brussels, 2006.

[BIM 06] BIMBOT R., *Histoire de la radioactivité. L'évolution d'un concept et de ses applications*, Vuibert, Paris, 2006.

[BIS 19] BISHOP I., XIAN S., FELLER S., "Robert A. Millikan and the oil drop experiment", *Phys. Teacher*, vol. 57, p. 442, 2019.

[BLA 25] BLACKETT P., MAYNARD S., "The ejection of protons from nitrogen nuclei, photographed by the Wilson method", *J. Chem. Soc. Trans. Ser. A*, vol. 107, no. 742, pp. 349–360, 1925.

[BLA 99] BLANC D., *Précis de physique nucléaire*, Dunod, Paris, 1999.

[BOH 39] BOHR N., WHEELER J.A., "The mechanism of nuclear fission", *Phys. Rev.*, vol. 56, p. 426, 1939.

[BOT 30] BOTHE W.H., BECKER H., "Künstliche Erregung von Kern–γ–Strahlen (Artificial excitation of nuclear γ rays)", *Z. Phys.*, vol. 66, p. 289, 1930.

[BRÖ 01] BRÖCKER B., *Atlas de la physique atomique et nucléaire*, Librairie Générale Française, Paris, 2001.

[BUS 13] BUSKULIC D., Notes de cours de PHYS 801. Introduction à la Physique Nucléaire, available at: https://lappweb.in2p3.fr/~buskulic/cours/PHYS801/PHYS801_Physique_Nucleaire.pdf, 2013.

[CAR 79] CARATINI R., *Physique Chimie*, Bordas-Encyclpédie, Paris, 1979.

[CHA 32a] CHADWICK J., "The existence of a neutron", *Proc. R. Soc.*, vol. 136, p. 692, 1932.

[CHA 32b] CHADWICK J., "Possible existence of a neutron", *Nature*, vol. 129, p. 312, 1932.

[COH 77] COHEN-TANOUNDJI C., DIU B., LALOE, F., *Mécanique Quantique 1*. Hermann Editeurs des Sciences et des Arts, Paris, 1977.

[CRO 79] CROOKES W., "On radiant matter", *Nature*, vol. 20, nos 419–423, pp. 436–440, 1879.

[CUR 32a] CURIE I., "Sur le rayonnement α du radiothorium, du radioactinium et de leurs dérivés ; complexité du rayonnement α du radioactinium", *J. Phys. Radium*, vol. 3, no. 2, pp. 57–72, 1932.

[CUR 32b] CURIE I., JOLIOT F., "Projections d'atomes par les rayons très pénétrants excités dans les noyaux légers", *C. R. Hebd. Séances Acad. Sci. Paris*, vol. 194, p. 876, 1932.

[CUR 32c] CURIE I., JOLIOT F., "Sur la nature du rayonnement pénétrant excité dans les noyaux légers par les particules α", *C. R. Hebd. Séances Acad. Sci. Paris*, vol. 194, p. 1229, 1932.

[DAU 99] DAUGAS J.-M., Etude de la structure et du mécanisme de production des états isomères aux énergies intermédiaires, PhD thesis, Université de Caen, Caen, 1999.

[DAY 53] DAY R.B., HENKEL R.L., "Neutron total cross section at 20 MeV", *Phys. Rev.*, vol. 92, p. 358, 1953.

[DEG 17] DEGRELLE D., Caractérisation numérique de la technique de spectroscopie gamma par simulation Monte-Carlo. Application à la datation d'échantillons environnementaux, PhD thesis, Université Bourgogne Franche-Comté, Besançon, 2017.

[DEL 03] DELAUNAY F., Structure des états du ^{11}Be excités par la réaction d(^{10}Be,p)^{11}Be, PhD thesis, Université Paris 6, Paris, 2003.

[DUM 15] DUMORA D., Physique Nucléaire, Lesson, Version 4.2.1, Université de Bordeaux, Bordeaux, September 26, 2015.

[ELL 21] ELLIS C.D., "The magnetic spectrum of the ß-rays excited by γ rays", *Proc. Roy. Soc.*, vol. A99, pp. 261–271, 1921.

[ELL 22] ELLIS C.D., "ß-spectra and their meaning", *Proc. Roy. Soc.*, vol. A101, pp. 1–17, 1922.

[EVA 61] EVAN D., *Le noyau atomique*, Dunod, Paris, 1961.

[FAL 87] FALCONER I., “Corpuscles, electrons and cathode rays: J.J. Thomson and the ‘Discovery of the Electron’”, *Br. J. Hist. Sci.*, vol. 20, pp. 241–276, 1987.

[FEE 47] FEENBERG E., “Semi-empirical theory of the nuclear energy surface”, *Revs. Mod. Phys.*, vol. 19, p. 239, 1947.

[FER 50] FERMI E.E., *Nuclear Physics*, University of Chicago Press, Chicago, 1950.

[FER 64] FERMI L., *L’histoire de l’énergie nucléaire*, Fernand Nathan Éditeur, Paris, 1964.

[FRI 49] FRIEDLANDER G., KENNEDY J.W., *Introduction to Radiochemistry*, John Wiley & Sons, Inc., New York, 1949.

[FOO 12] FOOS J., *Manuel de radioactivité*, Hermann Editeurs des Sciences et des Arts, Paris, 2012.

[GEI 09] GEIGER H., MARSDEN E., “On a diffuse reflection of the α-particles”, *Proc. Roy. Soc.*, vol. A82, p. 495. 1909.

[GER 07] GERBAUX M., Sources de particules de hautes énergies obtenues avec des lasers intenses pour applications à la physique nucléaire, PhD thesis, Université Sciences et Technologies – Bordeaux I, Bordeaux, 2007.

[GRO 85] GROSSETETTE C.H., *Relativité restreinte et structure atomique de la matière*, Ellipses Éditions Marketing S.A., Paris, 1985.

[GUY 03] GUYMONT M., *Structure de la matière. Atomes, liaisons chimiques et cristallographie*, Belin Sup, Paris, 2003.

[HAL 11] HALIOUA B., Histoire de la découverte de la radioactivité, available at: https://www.allodocteurs.fr/actualite-sante-histoire-de-la-decouverte-de-la-radioactivite 3709.html, 2011.

[KRI 19a] KRIVIT S.B., The world’s first successful alchemist (It Wasn’t Rutherford), available at: https://news.newenergytimes.net/2019/05/14/the-worlds-first-successful-alchemist-it-wasnt-rutherford/, 2019.

[KRI 19b] KRIVIT S.B., Rutherford’s reluctant role in nuclear transmutation, available at: https://news.newenergytimes.net/2019/05/18/rutherfords-reluctant-role-in-nuclear-transmutation/, 2019.

[LAL 11] LALLOUET Y., Différents aspects de la physique nucléaire depuis les basses énergies jusqu’aux énergies intermédiaires, PhD thesis, Université Claude Bernard – Lyon I, Lyon, 2011.

[LEP 56] LEPRINCE RINGER L., *Grandes découvertes du XXe siècle*, Librairie Larousse, Paris, 1956.

[LES 10] LE SECH C., NGÔ C., *Physique Nucléaire. Des quarks aux applications*, Dunod, Paris, 2010.

[MAR 69] MARMIER P., SHELDON E., *Physics of Nuclei and Particles*, Academic Press, Cambridge, 1969.

[MAR 11] MARQUET C., SIMARD L., "Le neutrino est-il identique à son antiparticule ?", *Reflets de la physique*, vol. 24, p. 11, 2011.

[MAT 46] MATTAUCH J., FLUEGGE S., *Nuclear Physics Tables and An Introduction to Nuclear Physics*, Interscience Publishers, Inc., New York, 1946.

[MAY 17] MAYET R., *Physique nucléaire appliquée*, 2nd edition, Edition De Boeck Supérieur, Louvain-la-Neuve, 2017.

[MIL 10] MILLIKAN R.A., "A new modification of the cloud method of determining the elementary electrical charge and the most probable value of that charge", *Philos. Mag.*, vol. 19, p. 209, 1910.

[MIS 96] MISDAQ M.A., *Structure du noyau atomique*, Editions Afrique-Orient, Casablanca, 1996.

[MOU 11] MOUMENE A., Étude des états liés des potentiels d'écrantage de type exponentiel coulombien, Physics Thesis, Université des Sciences et de la Technologie "Houari Boumediene", Bab Ezzouar, 2011.

[NES 17] NESVIZHEVSKY V., VILLAIN J., "The discovery of the neutron and its consequences (1930–1940) ; La découverte du neutron et ses conséquences (1930–1940)", *C.R. Physique*, vol. 18, pp. 592–600, 2017.

[NIC 10] NICOL S., Étude et construction d'un tomographe TEP/TDM pour petits animaux, combinant modules phoswich à scintillateurs et détecteur à pixels hybrides, PhD Thesis, Université de la Méditerranée – Aix-Marseille II, Marseille, 2010.

[OLS 91] OLSCHWANG D., DIOP C., *La Matière, de la réalité sensible au modèle formalisé*, Presses Universitaires de Dakar, Dakar, 1991.

[PAG 85] PAGELS H., *L'Univers quantique. Des quarks aux étoiles*, Inter Éditions, Paris, 1985.

[PER 82] PERRIN F., "La découverte de la radioactivité bêta positive", *Journal de Physique Colloques*, vol. 43, no. C8, pp. C8-431-C8-433, 1982.

[PER 95] PERRIN J., "Nouvelles propriétés des rayons cathodiques", *Comptes Rendus*, vol. 121, p. 1130, 1895.

[POL 35] POLLARD E.C., "Nuclear potential barriers: Experiment and theory", *Phys. Rev.*, vol. 47, p. 611, 1935.

[ROU 60] ROUSSEAU, P., *L'histoire de l'atome*, Librairie Arthème Fayard, Paris, 1960.

[ROY 06] ROY CHOWDHURY P., BASU D.N., "Nuclear matter properties with the re-evaluated coefficients of liquid drop model", *Acta Phys. Polon. B*, vol. 37, p. 1834, 2006.

[RUT 11] RUTHERFORD E., "The scattering of α and β particles by matter and the structure of the atom", *Philos. Mag. Ser.*, vol. 21, pp. 669–688, 1911.

[RUT 19a] RUTHERFORD E., "Collisions of alpha particles with light atoms III: Nitrogen and oxygen atoms", *Philos. Mag. Ser. 6*, vol. 37, pp. 571–580, 1919.

[RUT 19b] RUTHERFORD E., "Collisions of alpha particles with light atoms IV: An anomalous effect in nitrogen", *Philos. Mag. Ser. 6*, vol. 37, pp. 581–587, 1919.

[SEE 61] SEEGER P.A., "Semiempirical atomic mass law", *Nucl. Phys.*, vol. 25, pp. 1–135, 1961.

[SAK 10] SAKHO I., *Guides Pratiques du Lycéen, Physique Premières S*, Éditions Publibook, Paris, 2010.

[SAK 11] SAKHO I., *Histoire de l'atome, de l'insécabilité au modèle probabiliste*, Éditions Publibook, Paris, 2011.

[SAK 12] SAKHO I., *Mécanique Quantique 1 : Licence 2 MPCI, Exercices corrigés et commentés*, Hermann Editeurs des Sciences et des Arts, Paris, 2012.

[SAK 16] SAKHO I., *Guides Pratiques du Lycéen, Physique Terminales S*, Nouvelles Editions Africaines du Sénégal (NEAS), Dakar, 2016.

[SAK 18a] SAKHO I., "Electrodynamics calculations of the unit nuclear radius in agreement with the constant density model", *AASCIT J. Phys.*, vol. 4, pp. 26–44, 2018.

[SAK 18b] SAKHO I., *Electrostatique, magnétostatique & induction électromagnétique. Résumés de cours & 120 exercices corrigés*, Editions Ellipses, Paris, 2018.

[SAK 19] SAKHO I., *Introduction to Quantum Mechanics 1: Thermal Radiation and Experimental Facts Regarding the Quantization of Matter*, ISTE Ltd, London, and John Wiley & Sons, Inc, New York, 2019.

[SAK 20] SAKHO I., *Physique atomique : systèmes hydrogènoïdes et systèmes héliumoïdes.* Editions Ellipses, Paris, 2020.

[SAO 04] SAOUTER S., Possibilité d'observation par le télescope Antares des sources ponctuelles de rayons gamma observées par le détecteur Egret et étude d'un prototype, PhD thesis, Université Paris VI – Pierre et Marie Curie, Paris, 2004.

[SAR 33] SARGENT B.W., "The maximum energy of the β-Rays from uranium X and other bodies", *Proc. Roy. Soc.*, vol. A139, p. 659, 1933.

[SIV 86] SIVOUKHINE D., *Cours de physique générale, Tome V : physique atomique et nucléaire*, Éditions Mir, Moscow, 1986.

[SOD 11] SODDY F., "Radioactivity", *Chem. Soc. Annu. Rep.*, vol. 7, pp, 256–286, 1911.

[SOD 14] SODDY F., *Chemistry of the Radio-Elements, Part II*, Longmans, Green & Co. Ltd., London, 1914.

[STÖ 07] STÖCKER H., JUNT F., GUILLAUME G., *Toute la Physique*, Dunod, Paris, 2007.

[SUR 98] SURAUD E., *La matière nucléaire : des étoiles aux noyaux*, Hermann, Editeurs des Sciences et des Arts, Paris, 1998.

[THO 97] THOMSON J.J., "Cathode rays", *Proc. Roy. Inst.*, vol. 15, pp. 419–432, 1897.

[VUC 55] VUCCINO S., "Étude de la méthode des coïncidences appliquée à l'étalonnage absolu du ^{24}Na, ^{60}Co et ^{46}Sc", *J. Phys. Radium*, vol. 16, no. 6, pp. 462–468, 1955.

[WAP 58] WAPSTRA A.H., *Atomic Masses of Nuclides*, Handbuch der Physik, Springer, Berlin, 1958.

[WEI 51] WEISSKOPF V.F., "Radiative transition probabilities in nuclei, Letters to the Editor", *Phys. Rev.*, vol. 83, p. 1073, 1951.

Index

A, B

C, D

E, F

L, M

N, O

P, Q

R, S